The Gene Bomb

Does Higher Education and Advanced Technology Accelerate the Selection of Genes for Learning Disorders, ADHD, Addictive, and Disruptive Behaviors?

by

David E. Comings, M.D.

The Gene Bomb

Does Higher Education and Advanced Technology Accelerate the Selection for Genes for Learning Disorders, ADHD, Addictive and Disruptive Behaviors?

Published by **Hope Press** P.O. Box 188
Duarte, CA 91009-0188 U.S.A.

Other books by Hope Press:

Tourette Syndrome and Human Behavior
by *David E. Comings, M.D.*

Search for the Tourette Syndrome and Human Behavior Genes
by *David E. Comings, M.D.*

Teaching the Tiger
by *Marilyn P. Dornbush, Ph.D.* and *Sheryl K. Pruitt, M.Ed.*

RYAN and **What Makes RYAN Tick?**
by *Susan Hughes*

(to order these and others see back leaf forms)

Printed in the United States of America

Library of Congress Cataloging-in-Publication Data

Comings, David E.
The gene bomb : Does higher education and advanced technology accelerate the selection of genes for learning disorders, ADHD, addictive and disruptive behaviors? / David E. Comings.
p. cm.
Includes bibliographical references and index.
ISBN 1-878267-38-8. — ISBN 1-878267-39-6 (pbk.)
1. Personality disorders — Genetic aspects. 2. Antisocial personality disorders — Genetic aspects. 3. Behavioral genetics. 4. Natural selection. 5. Human population genetics. I. Title.
RC554.C66 1996
616.89'042 —dc20 96-8110
CIP

Dedicated to:

Mark,

Scott,

Karen,

Nicole,

Kim,

and all others who were born in the 20th century

and will be living in the 21st

Acknowledgements

Thanks to the National Atomic Museum, Albuquerque, N.M.
for the cover

Table of Contents

Introduction

A number of different studies suggest that the levels of many of the most disruptive behaviors affecting our society, such as alcoholism, drug abuse, crime, aggression, childhood hyperactivity, conduct and antisocial personality disorders, learning disabilities, major depression, suicide, and anxiety, are increasing at a rate that is significantly greater than can be accounted for by the increase in population *per se*. Most of us defend ourselves against these disturbing observations by either denying that such trends are real, or suggesting that if they are real they are the expected outcome of an increasingly complex and fast-paced society. Although both genetic and environmental factors play a role in each of these behaviors, it is taken for granted that the change in their frequency must be solely due to changes in the environment, since genes are stable and the genetic evolution of mankind supposedly stopped many generations ago.

As a human geneticist, I have been involved for the past sixteen years in the study of the role of genes in human behavior. Much of this time has been spent treating and studying two very similar hereditary childhood behavioral disorders – attention deficit hyperactivity disorder (ADHD) and Tourette syndrome (TS). It quickly became apparent to me that the frequencies of the behaviors listed above were dramatically increased in patients with ADHD and TS and in their relatives, and thus had a significant genetic component. Over the years, as I became more familiar with the evidence that the above behavioral problems were increasing in frequency, I also became convinced that the usual explanation for the increase was upside down and backward. Thus, *instead of these behaviors being the result of an increasingly complex society, I felt the increasingly complex society was selecting for the genes causing these behaviors.*

In its simplest form, I felt that, in many cases, alcohol and drug abuse, conduct, learning problems, and the other disorders listed above were due in large part to the chance coming together of a number of different genes. The effects of these genes were complemented by the environment, and individuals carrying them had "problem behaviors" consisting of impulsive, compulsive, oppositional, addictive, and cognitive disorders; were bored easily; preferred instant gratification to long-term goals; and tended to drop out of school early and become involved in teenage pregnancies, while individuals who did not carry these genes tended to stay in school, go to college, or develop useful skills before starting a family.

Since any living object that reproduces every fifteen to twenty-two years will increase its numbers faster than one that reproduces every twenty-three to thirty years, *this earlier age of first pregnancy tends to act as a powerful selective advantage for the genes involved.* The more complex the society and the longer the duration of education required to perform well in that society, the more these two rates will diverge. As a result, the rate of selection of these genes will progressively increase, and this can potentially destroy the species from within. Throughout the book I will refer to this as *the hypothesis*.

For a genetic disorder, especially one that is polygenic (i.e. caused by a number of genes), as the number of such genes people carry increases, the frequency of the disorder increases, and the age that the disorder first appears decreases. Thus, if there is reasonable evidence to support the suspicion that a given set of disorders – in this case a spectrum of addictive, impulsive, compulsive, mood, anxiety, and learning disorders – increases, or the age of onset of one or more of these disorders is decreasing, one explanation can be that the genes involved are increasing in frequency. This book reviews that evidence and concludes that the frequency of many of these disorders is increasing, the age of onset is decreasing, and at least part of this is due to an increasing frequency of the genes involved. It discusses the mechanisms by which this gene selection may be occurring and suggests a number of *voluntary* programs that could reverse this trend.

This issue is of critical importance because many sociologists explain away these behavior problems using a purely psychosocial model. This assumes that any social program that relieves the socioeconomic problems will eliminate the behaviors. This may not be the case. If many of these behaviors are genetically influenced, by the time such social programs are in place, it may be too late because the species has become a genetically different one than it used to be.

I first wrote up an account of the basic aspects of the hypothesis in 1989 and discussed it with some colleagues. They advised me not to publish it, fearing that it would be too controversial, and that I might be labeled a racist, a Nazi, or any of the many other appellations that tend to be applied to people who suggest that genetic factors play a role in social issues and behavior. As a result, I filed it away in a back drawer. In 1991, after we and others had published evidence suggesting that a *dopamine* receptor gene could be one of the genes involved, I dusted it off and submitted it for publication. It was rejected by journals ranging in specialties from psychiatry to genetics to public health. In large part, the rejections were for the same reasons my colleagues had warned me about. Once again it went back into the drawer.

As we continued our work on genetic factors in the disruptive and addictive behavioral disorders, we identified a role for several additional genes and described evidence that the genes for TS and ADHD produced a wide spectrum of different associated behavioral disorders. By this time the total number of studies supporting the hypothesis had grown to the degree that a single paper no longer seemed to be an adequate vehicle. Presentation in book form, for both a professional and general audience, seemed better.

In addition, I was becoming increasingly concerned that if the hypothesis was true and I didn't write about it, no one else would. I often thought of the book *And the Band Played On* by Randy Shilts.[239] He severely chastised politicians and the medical and scientific professions for failing to mount an early response to the AIDS epidemic before it had gotten out of control. In contrast to AIDS, where the problem is easy to see because people die of AIDS, the epidemic I was concerned about was invisible and far more subtle. Like the destruction of the civilization on Easter Island,[92] it could occur so gradually as to go unnoticed, until it was too late to correct. However, its eventual effect on the human race could be far more disastrous than all the microbial epidemics combined.

If I was wrong and this theory was a giant figment of my demented imagination, it didn't matter if it was never published. But what if I was right? Concerns that the human race would disappear in cloud of nuclear dust were rapidly diminishing. While overpopulation and environmental pollution continued to be a threat, they were receiving more attention and could possibly be managed. Outside of a collision with an asteroid, a runaway lethal ebola virus spilling out from a remote habitat of a disrupted rain forest, or, more likely, a worldwide AIDS epidemic,[138] it began to look like at least a portion of the human race would survive for some time.

However, life might become a miserable existence if genes for disruptive, addictive, depressive, and anxious behaviors were being selected for at an increasingly rapid pace. Was the human race doomed to more and more learning disorders, depression, anxiety and violence, bigger and better prisons, larger and more expensive police forces, greater disaffection with education, and increasing numbers of addiction and other treatment centers – eventually ending in a grotesque form of a species-specific genetic meltdown?

Because of the central role of higher education in the selection process, the initial title of the book was the present subtitle – *Does Higher Education and Advanced Technology Accelerate the Selection of Genes for Learning Disorders, ADHD, Addictive, and Disruptive Behaviors?* A more succinct and expressive title was needed, thus – *The Gene Bomb.*

The hypothesis is presented in six parts.

Part I. Evidence that Learning, Addictive, Disruptive, and Other Behavioral Disorders are Increasing in Frequency. I have presented relatively little descriptive material on the different behavioral disorders themselves. For this, the reader is referred to an earlier book, *Tourette Syndrome and Human Behavior*.[52]

Part II. Evidence These are Interrelated Genetic Disorders. While some of the evidence is presented here, for a much more detailed description of this subject the reader is referred to the companion book – *Search for the Tourette Syndrome and Human Behavior Genes*.[60]

Part III. Evidence that People with Learning Disorders and Addictive-Disruptive Behaviors have Children Earlier. This section includes the evidence that individuals and parents of individuals with these behavioral disorders tend to have children earlier. As a result of this increased rate of turnover or reproduction, these genes are preferentially selected for and increase in frequency in the general population. This part also reviews the evidence that individuals with these disorders have more children and come from larger families, which also selects for these genes.

Part IV. Causes. Assuming that genes for disruptive, addictive, and learning disorder genes are being selected for, this section describes some of the culprits and concludes that the major cause is the dramatic increase in number of individuals attending college and graduate school in the latter part of the twentieth century. These individuals tend to have the fewest number of genes for learning disorders, disruptive, and addictive behaviors. Since higher education has a powerful effect on delaying the onset of childbirth, there is a selection against the genes these individuals carry, and selection for the genes carried by those who drop out of school early.

Part V. Proving or Disproving the Hypothesis. This section responds to arguments about why the hypothesis may be wrong, as well as what needs to be done to prove whether it is correct.

Part VI. What to Do. The final part discusses what can be done to correct the problem, assuming the hypothesis is correct. It discusses why the hypothesis is not racist, and why it does not call for discredited coercive eugenic approaches. The best way to diffuse the bomb is by identification and verification of the problem through research, and the institution of *voluntary* programs that are beneficial to all members of society. Best of all, this would be a win-win situation in which these programs are reasonable and beneficial – even if the hypothesis turns out to be wrong.

Part I

Evidence that Addictive, Disruptive, and Other Behavioral Disorders are Increasing in Frequency

The following chapters present evidence that the frequency of depression, suicide, alcoholism, drug addiction, anxiety disorders, attention deficit hyperactivity disorder, autism, conduct disorder, criminal behavior, and learning disorders have been increasing since World War II, and IQ may be decreasing. In many cases the disorders are also affecting progressively younger and younger individuals.

Chapter 1

Increase in Depression

In his forward to the book *Psychiatric Disorders in America*, by L.N. Robins, Dr. D.X. Freedman stated:

> "Less than a month after President Carter's 1977 inauguration, Rosalyn Carter was authorized to assemble a commission to examine the nation's needs for mental health services and new knowledge about [these] disorders. Sitting with her in a basement office of the White House East Wing as the initial plans were made, one could not mistake Mrs. Carter's clear respect for and interest in sound information. Her questions were unerringly straightforward. How many are suffering from these illnesses, who are they, and how are they treated? Embarrassingly, equally straightforward answers could not be provided. The base of information about the scope and boundaries of mental illness was simply inadequate."[119]

There was clearly a need for both accurate psychiatric diagnoses and studies on their prevalence in the population.

The criteria for the different psychiatric diagnoses were defined in the Third Edition of the Diagnostic and Statistical Manual of Mental Disorders published by the American Psychiatric Association. This was termed the DSM-III, for short. However, in order to determine how common these different disorders were, it was necessary to design an instrument capable of making accurate assessments. Since studies in a number of cities would be required, the instrument had to be capable of producing reproducible and consistent results, regardless of where the tests were being given. In addition, to keep costs down, the tests would have to be administered by trained, but non-professional, interviewers.

The task of producing this instrument was given to Dr. Lee Robins and her colleagues at the Washington University School of Medicine in St. Louis. After years of development, testing, and refining, the end product was called the Diagnostic Interview Schedule, or DIS[227] for short. Eventually this structured psychiatric interview was given to over 18,000 subjects in five cities around the United States. This was called the Epidemiologic Catchment Area (ECA) study. The results for the prevalence of various psychiatric disorders were published in 1984.[228] (See Table 1)

Approximately one-third of the population had some type of psychiatric disorder. This frequency is higher than that for any other group of disorders. In the table, the results are also given for another survey entitled the National Comorbidity Survey

(NCS), performed ten years later.[149] Although the two were not identical surveys, they were similar enough to show the apparent increase in frequency of all of the disorders that were tested in both surveys. For example, the prevalence of any disorder was up from 32.8% to 48.0%, any mood disorder was up from 7.8% to 19.3%, anxiety disorders were up from 16.0% to 24.9%, and alcohol or drug abuse or dependence was up from 16.7% to 26.6%.

Studies in the United States

In the mid 1970s Dr. Gerald Klerman of the Massachusetts General Hospital in Boston, began writing about the apparent increase in the prevalence of depression in the United States, especially in young people.[152-154] He noted that the average age of onset of depression in large studies done after World War II was considerably younger than in studies done prior to the war,[152] that many of the patients being treated for depression were younger than the textbook description of the disorder, and that increasing attention was being drawn to depression in children and women.[155] Prior to the development of the DIS and similar instruments, it was difficult to determine if these observations were true or were simply due to changes in criteria for making the diagnosis of depression, or awareness of the disorder. In 1985, using the new structured instruments, Klerman and colleagues[156] examined the frequency of depression in 2,289 relatives of 523 patients with depression. The relatives were analyzed for frequency of depression by the decade of the age of their birth. These age groupings are called cohorts, and thus the studies were called cohort studies. This study showed a progressive increase in frequency, and an earlier age of onset of depression, for individuals in the younger cohorts. This is shown for female and male relatives in Figure 1.

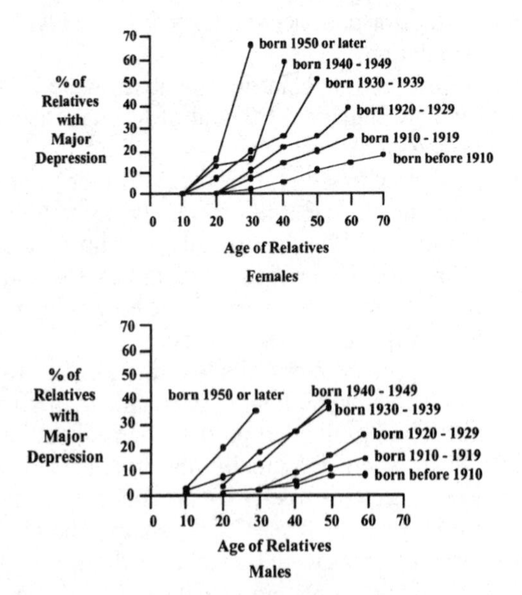

Figure 1. Cumulative probability of major depression in female (top) and male (bottom) relatives of patients with depression. Redrawn from Klerman et al., *Arch. Gen. Psychiatry* 42:689-693. Copyright 1985, Amer. Med. Assn.

For female relatives born before 1910, 18% had developed an episode of major depression by age 70. For those born in the 1910 to 1919 cohort, by age 60 26% had developed depression. In successive ten-year cohorts, the lifetime frequency of depression increased to 40, 52, 60, and 65%, with the latter high frequency occurring by 30 years of age. The rates for male relatives increased in a similar but less dramatic fashion, with 35% of those born after 1950 having depression.

Because of the strong genetic influences on depression and the fact that these studies were performed on the relatives of patients with depression, the rates are much higher than for the general population. However, studies on depression in the general population produced similar results. For example, the results of administering the DIS in five different cities provided a large body of data to examine the question of whether there was an increase in the frequency of depression by birth cohorts in the general population. A summary of these results is shown in Figure 2.[157]

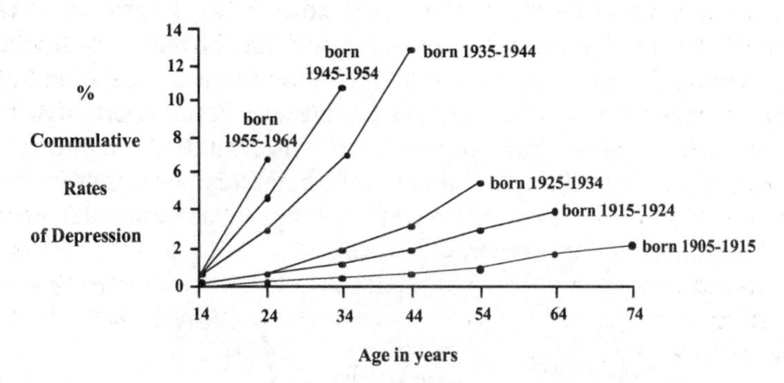

Figure 2. Lifetime prevalence of major depression from the National Institute of Mental Health and Epidemiologic Catchment study at five sites. Both sexes, Whites only. Redrawn from Klerman, G.J., *Journal of the American Medical Assn.* 261:2229-2235. Copyright 1989, Amer. Med. Assn.

While the cumulative lifetime rates of depression were lower than for the studies of relatives of patients with depression, the results were the same. For those born between 1905 and 1914, less than 2% developed depression, while for those born between 1935 and 1944, 13% had depression, and the onset was at an earlier age. While those born between 1945 and 1964 were too young to show such a high frequency of lifetime depression, the age of onset was still younger.

Several possible trivial explanations for these results have been entertained.[156,157] These included poor memory of earlier episodes of depression in older subjects, inability to interview older subjects with depression because they had died or moved away, decreased sensitivity to feelings of depression in older subjects, and changing diagnostic practices. None of these factors was considered to account for the magnitude of the results. The possibility of poor recall is discussed later.

A number of possible non-trivial explanations have also been proposed.[156,157] While it was recognized that depression was the result of a complex interaction between genes and the environment, only environmental explanations of the results were offered. These included nutritional factors, viruses, increasing urbanization, changes in family structure, greater geographic mobility with resultant loss of family attachments, alterations in the roles of women, and shifts in male-female occupational patterns.

The younger the age of onset of depression, the more likely it is that the relatives will also have a history of depression.[197a] This supports the idea that those with a younger age of onset have a higher degree of genetic loading than those with an older age of onset.

Manic-Depressive and Schizoaffective Disorders

Two other disorders related to depression are manic-depressive disorder and schizoaffective disorder. The latter is a category of depression with some schizoid or schizophrenic features. Elliot Gershon and colleagues at the National Institutes of Mental Health[123] (NIMH) studied 823 relatives of patients with manic-depressive and schizoaffective disorders to determine if the relatives showed the same increase in frequencies and age of onset of depression as had been observed for relatives of subjects with depression. They found that the frequency of manic-depressive, schizoaffective, or major depressive disorder was lowest in the relatives born before 1910, intermediate in the relatives born between 1911 and 1939, and highest in the relatives born after 1940. Unlike the above studies, the mood disorders were not significantly greater in females. The results for both sexes are shown in Figure 3.

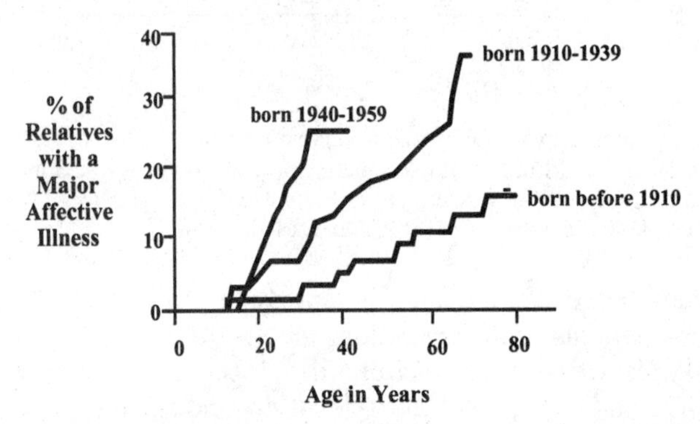

Figure 3. The lifetime frequency of major affective illness (manic-depressive, schizoaffective, or major depression) in relatives of subjects with manic-depressive or schizoaffective disorder, by age of onset in different birth cohorts. Redrawn from Gershon et al., *Arch. Gen. Psychiatry* 44:314-319. Copyright 1987, Amer. Med. Assn.

These results are significant because they show an increase in frequency and earlier age of onset of two additional entities – manic-depressive and schizoaffective disorder. Thus, the results were not restricted to simple depression. Some of the explanations for the possible causes of the increased frequency of depression, such as increased urbanization, the break-up of the family unit, or alterations in gender-related occupational roles, would be expected to potentially cause depression, but would seem less likely to cause mania or schizoaffective disorder. The greater the number of different disorders involved in the cohort studies, the more difficult it becomes to explain away the increases by a single environmental cause. In attempting to understand this phenomena, Gershon and colleagues stated:

> "Any hypothesis on the cause of the cohort differences must account for the increase in both manic and depressive disorders, the finding that sex is unrelated to the cohort differences, and the finding that the risk remains elevated in families of patients with affective illness. There is no obvious social or historical event or trend that offers such a hypothesis. No genetic change could have occurred so rapidly, but the observed phenomena could reflect a genetic-environmental interaction, such that an inherited vulnerability is more likely to be expressed in recently born cohorts. The cohorts with elevated rates of affective illness seem to have been born during wartime decades. Use of alcohol, tobacco, and illicit drugs, as well as nutritional practices and exposure to infectious diseases, have all changed during the past decades. However, there is no evidence that any of these events are related to the cohort effect, which is perhaps the most dramatic finding on risk differences yet reported in the epidemiology of affective illness. But the observation that rates of affective disorder can increase rapidly over a few decades suggests that once the cause of this increase becomes clear, an opportunity for powerful intervention in these disorders may present itself."[123]

They went on to conclude that:

> "When these data are combined with other reports, an ominous trend may be present, leading to an increase in prevalence of a broad spectrum of familial affective disorders in the coming decades."

Studies Worldwide

One possible explanation of the data was that it was unique to the United States, and we are just naturally crazier and more depressed than people in other countries. While several studies performed in other countries supported these findings,[157] differences in diagnostic criteria, time periods, and methods of analysis suggested that the results would be reliable only if the studies were done with a common plan, definitions, the same diagnostic criteria, and the same computer programs as used in the studies in the United States. This led to the formation of a Cross-National Collaborative Group of psychiatrists from the following countries: United States (five cities), Canada (Edmonton), Puerto Rico, Germany (Munich), Italy (Florence), France (Paris), Lebanon (Beirut), New Zealand (Christchurch), and Taiwan.[84] One of the U.S. cities (Los Angeles) included a non-Hispanic and a Mexican-American sample.

The U.S. results for the five sites are similar to those shown in Figure 2. The Los Angeles/non-Hispanic site showed an increase in rates of depression with the 1935-1944 cohort. The Edmonton results were very similar, except that the rate of depression showed the most dramatic increase in the 1925-1934 cohort, ten years earlier than the Los Angeles sample. The results were negative in the Puerto Rican, Los Angeles/Mexican-American, and Taiwan samples, where the rates of depression among the older birth cohorts was approximately equal to or less than the rates for the younger cohorts. This is discussed in more detail under exceptions (see below).

Except for these three groups, the results with all of the other sites were similar, with increases in the rate of depression being observed in the cohorts born between 1924 and 1944. The results for Beirut were unique in that they also demonstrated period effects – that is, the effect of environmental factors operating during a specific time period to cause depression; thus, there was a dramatic increase in the rate of depression during the two periods of war and chaotic political change in the area. No other site showed such dramatic time-related effects.

The Cross-National Study also examined the frequency of depression in the relatives of depressed patients in three different cities sites – the NIMH collaborative study, the New Haven, Connecticut, and Mainz, Germany studies. These results were similar to those shown in Figure 1. These studies suggested that "the increase in rates in the family studies is gradual and less likely to be due to the occurrence of specific [environmental] events."

The authors of the Cross-National Study concluded that, "It is clear that major depression occurs across a broad range of cultures and that *more recent generations are at increased risk.*"[84]

Exceptions

As mentioned above, there were several studies in less technologically complex populations that showed trends opposite to those in the more complex countries. These are very important exceptions, since certain theories could not account for these results.

The studies in Puerto Ricans[32,84] showed a trend that was in the opposite direction to that seen in the studies of non-third world countries, in that there was a progressive increase in the prevalence of a wide range of psychiatric disorders with increased age of cohorts of 18-24, 25-44, and 45-64 years of age, both in lifetime and six-month rates (see Figure 4, next page).

A study of Koreans also showed no increase in psychiatric disorders in younger cohorts.[165] Karno and colleagues[148] compared the prevalence rates of psychiatric disorders in Mexican-Americans, most of whom were immigrants, to non-Hispanic Whites who had spent their lives in the U.S. Both groups lived in Los Angeles. Figure 5 shows the results in 18- to 39-year-old versus 40+-year-old cohorts. The figures are the percentage increase in rates for the younger versus the older cohorts, comparing the Mexican-American versus non-Hispanic Whites.

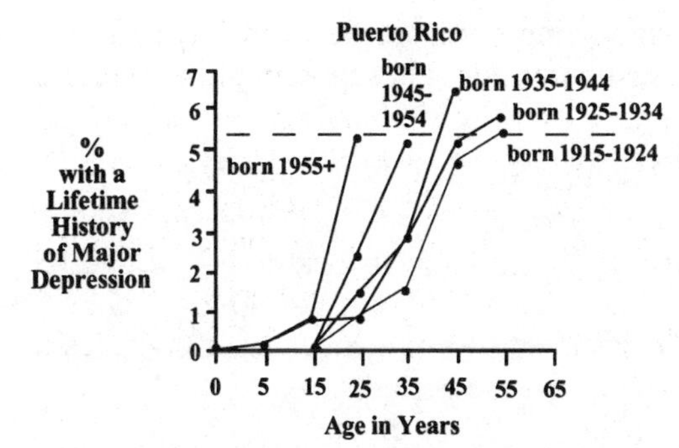

Figure 4. The lifetime prevalence of major depression in Puerto Rico from the Cross-National Collaborative Group study showing both sexes. Redrawn from the *Journal of the American Medical Assn* 266:3098-3105. Copyright 1992, Amer. Med. Assn.

Note the high rates of depression even in the older cohorts (above the dotted line).

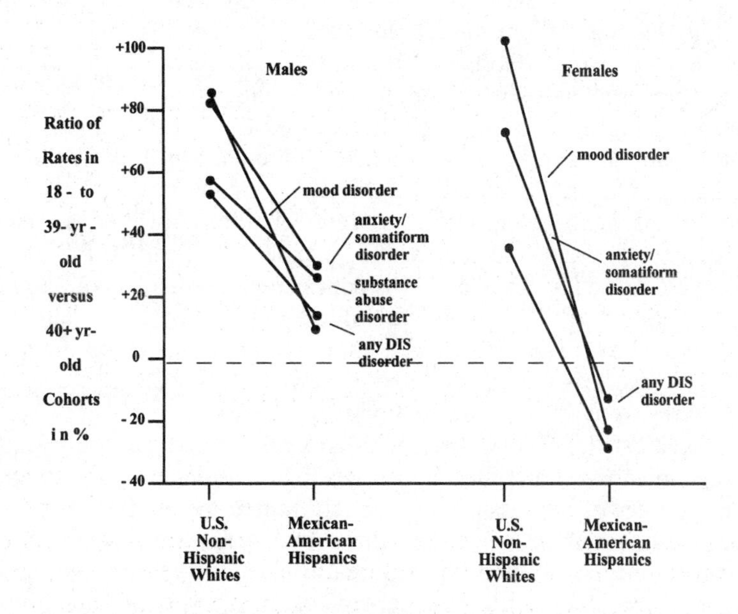

Figure 5. Percentage of the rates in the 18- to 39-year-old cohorts versus the 40+-year-old cohorts for male and female U.S. Whites compared to Mexican-American immigrants, both living in Los Angeles.

For example, for White (non-Hispanic) Los Angeles females, the frequency of a mood disorder in the 18- to 39-year-old cohort was 103% greater than for the 40+-year-old cohort. By comparison, for Mexican-American Los Angeles females, the frequency of a mood disorder in 18- to 39-year-old cohort was 25% less than in the 40+-year-old cohort. The results, especially for females, illustrated the dramatic differences in cohort rates for several psychiatric disorders between the Whites living in west L.A. versus the Mexican-American immigrants living in east L.A.

These exceptions are important to the concept that the selection of genes for addictive-disruptive disorders is driven by two aspects of technologically advanced societies – effective birth control and a need for, and wide access to, higher education. Those countries where these factors were less prominent, at least at the time the younger cohorts were being born, would be anticipated to be exceptions to the otherwise worldwide secular trends. This appears to be the case.

Are the Results Simply Due to Poor Memory in the Older Subjects?

In an article entitled "Reevaluation of Secular Trends in Depression Rates," Simon and VonKorff at the University of Washington in Seattle,[243] suggested that the apparent increase in lifetime rates of depression in the younger cohorts was simply due to the failure of the older subjects to remember prior episodes of depression. To support this, they examined the time period since the first onset of major depression and claimed this period tended to cluster around ten years regardless of the age of the subjects. However, their figures showed that the percentage of time this interval clustered at ten years dropped from over 70% for the younger 18- to 34-year-old cohorts to 30% for the older-than-55-year-old cohorts, and that much longer intervals were common in the older subjects.

They also examined the secular trends for several other diagnoses, including alcohol abuse, phobia, panic attacks, and schizophrenia. Since all of these disorders showed a trend for increased rates in the younger cohorts, they suggested this indicated that none of the cohort results were reliable. However, as discussed later, these are exactly the results that would be expected if many different psychiatric disorders are caused by similar sets of genes and thus share a common cause.

In a subsequent report, Simon and coworkers reported on a new international study that also showed a worldwide increase in depression in younger cohorts.[243a] However, they presented and reviewed evidence that much of these apparent increases were due to poor memory.

Evidence against the theory that the cohort results were entirely due to an artifact of poor memory is provided by studies that included only the longest and most severe episodes of depression, and which limited the studies to individuals 50 years of age or younger.[164] When these two precautions were taken, the results were the same. In addition, the failure of the studies in third world populations to show a decrease in lifetime depression in older cohorts suggests the study methods were reliable. It seems unlikely that older Mexican immigrants, Puerto Ricans, and Koreans would have much better memories than older subjects in the rest of the world.

One especially relevant study showing that the cohort studies could not be explained away by poor memory was reported by Coryell and colleagues from the University of Iowa.[83] They were able to eliminate recall artifacts by performing a prospective study – that is, by examining the same group of subjects at two different times. They examined 965 subjects who had never had an affective illness and then restudied them six years later. By that time, 11.8% had developed at least one episode of of major depression. Subjects younger than 40 years of age were three times more likely than older subjects to develop depression. Of additional interest, individuals with pre-existing disorders such as alcoholism, drug abuse, or phobic or panic disorder developed major depression at much higher rates, ranging from 27.9 to 42.9%, than those without such pre-existing problems. This is relevant to the concept, presented later, that these disorders are part of a genetic spectrum of interrelated disorders.

Summary – Studies of major depression in the relatives of patients with depression, and in the general population, suggest that the frequency of a lifetime history of depression is significantly higher, and the age of onset of depression is significantly earlier, in younger people. This is a worldwide trend with the only exceptions occurring in countries that, at least until recently, were less technologically advanced. While some of these results may be due to trivial explanations, such as poorer recall in older subjects, they do not explain away all of the findings.

Table 1. Lifetime Prevalence Rates for Psychiatric Disorder for U.S. Males and Females

Epidemiologic Catchment Area (ECA): Rates for DSM-III Disorders, from Robins et al., *Archives of General Psychiatry* 41:949-958. Copyright **1984**[228] Amer. Med. Assn. (N = 9,543)

National Comorbidity Survey (NCS): Rates for DSM-III-R Disorders, from Kessler et al., *Archives of General Psychiatry* 51:8-19. Copyright **1994**[149] Amer. Med. Assn. (N = 8,098)

Disorder	ECA:1984 %	NCS:1994 %
Any disorder covered	32.8	48.0
Any disorder except phobias	24.9	
Any disorder except substance abuse	22.8	
Substance use disorders	16.7	26.6
Alcohol abuse/dependence	13.6	23.5
Drug abuse/dependence	5.6	11.9
Schizophrenia/schizophreniform	1.7	
Schizophrenia	1.5	
Schizophreniform	0.2	
Affective Disorders	7.8	19.3
Manic episode	0.9	1.6
Major depressive episode	5.2	17.1
Dysthymia	3.0	6.4
Anxiety/somatiform disorders	16.0	24.9
Phobia	14.0	24.3
Panic	1.4	3.5
Obsessive-compulsive	2.5	
Somatization	0.1	
Eating disorders – anorexia	0.1	
Antisocial Personality Disorder	2.7	3.5
Cognitive impairment (severe)	1.2	

Chapter 2

Increase in Suicide

In 1980 Solomon and Hellon reported that suicide rates in Alberta, Canada were increasing, especially in young people.[247] Figure 1 shows their results for subjects 15 to 19 and 20 to 24 years of age.

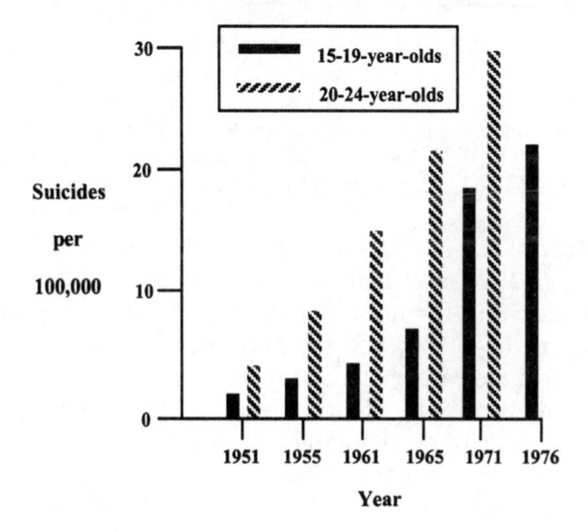

Figure 1. Progressive increase in the suicide rate for Canadian adolescents and young adults from 1951 to 1976. Data from Solomon and Hellon, *Archives of General Psychiatry* 37:511-513. Copyright 1980, Amer. Med. Assn.

For example, the suicide rate for young adults in the 20 to 24 age range increased from 2 per 100,000 in 1951 to 30 per 100,000 in 1976. There was no comparable increase for older subjects.

In a companion paper in the same issue, Murphy and Wetzel[197] examined similar data from a much larger database in the United States. Except for lower overall rates of suicide, the results were virtually identical to those of the Canadian study (Figure 2, next page).

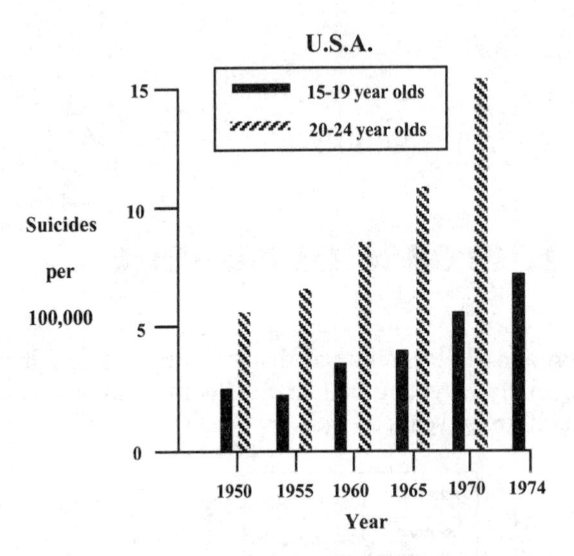

Figure 2. Progressive increase in the suicide rate for U.S. adolescents and young adults from 1949 to 1974. Data from Murphy and Wetzel, *Archives of General Psychiatry* 37:519-523. Copyright 1980, Amer. Med. Assn.

For the 20- to 24-year-olds, there was an increase in suicide rate from 5.9 per 100,000 in 1950 to 15.1 in 1974, and again there was no comparable increase or even a decrease in suicide rate for the older cohorts. Studies in Australia,[126] Zealand[245] and in England[26,252] showed the same trend. A more recent 1990 study in Canada[178] verified the earlier one and indicated that the greatest change in suicide rates at that time was in the 15- to 19-year-old age group. Suicide was the second most common cause of death in this age group.

While Figure 2 covers the years from 1950 to 1974, Figure 3 (next page) illustrates increased rates of suicide from 1965 to 1991 and compares Whites to Blacks.[237] The suicide rate in Whites is greater than that in Blacks. Between 1973 and 1986 the rate in Blacks was relatively stable, while the rate in Whites continued to increase; however, since 1986 the rate in Black youths increased more rapidly than in Whites.

The Causes of Suicide

Psychiatric problems are the major cause of suicide in the young. In a study in San Diego,[117] 53% of 133 consecutive young suicides had a principal psychiatric diagnosis of substance abuse. Twenty-four percent had an additional diagnosis of atypical depression, atypical psychosis, or adjustment disorder with depression. Typically, the substance abuse was a chronic condition present for nine years. In a study of young suicide attempts in Zurich,[10] there was an increase in a history of a disturbed childhood and sexual abuse, and subjects scored higher on scores of neuroticism and aggression. In later years they had increased rates of depression, anxiety, substance abuse ,and sociopathy.

A study of suicide in young men in Scandinavia[7,8] was based on a longitudinal study of 50,465 conscripts into the Swedish Army. The suicide rate was elevenfold greater for those who had ever required inpatient psychiatric treatment, and thirteenfold higher for those with a diagnosis of schizophrenia. Other risk indicators were substance abuse, personality disorder, a history of deviant behavior, and a poor social background.

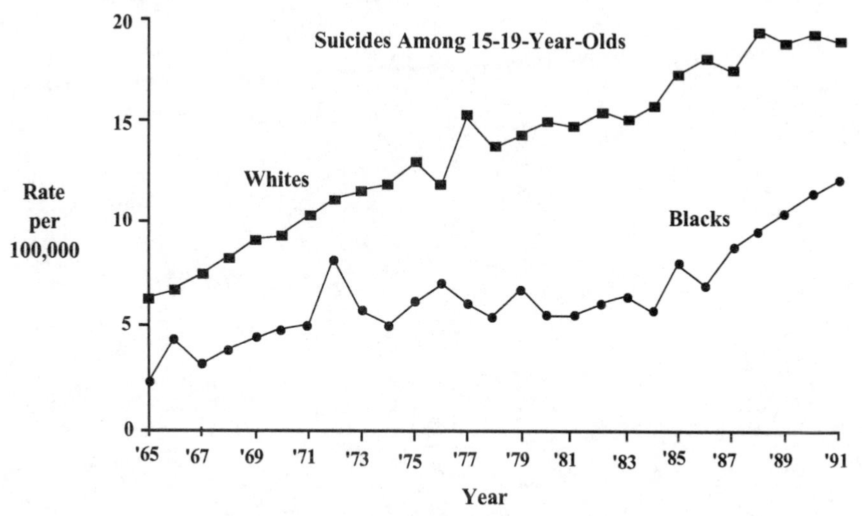

Figure 3. Increased suicide rates in White and Black youths and adolescents from 1965 to 1991. Redrawn from Shaffer et al., *American Journal of Psychiatry* 151:1810-1812, 1994. Copyright 1994, American Psychiatric Association, by permission.

Summary – Studies in many countries have shown a progressive increase in the suicide rate in young people since 1945. These are highly correlated with the presence of psychiatric disorders, especially substance abuse, deviant behaviors, sociopathy, depression, and schizophrenia. These results are consistent with those in the previous chapter showing an increase in frequency and a decrease in age of onset of depression, and with subsequent chapters showing a similar increase for other psychiatric disorders.

Chapter 3

Increase in Alcohol and Drug Abuse

Alcoholism

The frequency of alcoholism in the U.S., based on the 1980 ECA studies, was 13.6% for all ages combined (see Table 1, Chapter 1), and 21.8% for males 25 to 44 years of age.[46] In a study of the frequency of alcoholism in *relatives of alcoholics*, Dr. Robert Cloninger and colleagues at the Washington University School of Medicine in St. Louis, reported that the risk of developing alcoholism by age 25 progressively increased from 26% in men born before 1924 to 67% for those born after 1954.[46] The results for all the cohorts is shown in Figure 1.

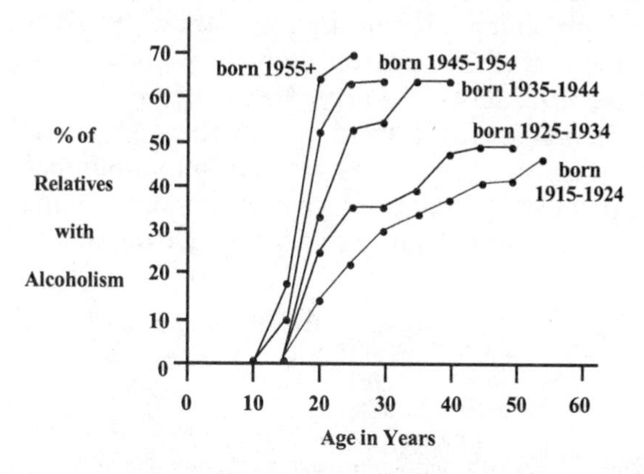

Figure 1. Increasing frequency of alcoholism in the younger relatives of alcoholics. Redrawn from Cloninger et al., *Alcoholism: Origins and Outcome*, R.M. Rose and J. Barrett Eds, by permission, Raven Press, New York, 1988.

They stated, "The high risks in the younger birth cohorts are especially frightening when it is realized that they have yet to pass through the full period of risk."

In a follow-up study, these workers confirmed that younger cohorts had an increased lifetime prevalence of alcoholism and a decreased age of onset.[219] As with the cohort studies on depression, the frequencies are lower when the studies are done on the general population rather than on relatives of affected individuals, but the trends are the same. This is shown by the analyses of the ECA data by Simon and VonKorff[243] of psychiatric disorders other than depression. Their results for alcohol abuse and dependence are shown in Figure 2.

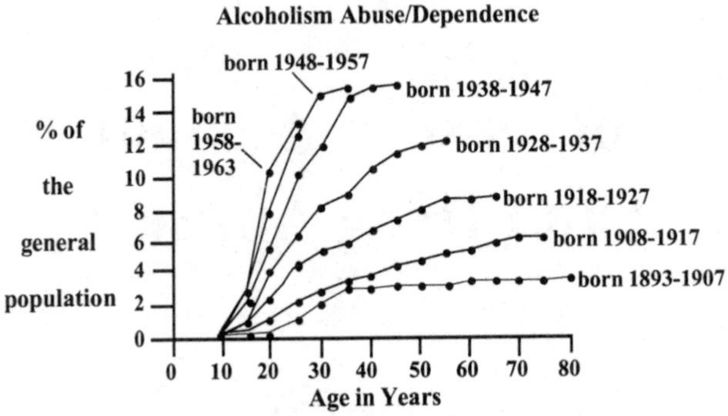

Figure 2. Percentage of the general population with alcoholism based on age cohorts showing a increase in frequency and earlier age of onset of alcoholism in younger cohorts. Redrawn from Simon and VonKorff, by permission, *American Journal of Epidemiology* 135:1411-1422, 1992.

Instead of reporting the results in terms of the frequency of alcoholism by birth cohort, Kimberly Burke and colleagues[30] from the National Institute of Mental Health examined the issue of the apparent increase in alcoholism in younger subjects by combining a hazard rate with the cohort approach. The hazard rate represents *the probability that an individual who enters a time interval, and who is free of the disorder at the beginning of the interval, will develop the disorder during that time interval*. Hazard rate numbers are much lower than lifetime frequencies because the frequencies are split into a number of shorter time periods. Viewed in this fashion, the results are even more striking. Figure 3 illustrates such an analysis for alcohol abuse or dependence based on the ECA data.

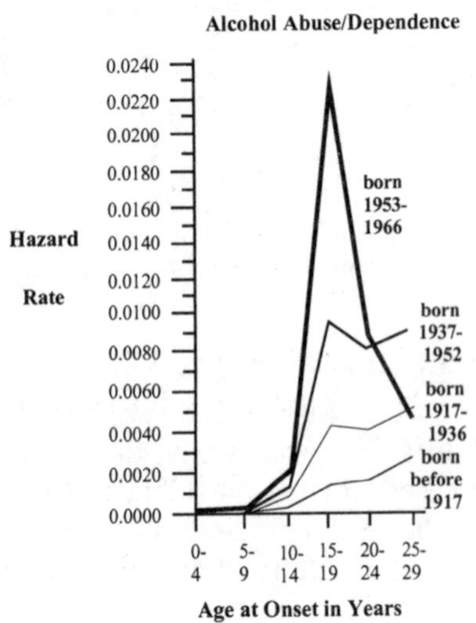

Figure 3. Hazard rates of alcohol abuse or dependence for six time intervals from birth to 29 years of age, for different age cohorts, showing a dramatic increase for the 15- to 19-year-olds of the most recently born cohort. Redrawn from Burke et al., *Archives of General Psychiatry* 48:789-795, 1991.

There was a marked increase in the rate of onset of alcohol abuse and dependence at age 15 to 19 years of age, for the cohort born between 1953 and 1966.

Fetal Alcohol Syndrome

Fetal alcohol syndrome is one of the most devastating aspects of alcoholism, because instead of affecting the female alcoholic herself, it affects her innocent children, producing lifelong physical and mental disabilities. The physical aspects include abnormal facial features, widening of the base of the nose, narrowing of the upper lip, and other features. The mental disability includes mental retardation. The Centers for Disease Control reported an increase in the frequency of reported cases of fetal alcohol syndrome between 1979 and 1992. The report was based on data from hospital discharge diagnoses of "Noxious influences affecting the fetus via placenta or breast milk, specifically alcohol, including fetal alcohol syndrome." The reported rate was 6.7 per 10,000 in 1993, versus 1.0 per 10,000 in 1979, a greater than sixfold increase. Since virtually all women are aware of the dangers of consuming alcohol when they are pregnant, fetal alcohol syndrome represents an addiction so severe that, despite this knowledge, they persist in their drinking.

Drug Abuse

An intriguing study that serves as a good transition into the epidemiology of drug abuse comes from the Washington University study discussed above. Here, instead of examining the lifetime frequency of alcoholism in the *relatives of alcoholics*, the lifetime frequency of drug abuse was studied instead. This showed a similarly dramatic increase.[93] For example, for males age 41 or greater, the frequency of probable or definite drug dependence was 1.8%. This increased to 26.7% for those 31 to 40 years of age, and to 50.0% for those 17 to 30 years of age. For female alcoholic relatives, the comparable figures were 13.0%, 42.9%, and 57.9%. This compares to a lifetime frequency of drug dependence of 5.6% for all ages, based on the ECA studies (see Table 1, Chapter 1). The dramatically higher rates of drug dependence in relatives of alcoholics indicates that both alcoholism and drug abuse have strong genetic factors, that the genetic factors for drug abuse are similar to those for alcoholism, and that, as with the cohort studies on depression and alcoholism, there is both an increase in the frequency of drug abuse and a trend for an earlier age of onset.

The studies by Burke and colleagues[30] also showed that the hazard rates for drug abuse and dependence were virtually identical to those for alcoholism. These are shown in Figure 4 (next page).

In a sense, these results are even more dramatic for drug abuse because of the virtual absence of significant drug abuse in the older cohorts. While it is clear that the increased availability of drugs plays a role in these figures, this is not true of alcohol, and yet both hazard rates are very similar.

Other Epidemiologic Studies

When it comes to drug abuse, the statistics that most of us are accustomed to are diagrams showing a dramatic increase in drug abuse and its consequences in recent years. Two such graphs are shown in Figures 5 and 6 (next page).

The first graph compares the number of subjects admitted for treatment of methamphetamine abuse in 1982-83 versus 1987-88, by different methods of administration.

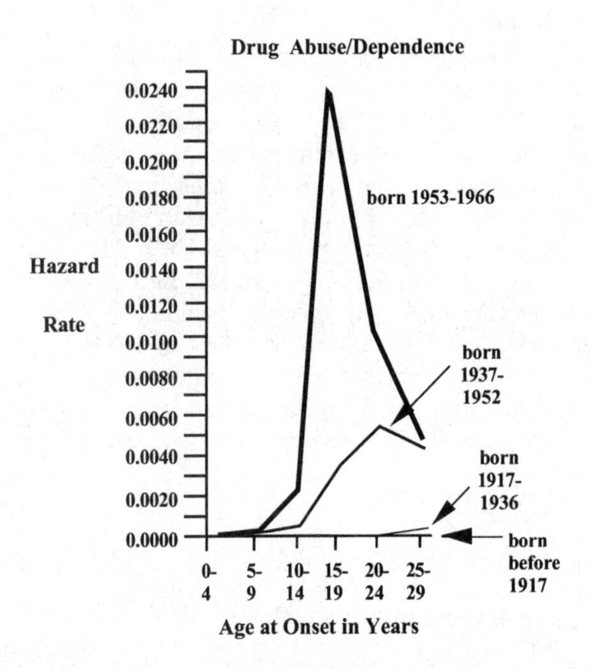

Figure 4. Hazard rates for drug abuse or dependence for four time intervals from birth to 29 years of age, for different age cohorts, showing a dramatic increase for the 15- to 19-year-olds in the most recently born cohort. Redrawn from Burke et al., *Archives of General Psychiatry* 48:789-795, 1991.

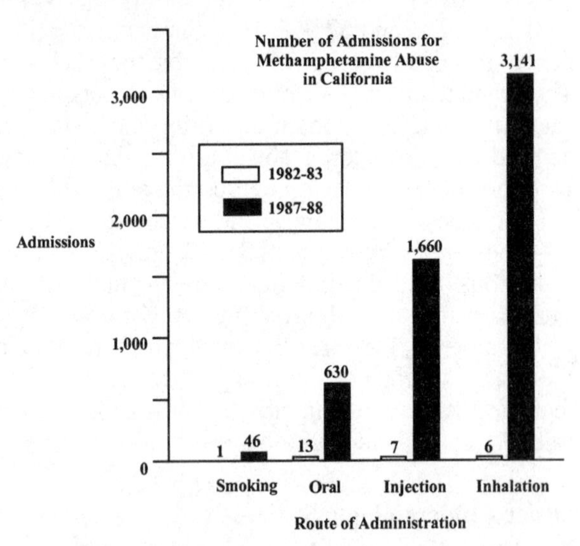

Figure 5. Number of admissions for treatment of methamphetamine abuse, by method of administration, for 1982-83 versus 1987-88. From Helsch-ober and Miller, *Methamphetamine Abuse in California*, in Miller and Kozel; *Methamphetamine Abuse: Epidemiologic Issues and Implications*, Research Monograph 115, 1991, Nat. Inst. Drug Abuse, Rockville, MD.

While this type of dramatic increase is predominantly due to increased availability of drugs and psychosocial factors, biological factors may play a role.

The second graph illustrates the number of deaths from drug overdose in Los Angeles County from 1984 to 1992.

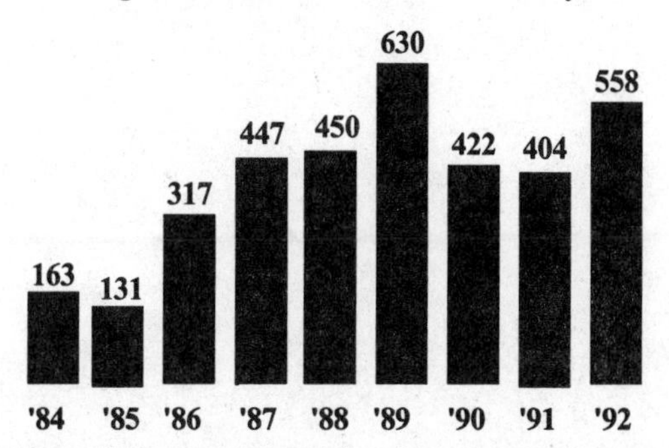

Figure 6. Number of deaths from drug overdose (cocaine and crack) in Los Angeles County from 1984 to 1992. From U.S. Dept. of Health and Human Services, California Dept. of Justice, UCLA Drug Abuse Research Center and *L.A. Times* 12/18/94.

While changing social values have resulted in a recent decrease in the amount of alcohol consumption in the U.S., the frequency and early age of onset of alcoholism has not changed.[218a]

Summary – Studies of alcoholism and drug abuse show that the lifetime prevalence of both has dramatically increased in recent years in the younger age groups.

Chapter 4

Increase in Anxiety

The anxiety disorders include panic attacks, simple phobias (fear of animals, snakes, insects, blood, heights, closed places, air travel), social phobias (fear of public speaking, being in a crowd, eating in public, meeting new people), agoraphobia (fear of leaving the house), general anxiety disorder, and obsessive-compulsive disorder. Years ago, in a more psychoanalytic era, these were termed *the neuroses*. This terminology has been replaced by the above more accurate and less pejorative classification.

As with other disorders, using the ECA data, Simon and VonKorff[243] performed a cohort analysis of phobic disorder and panic disorder. These are shown in Figures 1 and 2.

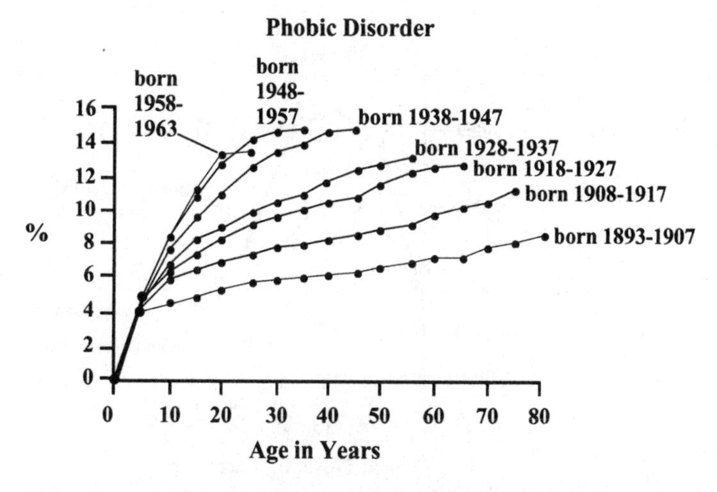

Figure 1. A cohort analysis of phobic disorder, based on the ECA data, showing an increase in frequency and earlier age of onset in younger cohorts. Redrawn from Simon and VonKorff, *American Journal of Epidemiology* 135:1411-1422, 1992.

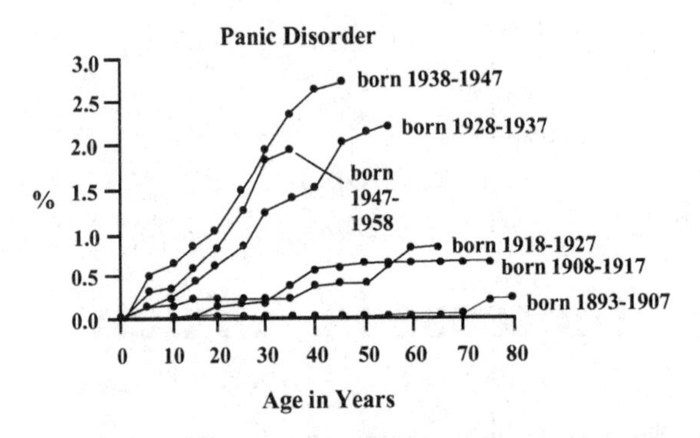

Figure 2. A cohort analysis of panic disorder, based on the ECA data, showing an increase in frequency and earlier age of onset in younger cohorts. Redrawn from Simon and VonKorff, by permission, *American Journal of Epidemiology* 135:1411-1422, 1992.

For both phobic and panic disorder there was a trend toward an increase in frequency and an earlier age of onset of both disorders. For panic disorder there was a very slight reversal of the trend for the latest cohort born in 1947 to 1958.

Burke and colleagues,[30] using the ECA data, also examined several of the anxiety disorders using the hazard rate technique. The results for agoraphobia, simple or social phobias are shown in Figure 3.

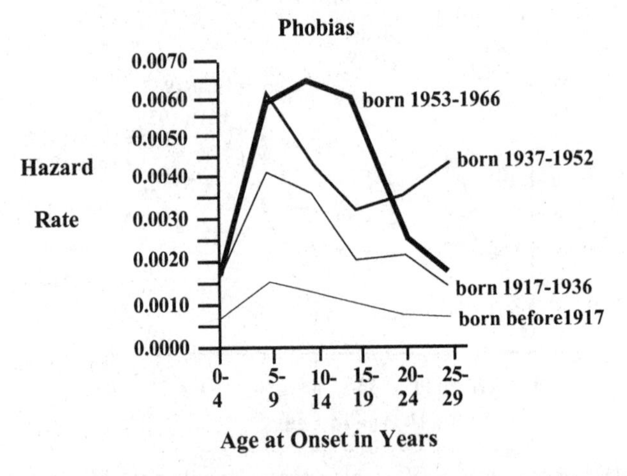

Figure 3. Hazard rate analysis of the ECA data for agoraphobia, simple and social phobias showing an increase in rate for the younger cohorts. Redrawn from Burke et al., *Archives of General Psychiatry* 48:789-795, 1991.

The hazard rate shows a progressive increase with younger cohorts over multiple time intervals for an onset of symptoms involving subjects 5 to 29 years of age. For example, at the age of 5 to 9 years, the hazard rate increased from .0015 for the 1917 cohort to .0060 for the 1937-1952 and 1953-1966 birth age cohorts, a fourfold increase. The results for panic disorder are shown in Figure 4.

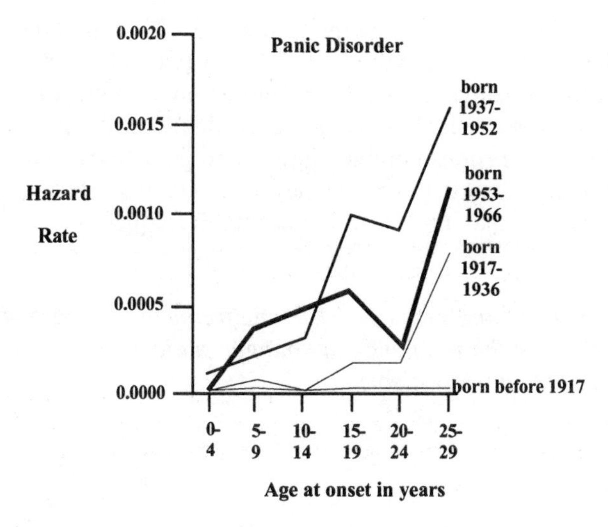

Figure 4. Hazard rate analysis of the ECA data on panic disorder showing a increase in rate and earlier age of onset for the younger cohorts. Redrawn from Burke et al., *Archives of General Psychiatry* 48:789-795, 1991.

While the curves for the 1953-1966 and the 1937-1952 cohorts tend to intersect, both show a dramatic increase in frequency and earlier age of onset for panic disorder than for the comparable ages in the 1917-1936 cohorts. The results for obsessive-compulsive disorder are shown in Figure 5.

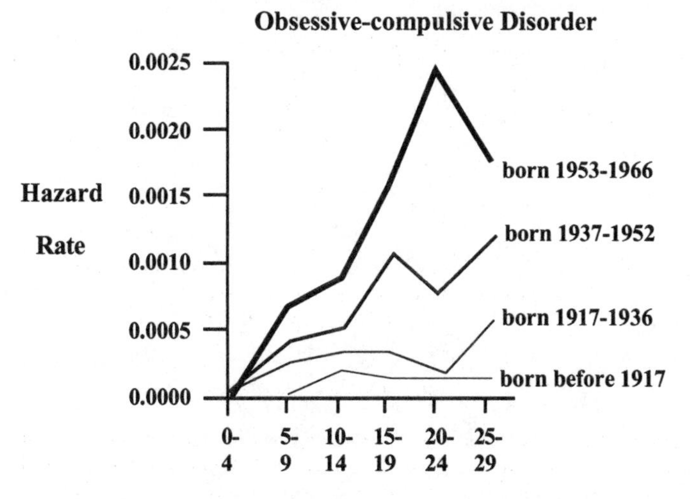

Figure 5. Hazard rate analysis of the ECA data on obsessive-compulsive disorder showing an increase in rate and earlier age of onset for the younger cohorts. Redrawn from Burke et al., *Archives of General Psychiatry* 48:789-795, 1991.

Here the most dramatic differences were seen for the onset of symptoms at age 20 to 24. The individuals in the youngest cohort, born in the years 1953 to 1966, would have been 20 to 29 years of age at the time of the study. Their hazard rate for the 20 to 24 years of age interval was .0025, compared to the individuals born in the 1917 to 1936 cohort of .0002, a twelvefold increase. In a later chapter, I will discuss relevant aspects of Tourette syndrome, a hereditary behavioral disorder. While no reliable studies are available on the possible increase in frequency of TS, there is considerable clinical and genetic overlap between it and obsessive-compulsive disorder. These results are one of the best confirmations of the clinical impression that TS is also increasing in frequency.

Summary – Cohort analyses using lifetime frequency or hazard rates suggest there is an increase in the frequency in younger subjects and often a decrease in the age of onset of the anxiety disorders.

Chapter 5

Increase in Attention Deficit Hyperactivity Disorder and Conduct Disorder

ADHD is a hereditary childhood disorder that affects 4-8% of boys and 3-5% of girls. It is characterized by the presence of problems with inattention, impulsivity, and hyperactivity. It is, at least in part, due to a genetic defect in genes that control dopamine, a neurotransmitter in the brain that plays a role in muscle (motor) activity, attention, motivation, and emotion. It is most often treated with medications that enhance dopamine activity, such as Ritalin or Dexedrine. In about half of the cases the symptoms persist into adulthood.

Approximately one-third to one-half of children with ADHD also have problems with conduct disorder or oppositional defiant disorder, and children with ADHD are at significantly greater risk of developing problems with alcoholism, drug abuse, criminal activity, and antisocial behavior as adults than are children without ADHD.

Another hereditary childhood disorder virtually identical to ADHD is Tourette syndrome. In addition to the above, TS children have chronic motor tics (eye blinking, facial grimacing, shoulder shrugging, head jerking, and others) and vocal tics (throat clearing, sniffing, snorting, barking, and others). Many more details about the symptoms, cause, and treatment of ADHD, TS, and conduct disorder are presented elsewhere.[52] The purpose of this chapter is to examine the question of whether these disorders are increasing in frequency.

Since none of these disorders were included in the DIS or inquired about in the epidemiological studies discussed in previous chapters, it is necessary to turn to other evidence to determine if they are increasing in frequency. Most clinicians who deal with these disorders have an intuitive feeling that they are becoming more common, but intuitive feelings are not quantifiable facts.

One indirect approach is to ask whether more children are being treated for ADHD in recent years than previously. This data can be obtained from records on the amount of Ritalin sold each year. Figure 1 shows this information.

Figure 1. Production of Ritalin (methylphenidate) by year in the U.S. From the U.S. Drug Enforcement Agency.

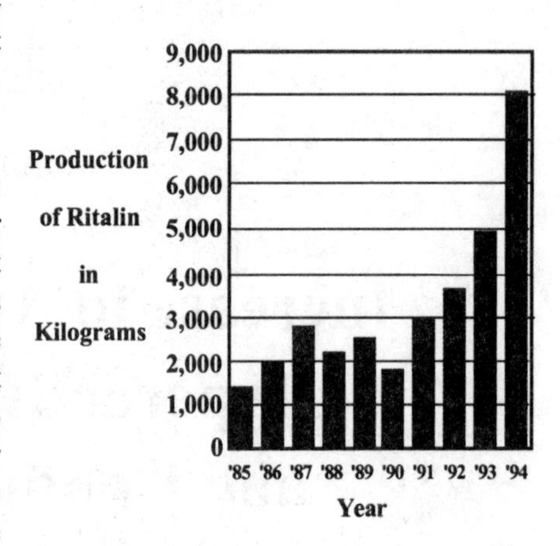

As can be seen, there was a dramatic increase in the amount of Ritalin used over the years from 1985 to 1994. This does not provide proof that the disorder itself is increasing in frequency, since the increased use could simply reflect increased awareness of the disorder due to a number of reports on TV and popular books on the subject.[128] This is likely to have contributed to the dramatic rise from 1990 to 1994. However, it is not unreasonable to conclude from these figures that the frequency of ADHD might have increased some from 1985 to 1994. As mentioned in the previous chapter, approximately 50% of children with TS also have obsessive-compulsive disorder, and there is evidence from the hazard rate analyses of Burke and coworkers[30] that the frequency of obsessive-compulsive disorder is increasing in frequency and the age of onset is decreasing. ADHD is closely associated with TS. It is present in 50-80% of TS cases that seek medical care.[52,61,62]

The Achenbach Study

The single most reliable method of determining if a disorder is increasing in frequency is to administer the identical test instrument to a large number of children in the general population at two different periods of time that are sufficiently far apart to detect a trend, if it exists. One of the most widely used standardized tests of childhood behavior was developed by Thomas Achenbach from the Department of Psychiatry at the University of Vermont. This is called the Child Behavioral Checklist (CBCL)[5] and it is completed by parents. There is a companion instrument for teachers called the Teacher's Report Form (TRF).[4] An extensive sample of the general population was given the tests in 1976 in the District of Columbia, Maryland, and Virginia. A second study was done in 1981 to 1982 in Nebraska, Tennessee, and Pennsylvania. A third nationwide sampling was done in 1989.[6] To make the earlier two samples geographically compatible, the subjects in the 1989 sample were drawn from the same general regions as in the 1976 and 1981-1982 studies.

One hundred and twelve specific behaviors were compared in the different samples. The scores on forty-two of these were higher in the 1989 survey than in the 1976 survey (Figure 2).

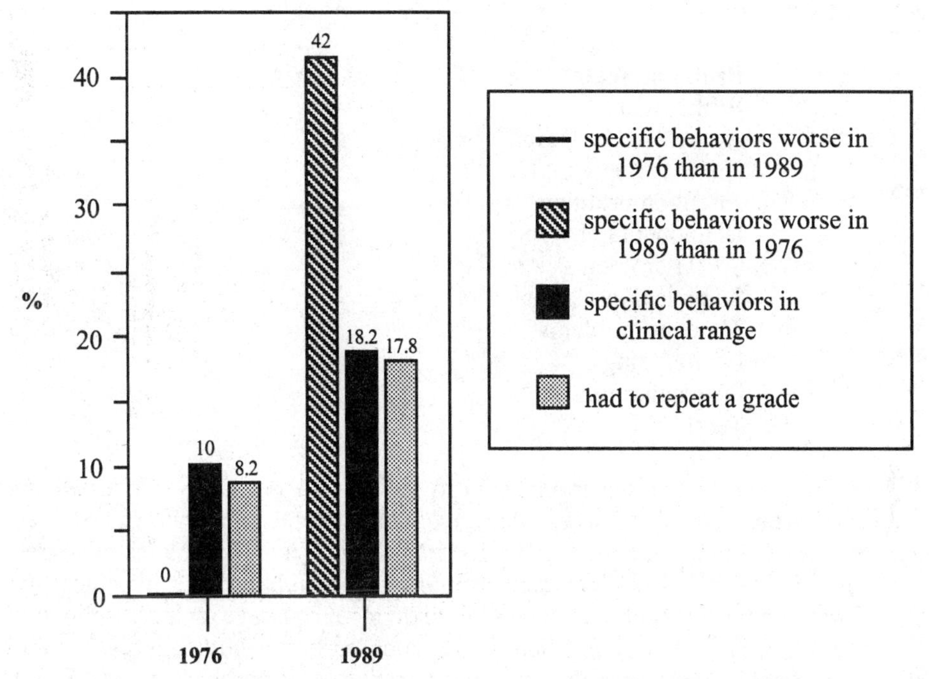

Figure 2. Comparison of results of Achenbach Child Behavioral Checklist in 1976 versus 1989. Data from Achenbach and Howell, *J. Am. Acad. Child. Adolesc. Psychiatry* 32:1145, 1993.

By dramatic contrast, *none* of the 112 scores were higher in the 1976 survey. The largest significant difference was the variable *stubborn, sullen, or irritable.* The items showing the greatest percent increase were *hangs around with others who get into trouble* (+120%), *underactive* (+110%), *whining* (+109%), *destroys things belonging to others* (+109%), and *poor in schoolwork* (+101%). The items were also examined for moving up in rank or moving down in rank in 1989. Of the four problems that moved down more than fifteen rank points, none was associated with any syndrome. By comparison, four of the five problems that increased in rank were associated with syndromes of aggression, inattention, and withdrawal. The variable of *ever had to repeat a grade* more than doubled, increasing from 8.2% in 1976 to 17.8% in 1989.

In scoring of the CBCL, the 112 variables were collapsed into a series of behavioral problem scores. All of these scores were significantly higher in 1989 than in 1976. These are shown in Table 1 (next page), ranked by percent of the total variability (variance) accounted for by each variable.

Table 1. Problem scores that were significantly worse in 1989 than in 1976, ranked by % of variance.

Problem Scale	% of Variance
Withdrawn	3
Aggressive behavior	2
Anxious/depressed	2
Attention problems	2
Delinquent behavior	2
Social problems	1
Somatic complaints	1
Thought problems	1
Internalizing	3
Externalizing	3
Total	4

The total score increased from 18.00 in 1976 to 24.44 in 1989. Internalizing refers to those variables and scores that involve symptoms of withdrawal, depression, poor motivation, and hypoactivity. Externalizing refers to symptoms of aggression, oppositional behaviors, conduct disorder, anger, fighting, and hyperactivity. These are not mutually exclusive. The same individual can have an increase in both internalizing and externalizing behaviors, such as being both aggressive and depressed. These scores describe significant increases in problems with ADHD, conduct disorder, and oppositional behavior, as well as anxiety and depression.

The teacher's report scores gave similar results, with seventeen scores being significantly higher in 1987 than in 1981-1982, and *none* higher at the earlier times. The total TRF score increased from 19.39 in 1981 to 22.42 in 1989. The largest difference was in the variable *apathetic or unmotivated.* The two items that increased in rank by more than fifteen items were *unhappy, sad, or depressed* and *dislikes school.*

When certain scores were high enough, they were considered to be diagnostic of specific entities. In 1989, 18.2% of the subjects had scores that were in the range for a clinical diagnosis, versus 10.0% of subjects in 1976 – almost double. The increase in attentional, delinquent, aggressive, and withdrawn behavior problems was particularly significant.

While the authors offered no explanation for these increases, they were able to determine that whatever it was, the groups did not differ by age, gender, ethnicity, or socioeconomic status.

Summary: The best way to determine if certain behaviors are increasing in frequency is to test the general population with the same instrument at two different time intervals. This was done using the well-standardized Achenbach Child Behavioral Checklist. Of 112 specific behaviors, forty-two were worse in 1989 than in 1976, while none were worse in 1976. In 1989 18.2% of the scores indicated clinically significant problems, compared to 10% in 1976. The increases specifically involved problems with poor attention, poor school performance, aggression, and depression.

Chapter 6

Increase in Autism

With the exception of the studies Achenbach described in the previous chapter, it is difficult to obtain reliable data on whether there has been an increase in the disruptive childhood behavioral disorders such as ADHD, TS, and conduct and oppositional defiant disorders. However, there are two indirect lines of evidence that this apparent increase is real.

Autism

The first comes from studies by Gillberg and colleagues in Sweden[124] indicating a significant increase in the frequency of autism over the past ten years. They entitled their report, "Is autism more common now than ten years ago?" The following figure shows the progressive increase in autism over three time periods in the 1980s.

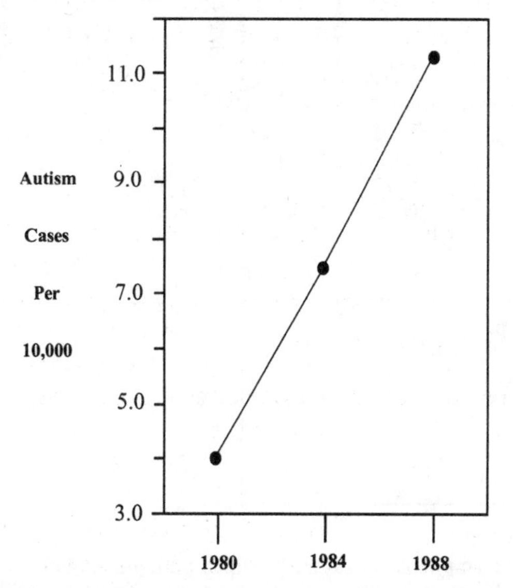

Figure 1. Progressive increase in the frequency of autism from 1980 to 1988, in Sweden. Data from Gillberg et al.[124]

This showed an almost threefold increase, from 4.0/10,000 in 1980 to 11.6/10,000 in 1988. The feature of this report that adds credibility to these figures is that the studies were done in the same location each time (western Sweden) and by the same investigators. Recent studies in England,[90] showing an autism prevalence of 9/10,000, are also consistent with this increase. The latter study also emphasized the progressive increase in autism and decrease in IQ associated with learning disorders. A study in Missouri[132a] showed a tenfold increase in the incidence of autism, from 50 cases diagnosed in 1988 to 720 cases in 1995.

Autism is a severe childhood disorder characterized by delayed onset of speech and other speech problems, severe social isolation, and repetitious ritualistic behaviors.[91] We[71] and others[28,253,254] have observed numerous young patients with autism who developed classic symptoms of TS when they grew older. In addition, it is not unusual for autism to be present in some of the relatives of TS patients. The following is a pedigree that illustrates this.

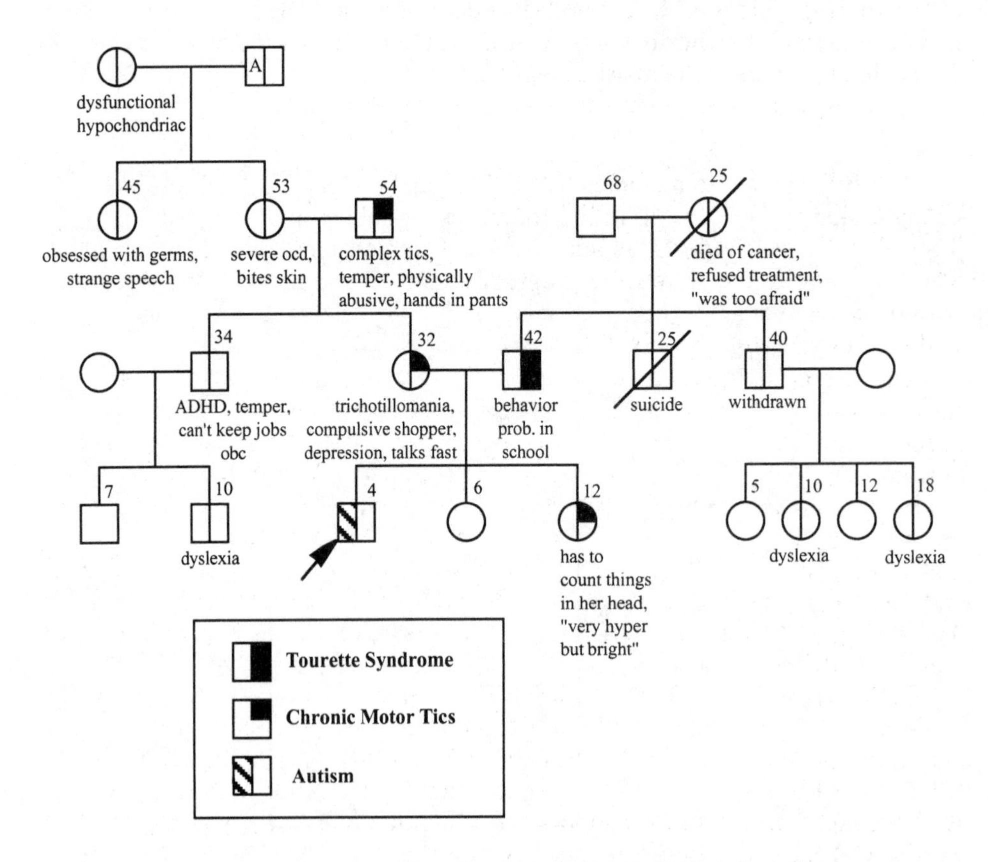

Figure 2. An example of a pedigree in which multiple individuals have Tourette syndrome or chronic motor tics and other behavioral disorders, including autism.

This pedigree shows that a 42-year-old father with TS and behavioral problems in school, and a 32-year-old mother with chronic motor tics and multiple obsessive-compulsive behaviors, including trichotillomania (hair pulling), had three children. The oldest had chronic motor tics and some obsessive-compulsive behaviors, and the youngest was autistic. These and many other pedigrees, including patients with both autism, pervasive developmental disorder (PDD, an autism-like disorder), and TS [29,71,253,254,256] suggest that these three disorders share genes in common. Molecular genetic studies (see Part II) tend to verify this.

Many cases of PDD are present with severe behavioral disorders, including aggression, psychosis, depression, inappropriate sexual behavior, firesetting, cruelty to animals, and a range of bizarre criminal behaviors.[256] In the introduction to a supplement on *Developmental Delay and Psychopathology in Young Children*, Kaminer and Cohen[142] commented on disorders similar to PDD and stated:

> "we are seeing many young children with different characteristics than those we saw ten years ago. Particularly, we are seeing young children whose developmental delays are somewhat milder than in the past, yet their behavioral deviations are more severe and consist of problems in relatedness, communication, and modulation of emotional response."

While they attributed this to social and demographic changes, one of the characteristics was an increase in a spectrum of psychiatric behavioral disorders which could be, in part, an effect of an increase in the frequency of multiple behavioral genes.

Sudden Infant Death Syndrome

The second indirect line of evidence comes from reports by us[72] and Dr. Jeffrey Sverd[255] indicating that the TS genes also seem to predispose infants to sudden infant death syndrome (SIDS). These studies suggested that SIDS is two to four times more common in TS and ADHD families than in the general population. The immediate cause of SIDS appears to be related to the presence of sleep apnea. The relationship with TS and ADHD may be that serotonin and dopamine play a role in the regulation of sleep, and genes affecting serotonin and dopamine metabolism are involved in causing TS and ADHD (see Part II). The pedigree on the next page shows the presence of SIDS in relatives of patients with TS.

In this family, as with the autism pedigree above, both parents of the SIDS and near-SIDS cases had TS or chronic motor tics. ADHD and alcoholism were also common in this family. This finding suggests that the genes predisposing to SIDS are related to the genes predisposing to TS and ADHD. A study in Finland showed that the annual incidence of SIDS in the period from 1969 to 1974 averaged 0.3/1,000 live births, and increased to 0.5/1,000 in the period from 1975 to 1980.[224] In Minnesota the incidence rate increased from 0.55 per 1,000 in 1950-1953 to 1.28 in 1990-1992.[185a] These increases provide indirect support to the possibility that these genes are increasing in frequency.

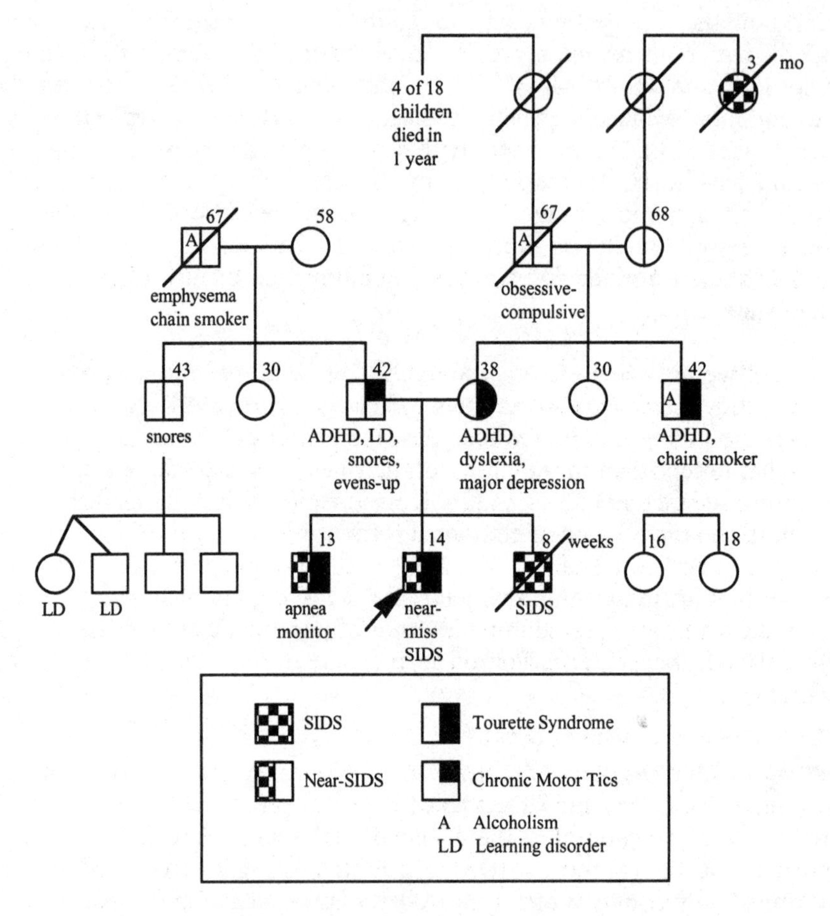

Figure 3. The presence of SIDS and near-SIDS in patients with Tourette syndrome and their relatives. From Comings, D.E. and Comings, B.G. *J. Dev. Physical Disabilities* 5:265,1993.

Summary – Autism, TS, ADHD, conduct disorder, and other disruptive childhood disorders share genes in common. The almost threefold increase in the frequency of autism in western Sweden lends support to the proposal that the genes for these disorders are increasing in frequency. Sudden Infant Death Syndrome also appears to be related to these genes, and studies in Finland and Minnesota suggest SIDS is also increasing in frequency.

Chapter 7

Increase in Learning Disorders

The ability to learn is a critical skill for a technologically advanced society. Children with learning disabilities are placed at a distinct disadvantage throughout their entire lifetime. If recognized, they are placed in special programs ranging from one to ten hours a week of tutoring in resource classes to full-time special education classes. Children placed in such programs are often discriminated against by their peers and taunted with viciously hurtful names ranging from "dummy" to "retard" to "stupid." Even when not recognized and left to struggle in regular classes, the taunts can be just as bad. Children with learning disorders are more likely to give up and drop out of school, and once out they are still at a significant disadvantage in the workplace.

The Prevalence of Learning Disorders

To obtain information on the prevalence of learning and related disorders in the U.S., the National Health Interview Survey of Child Health was conducted by the National Center for Health Statistics in 1981 and 1988. This consisted of in-person interviews carried out by interviewers trained by the U.S. Bureau of the Census. Adult family members were asked about the presence of delays in growth or development, learning disabilities, and emotional or behavioral problems. Positive responses were followed by more in-depth questioning. The emotional or behavioral problems had to have lasted more than three months and required psychological help. These included ADHD, aggressive behavior, conduct disorder, phobias, anxiety, and depression. The results of the 1988 survey, based on 63,569 children[1] 17 years of age or less, is summarized in Table 1 (next page).

Extrapolation of these numbers based on the total population of children indicates that 2.5 million children had developmental delays, 3.4 million had a learning disability, and 7 million had an emotional or behavioral problem. Combined, one in four children, or 10.2 million, had a developmental delay, learning or emotional problem. This exceeds the total combined number of children with asthma, dermatitis, orthopedic problems, or heart murmurs. These mental and emotional disorders are the most prevalent health conditions of modern childhood.[1]

Table 1. Data from 1988 U.S. National Health Interview Survey of Child Health

Characteristic	% of population affected		
	Total	Males	Females
Developmental delay	4.0	4.2	3.8
Learning disability	6.5	8.6	4.4
Behavioral disorder	13.4	15.4	11.3
Total	19.5	22.9	16.0

As would be expected, the prevalence of learning disabilities increased the longer children were exposed to school, progressing from 1.6% at 3-5 years of age, to 6.8% in the elementary school ages, and to 8.8% for students in junior and senior high school. As shown in Table 1, the prevalence of all three disorders is higher in males than females. This is especially true of learning disorders, which were almost twice as common in males.

Learning Disorders are Increasing in Frequency

Since the definition of "learning disability" can vary from state to state, district to district, and year to year, it can be difficult to accurately assess whether the rates are increasing or decreasing over time. The rates could be artificially increased with increasing awareness of the problem, or artificially decreased if school districts keep making the criteria more and more restrictive so there are a constant number of children so diagnosed each year.

Despite these difficulties, based on the above Survey of Child Health, of the total number of public school children, 4.8% were receiving special educational services for learning disabilities in 1987 – more than double the 1.8% recorded in 1977.[1] It is also relevant that there was a 50% increase in the number of children receiving treatment for emotional or behavioral problems, rising from 6.5% in 1981 to 10% in 1988.[1]

The U.S. Department of Education has also been keeping track of the number of children in the country who are served under Part B of the Individuals with Disabilities Education Act (IDEA), Chapter 1 of the Office of Special Education Programs (OSEP) Elementary and Secondary Education Act, and State Operated Programs (SOP), from 1977 to the present. They also track the percent of these students who have learning disabilities as opposed to other problems, such as mental retardation, hearing, visual, orthopedic, or other impairments. This allows an estimate of the absolute number of children with learning disorders served under these programs over this time period. The results are shown in Figure 1 (next page).

In the 1976-1977 school year, there were 890,000 children with learning disorders served under these three programs. In the 1992-1993 school year, there were 2,714,000 children with learning disorders served. This was a 305% increase, much greater than the increase in the general population.

Learning disorders and retardation do not exist as solitary entities. There is an increased frequency of psychiatric disorders in children with learning disorders and retardation.[43,218,234,269] In a population-based study by Dr. Michael Rutter and colleagues in England,[229] based on parent questionnaires, 30% of children with mental retardation had other psychiatric disorders, compared to 8% in the controls. Using teacher questionnaires, the rates were 40% versus 10%.

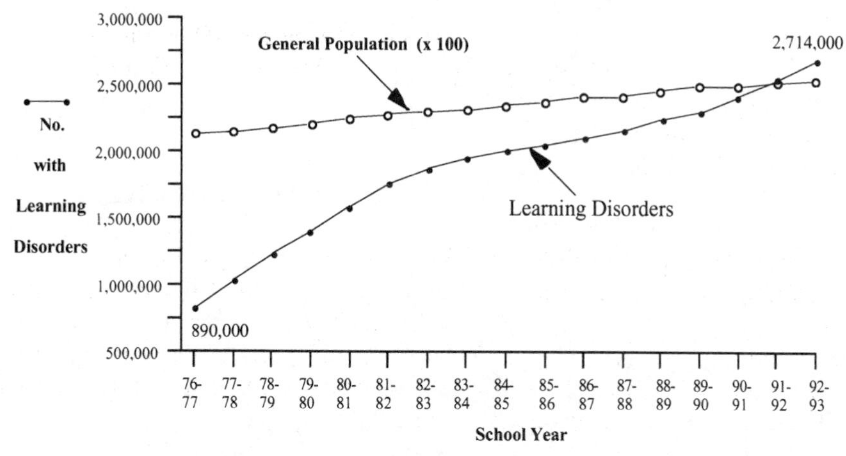

Figure 1. Number of children with learning disabilities served under IDEA, OSEP, and SOP from the school years 1976-1977 to 1992-1993. From the Sixteenth Annual Report to Congress of the Implementation of the Individuals with Disabilities Education Act, 1994.[1]

Socioeconomic Status

While there is a correlation between income and the prevalence of learning disorders, there is also a correlation between income and the prevalence of ADHD, conduct disorder, and emotional disorders; for example, a Canadian study[216] examined 3,292 children born in Ontario during a specific time period. Poor school performance, as defined by failing a grade or requiring special classes, was three times more common among families with an income of less than $10,000. However, it was of interest that psychiatric disorders such as ADHD, conduct disorder, somatization, depression, and anxiety were 2.4 times more common in the low income group, and social impairment consisting of constant problems in getting along with peers, teachers, or family members was 3.2 times more common. The 1988 U.S. National Health Interview Survey of Child Health described above gave similar, but less marked, results. Here, of families with an annual income of less than $10,000 U.S. dollars, 8.4% of the children 17 years of age or less had a learning disorder, versus 6.4% of children in families with higher incomes, a 1.3-fold increase.

Thus, while socioeconomic factors can play a role in learning problems, it must be kept in mind that genetic factors play a major role in learning disabilities (Chapter 13) and IQ (Chapter 8), and IQ plays a major role in learning disabilities and in socioeconomic status.[184,272]

Summary – From 1977 to 1993 there was a 305% increase in the number of children served by the programs for students with learning disorders.

Chapter 8

Decrease in IQ?

Humans evolved from the lower primates. If we take for granted that we have a significantly higher IQ than other primates, an assumption some may challenge, then that higher IQ appeared to require over 100,000 years to evolve. We may blithely discuss evidence that the frequency of depression, suicide, anxiety, aggression, addictive, and other behaviors are increasing in frequency, without having these thoughts evoke a distressing, visceral reaction. However, mention of the possibility that the IQ of the human race is beginning to turn a corner and evolve backward to lower levels strikes a raw nerve. Our IQ is like the core of our essence, analogous to masculinity or femininity. Is the mean IQ of the human race continuing to evolve upward, has it stabilized, or is it beginning to evolve downward? Before addressing these questions, it is necessary to determine if intelligence is controlled by genes.

Heritability of IQ

The subject of the inheritance of IQ is very controversial, as is the subject of how well IQ tests measure intelligence. However, most of the controversy revolves around the degree to which IQ is controlled by genes, not whether genes do or do not play a role. Virtually all studies concede that a portion of the IQ is inherited. These studies examine a variable called heritability. A heritability of 0.0 means genes are not involved at all. A heritability of 1.0 means a trait is totally controlled by genes. A heritability of 0.5 indicates a trait is half-genetic, half-environmental. A heritability of 0.5 is the perfect compromise, an even split that leaves both the environmentalist and the geneticist involved, if not totally happy.

The most controversial claims are those that suggest the heritability of IQ is significantly greater or lesser than 0.5. The range of controversy runs from heritabilities of 0.4 to 0.8, and most of the smoke and fire comes when estimates reach higher levels, indicating genes are more important than the environment. Thus, though arguing whether a figure is 0.7 instead of 0.5 may seem too trivial to generate so much heat, it has.[246]

One of the most reliable techniques of identifying the role of the environment versus genes in intelligence is the use of twin studies, especially studies comparing the IQs of identical twins reared apart. This is valuable because identical twins share 100% of their genes and, when raised apart, close to 0% of their environment. The more recent of these studies suggest heritabilities for intelligence of 0.70.[22,23] While a heritability of 0.5 generates the least amount of controversy, an estimate of 0.6 ± .2 is the most realistic[131] and covers a very wide range of estimates.

A very direct way to illustrate the importance of genetics in determining IQ is to use the National Longitudinal Study of Youth (NLSY) data (Chapter 19) to examine the relationship between the IQ of the mother and the IQ of her children. The following shows the results:

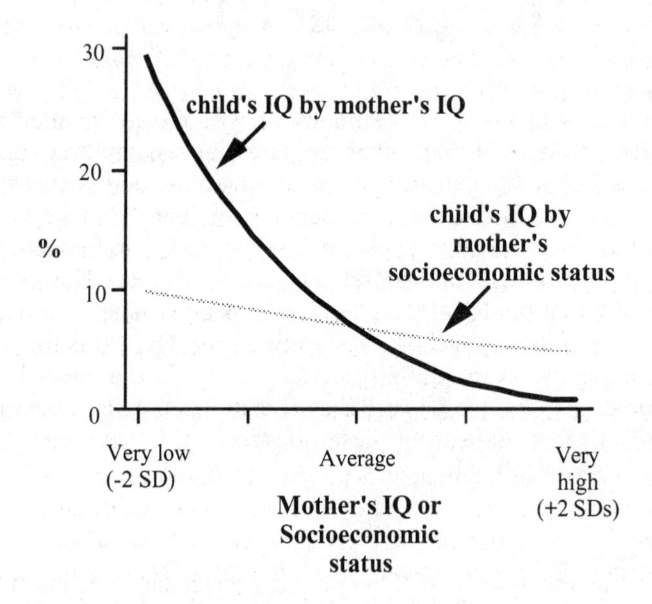

Figure 1. Relationship between a child's IQ and his/her mother's IQ or socioeconomic status. Redrawn from R. J. Herrnstein and C. Murray, *The Bell Curve*, The Free Press, 1994, p 231.

This shows a marked correlation between the mother's IQ and her child's IQ, but a relatively minor correlation between a child's IQ and his/her mother's socioeconomic status.

I have the luxury that whether the true heritability of intelligence is 0.4, 0.5, 0.6, or 0.8, it is irrelevant to the theme of this book. All that is important is that genes do play a significant role, and this is true of all of these heritabilities. Since there is almost universal agreement that the heritability of intelligence is 0.4 or greater,[131,246] those aspects of human behavior that relate to intelligence would also be genetically controlled to a similar degree. These are discussed later.

Evidence for Increasing IQ

One does not have to have an advanced degree in genetics or epidemiology to be concerned about the apparent inverse relationship between the number of children and the education level or socioeconomic status (SES) of a family. To the extent that SES and IQ are correlated, if individuals of lower education level or SES have significantly more children than those of higher education level and higher

SES, there will be a selection for the genes associated with the lower levels. In the 1930s several investigators suggested this effect was so strong that intelligence was plunging at the alarming rate of one to four IQ points per generation.[38,39,82,221]

In the face of these concerns several studies seemed to be very reassuring. For example, in 1984 James Flynn published a paper entitled "The mean IQ of Americans: Massive gains from 1932 to 1978."[114] If this was not enough, in 1987 he published another paper entitled "Massive IQ gains in 14 nations: What IQ tests really measure."[115] This reported that the gains noted in the United States were also seen in fourteen other countries worldwide. However, these reports are almost too good to be valid. The problematic aspects of these figures were reviewed by Herrnstein and Murray[131] and can be summarized as follows:

• If these trends were constant, it would imply that the mean IQ of Americans in 1776 was 30 and will be 150 in 2095. This is clearly ludicrous.

• There was a strong tendency for IQ scores to drift upward as a function of the number of years since a given test was standardized. This suggests some unknown aspect of the test is responsible for the rise, not a real increase in IQ.

• In several studies the rise was entirely due to an increase in scores in the bottom half of the IQ distribution.[175,260] This may simply reflect a tendency for the lower tail of the distribution to draw closer to the mean, with more universal access to education or other social programs.

• Flynn, himself, suggested the increases might be more related to improvements in test-taking skills than intelligence *per se*.[115]

• Most of the change has been concentrated in the non-verbal portions of the test. The relative stability of verbal IQ again suggests the changes are more related to some aspect of test-taking rather than a real increase in cognitive skills.

• In some countries there appears to be a leveling off of this trend.[175]

In toto, it appears that the apparent trend to higher IQs may represent some poorly-understood aspects of test-taking rather than a true increase in intelligence and cognitive ability. Herrnstein and Murray termed this the *Flynn effect*.[131]

Another optimistic assessment of IQ trends was published in 1962 by Higgins, Reed, and Reed.[132] In a paper entitled "Intelligence and family size: A paradox resolved," they suggested it was possible to have a negative relationship between family size and intelligence and yet still have an upward evolving trend in IQ. This was possible, they suggested, because people who had no children had been excluded from the earlier studies, and these low-fertility, often unmarried individuals actually had a lower average intelligence than fertile couples. They concluded that intelligence was rising slowly, despite a lower average IQ among larger families. I first read this paper in 1964 during my fellowship in genetics and still recall the great sense of relief I felt that this previously ignored aspect of the data appeared to have rescued the human race from a gradual decrease in IQ.

Evidence for Decreasing IQ

Unfortunately, this sanguine outlook soon began to erode.[131] To begin with, it did not take into account the Flynn effect. Second, most of the sites studied were based on a highly unrepresentative sample of all-White residents from the upper Midwest.[40,211] In a more representative study of 250,000 children in Georgia, Osborne[211] found a significant negative correlation between mental ability and fertility. All effects were dysgenic, i.e. consistent with a decrease in IQ. They also noted that per capita expenditure did not have a notable positive effect on school achievement at any grade level.

A third problem was that the studies were performed during a period of increasing fertility rates (the post-war baby boom).[131] Based on the National Longitudinal Survey studies (p.220), Daniel Vining Jr. proposed that in periods of rising fertility, individuals with higher-than-average intelligence tend to have as many or more children than those with lower intelligence.[268] He suggested that as this brief burst in fertility rates began to subside, the selection for higher intelligence would also subside. These results have particular validity since, unlike most studies of IQ and fertility, they were based on a representative sample of the population. A summary of Vining's results for Caucasian women is shown in Figure 2.

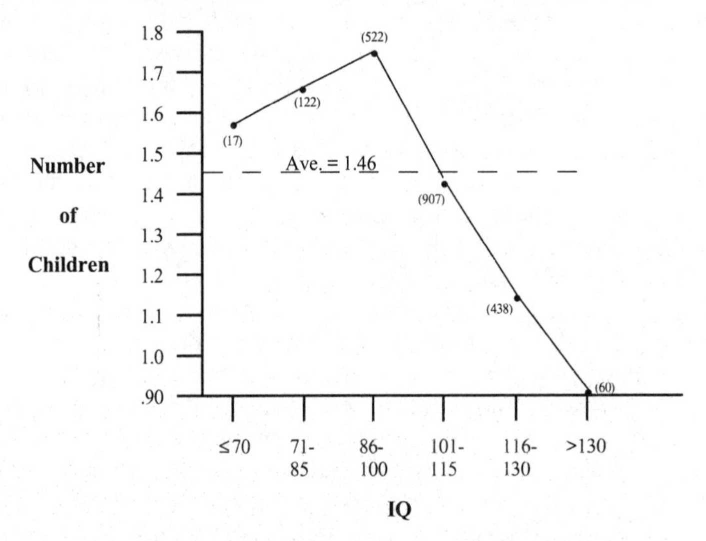

Figure 2. Diagram of the fertility rate of Caucasian women by IQ. From Vining, D.R. Jr., *Intelligence* 6:241-264,1982. N = ( ).

Vining's results were consistent with a decrease of 1 1/3 IQ points per generation for Caucasians and Blacks combined. The proposal that intelligence may actually increase during periods of raising fertility would explain why humans have a high IQ to begin with. Until modern times, the human race has been characterized by high fertility rates during most of its existence. The decline in birth rates in the West began in the eighteenth century.[275,276] In the United States, there has been a progressive downward trend in the number of children per family, from eight before the Revolutionary War to less than two by the 1970s.

A fourth problem is that Van Court and Bean[264] suggested that not even the optimistic suggestion that intelligence increased during the baby boom was correct. Based on studies of fifteen birth cohorts covering five-year birth intervals from 1894 to 1964, they showed a negative correlation between IQ and number of offspring for all fifteen cohorts, with all but three being significant. The correlation coefficients for Caucasians, corrected for attenuation, or loss of cases, are shown in Figure 3.

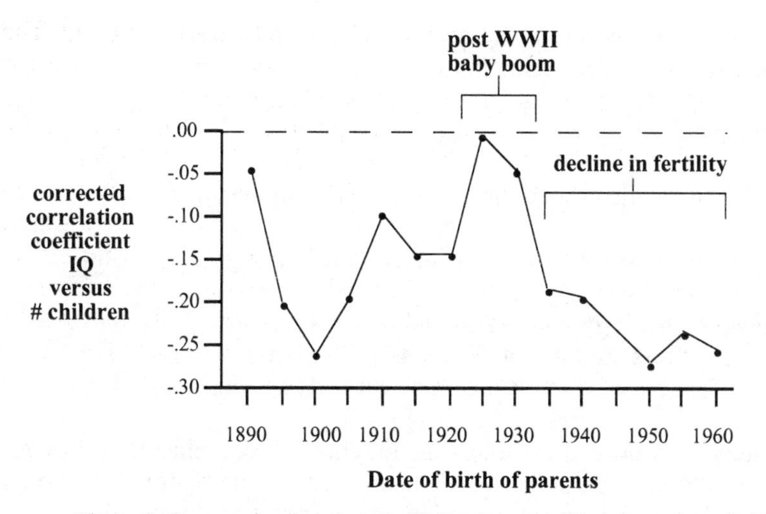

Figure 3. Corrected correlation coefficient between IQ and number of children over fifteen for birth cohorts in five-year intervals from 1890 to 1965. From Van Court and Bean, *Intelligence* 9:23-32, 1985.[264]

The only cohort that was not negative was for the post-World War II baby boomers (r=.00). There is a significant decrease in the correlation coefficient for families having children after 1965 (date of birth of parents after 1935), when the national fertility rate began to drop. Van Court and Bean also provided evidence that individuals with no children actually had slightly higher IQs than people with children, and suggested a slight decrease in the IQ rates has been occurring throughout the twentieth century.

A fifth concern is the progressively declining Scholastic Aptitude Test (SAT) results. The SATs could be considered the equivalent of an IQ test given each year over a period of many years. It has the advantage that it is standardized and is given to a group in a very narrow age range, i.e. high school graduates. The mean of the SAT distribution fell by about .5 of a standard deviation between 1963 and 1980.[131,167,281] While the family environment, a broader base of adolescents taking the test, and the educational establishment itself are often blamed, a gene-based progressive decline in IQ and increase in learning disorders could be a contributing factor.

In combination, studies and all the above factors suggest a modest, but real, decrease in IQ has been occurring in the twentieth century. A number of recent studies suggest the rate is at least 0.8 IQ points per generation.[131,221] In commenting on Vining's[267] report on this trend, Eibl-Eibesfeldt[99] stated:

> "We might comfort ourselves with the thought that the trend found by Vining could be temporary and that for the time being the range of modifiability by education leaves enough room to compensate for the genetic change of our intellectual capacity. The trend might also change in time. There are, in fact, indications that intelligent women are beginning to value motherhood again. But what if the trend continues? One IQ point per genera-

> tion is a lot, in evolutionary terms. To wait passively for the mechanisms of natural selection to operate – and hopefully correct this – may prove disastrous to [*Homo*] *sapiens*. We must face the problem."

In a lighter vein, he also stated that a cynic might suggest that:

> "...man, in spite of all his intelligence, imagination, and ingenuity is about to endanger his further existence by his advanced technology of warfare and by overexploiting his environment. A less intelligent *Homo* is perhaps the solution. He might be less harmful to his environment and less suicidal."

Unfortunately, this only emphasizes intelligence. If, coincidentally, because of the selection for other genes, the *Homo* also became more aggressive, violent, impulsive, compulsive, and addiction-ridden, the decrease in IQ would hardly be a "solution."

Summary – The study of genetic factors in intelligence is even more controversial than the study of the genetics of behavior in general. Despite this, there is almost universal agreement that genes account for between 40 to 80% of our level of intelligence. The fact that families at the lower end of intelligence tend to have more children than families at the higher end has been a subject of considerable concern. Since all agree that genes play some role, this would tend to select for those genes contributing to a lower intelligence level. Whether this is actually happening has been a subject of much discussion. All factors considered suggest that in the twentieth century there has been a slight downward trend per generation in IQ and that this trend may accelerate as fertility rates in advanced countries continue to decline.

Chapter 9

Increase in Crime

One could argue that while the thought of a decrease in IQ was bad enough, a progressive increase in crime is even worse and is an even greater threat to the status quo. Despite significant regional variations in crime frequency, an increasingly violent and crime-ridden society has an impact on all members, regardless of where they live or what their socioeconomic status. The increase in crime also carries with it the greatest level of visceral fear and the greatest likelihood that those fears will be translated into a variety of self-protective political agendas.

There are three major sources of information on crime rates over time – the National Crime Survey, the Uniform Crime Reports, and the National Center for Health Statistics. The National Crime Survey is based on interviews of all subjects 12 years of age and older in a national sample of households asked to recall and describe recent non-fatal victimizations. The Uniform Crime Reporting System records information reported to police, plus additional information about homicides, and direct information on individuals arrested. The National Center for Health Statistics provides data on homicides based on death certificates. For a variety of reasons each of these gives somewhat different results.[220]

Is the United States more violent now than in the past? The answer to this depends upon the definition of violence. If defined on the basis of the homicide rates, the answer is no, not much. Figure 1 (next page) summarizes data on the rates of various crimes over the period from 1960 to 1992. The homicide (murder) rate has not changed much. A first major peak was in 1930 (not shown), with a second peak in the early and late 1970s. By contrast, on the basis of these statistics, the rate of total violent crime (homicide, aggravated assault, forcible rape, etc.) has increased dramatically. Specifically, the data show a 3.5-fold increase in the rate of violent crime (160.9 per 100,000 in 1960 versus 757.5 per 100,000 in 1992). The increase was 5.1-fold for aggravated assault (86.1 vs 441.8), and 4.5-fold for forcible rape (9.6 vs 42.8).

There are marked differences in the rates of increase of violent crime and aggravated assault, depending upon the population of the towns and cities involved. Figure 2 shows the rates of total violent crime for several representative-sized communities.

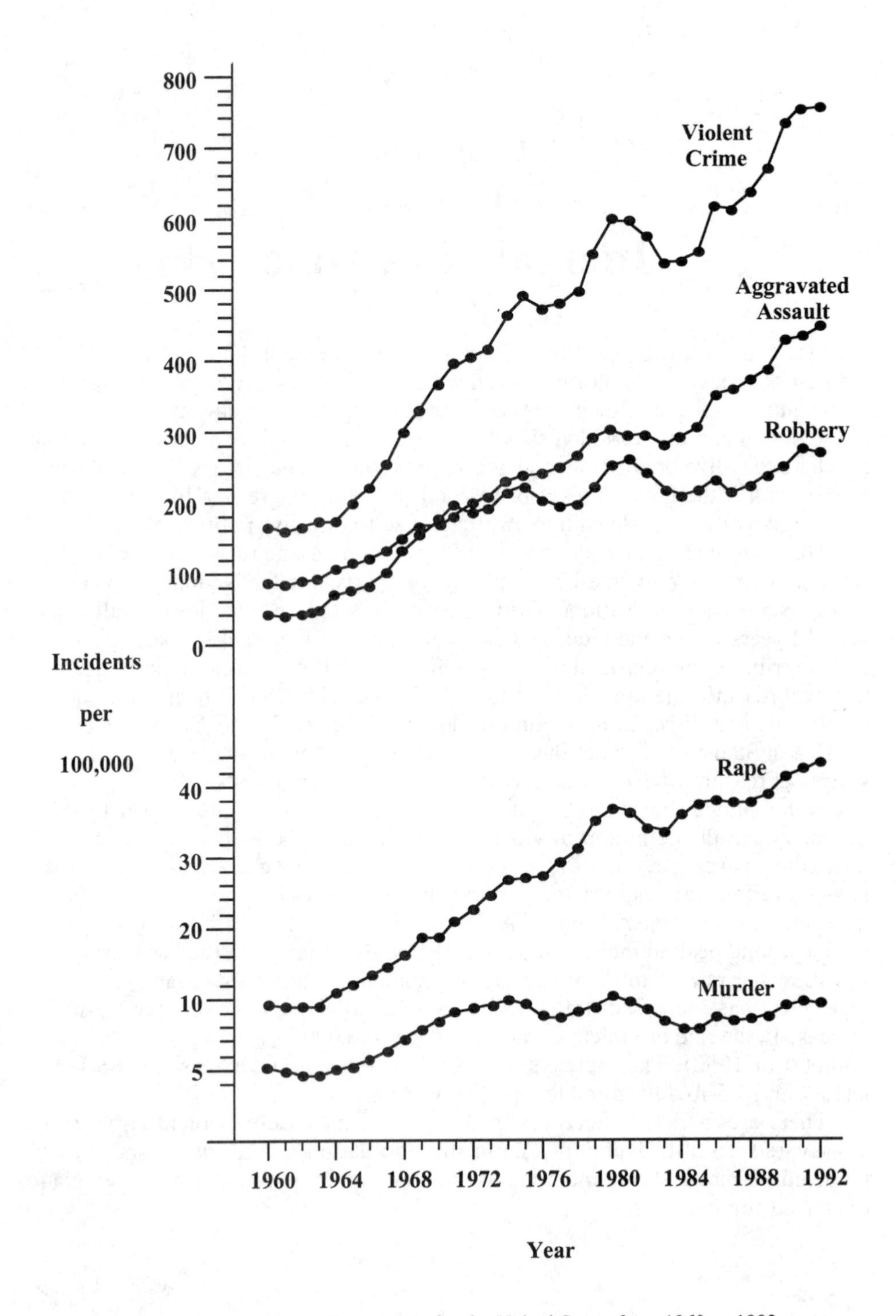

Figure 1. Crime Index offense totals for the United States from 1960 to 1992. From the Federal Bureau of Investigation, U.S. Dept. of Justice.

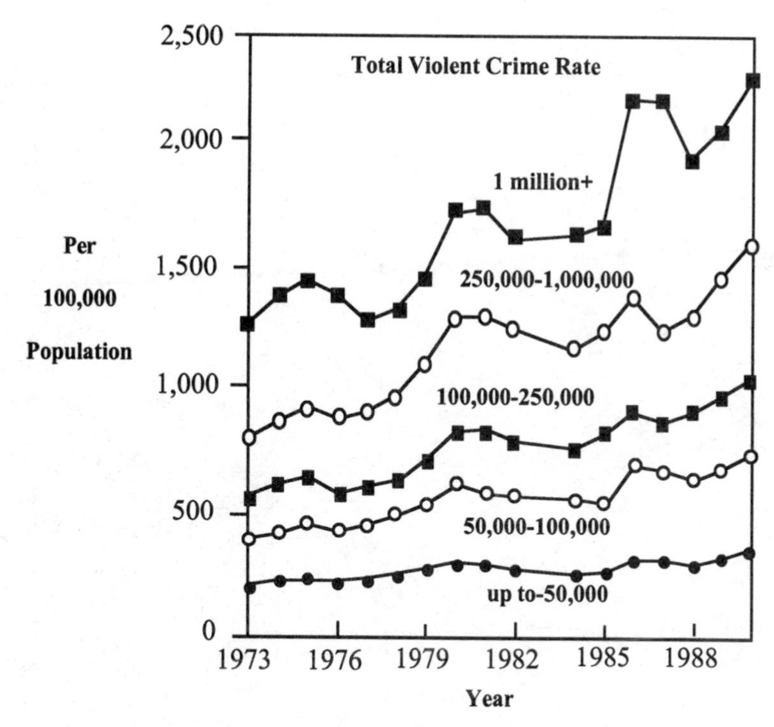

Figure 2. Total violent crime rates by city size. Redrawn from Reiss, A.J. Jr. and Roth, J.A., *Understanding and Preventing Violence.* Copyright 1993 by the National Academy Press, Washington D.C. p81.

This shows that both the rate of violent crime per 100,000 population and the rate of increase from 1973 to 1990 are higher in the larger metropolitan areas. For cities of more than one million, the rate of violent crime increased almost twofold in only seventeen years.

Summary – The rate of total violent crime based on incidents per unit of population has increased 3.5-fold from 1960 to 1992.

Part II

Evidence these Are Interrelated Genetic Disorders

If all of the disorders and entities reviewed in Part I were purely the result of learned behavior or psychological factors, there would be no sense in writing this book. The increases in frequency and occurrence at younger and younger ages would be attributed entirely to environmental factors; however, there is burgeoning evidence that genetic factors play a major role in all of these entities. A more rapid reproduction rate of these genes will take place if individuals carrying them have children at an earlier age, have more children, or come from larger families than those without these genes. It is beyond the scope of this book to review all of the evidence for these genetic factors. For a more detailed presentation of this material the reader is referred to the companion books, *Search for the Tourette Syndrome and Human Behavior Genes*[60] and *Tourette Syndrome and Human Behavior*,[52] as well as other books.[269,282] The following chapters provide a brief review of some of the material given in these other sources.

Chapter 10

Tourette Syndrome – A Hereditary Spectrum Disorder

Tourette syndrome is a hereditary disorder characterized by the presence of chronic motor tics such as eye blinking, facial grimacing, shoulder shrugging, neck jerking, and others, and vocal tics such as throat clearing, grunting, barking, spitting, and others. When we first began to see patients with TS, it quickly became apparent that the probands – that is, the first member of the family to be seen in the clinic – had a wide range of behavioral disorders. To study this in more detail we examined a series of 247 TS probands and 47 controls.[50,51,62-66] These results are shown in Figure 1.

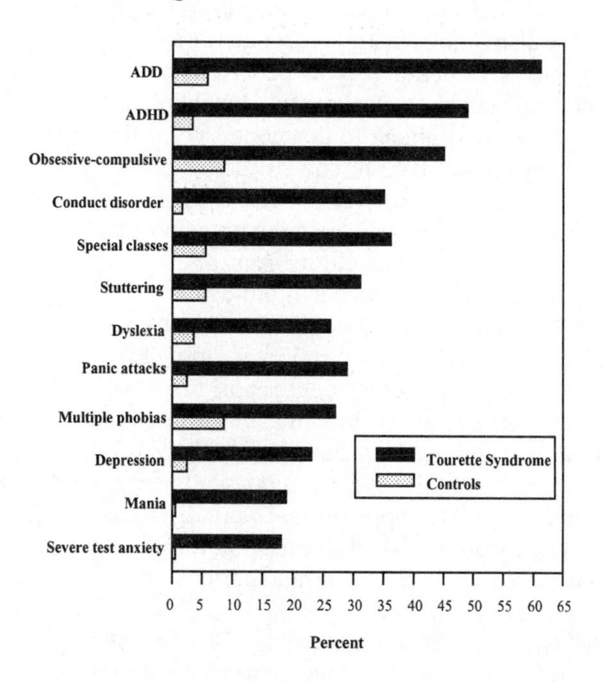

Figure 1. Prevalence of a range of behavioral disorders in TS probands versus controls. From Comings and Comings.[50,51,62-66]

This shows that the frequency of a wide range of disorders, including ADHD, obsessive-compulsive behaviors, conduct disorder, dyslexia, anxiety, phobias, depression, mania, and sleep disorders was much higher in patients with TS than in controls. This led us to suspect that TS (*Gts*) genes were causing a complex neuropsychiatric spectrum disorder. However, some [213,214] suggested that these associated behaviors were present in the probands by chance (ascertainment bias) and were not caused by the *Gts* genes. To distinguish between these two possibilities, we examined relatives of TS probands. If the associated behaviors were due to the *Gts* genes, they should be significantly more common in the relatives with TS than in relatives without TS. If the associated behaviors were due to ascertainment bias, they would be present with approximately equal frequency in relatives with and without TS. In many studies we found that the frequency of the same wide range of behaviors present in the TS probands were also present in the relatives with TS, but not in the relatives without TS,[54-56,58,59,68-70,73,156,158] indicating the *Gts* genes were causing a wide spectrum of psychiatric disorders.

The best way I know to briefly introduce the essence of TS, and with it the role of genes in behavior, is to present two cases from our clinic. For this purpose I have chosen two families seen in the week before writing this chapter. The reader should note that most TS cases are not this severe.

Family 1 – When A.J. was in nursery school at the age of 3, his teacher complained he was very distractible, disruptive, couldn't focus, and was constantly kicking and fighting with other children. This pattern of behavior continued the rest of his life. In first grade he began to display rapid eye blinking, head jerking, and facial grimacing tics. By the third grade he developed vocal tics consisting of constant grunting and barking noises. By the end of the year, a diagnosis of ADHD was made, and he was treated with Ritalin. By the fifth grade, a diagnosis of Tourette syndrome was made, and he was also treated with haloperidol, a medication for tics.

A.J. has been followed in our clinic from age 11 to his present age of 17. During this time, he has demonstrated symptoms of virtually every major psychiatric disorder, including anxiety, panic attacks, severe obsessive-compulsive disorder, motor and vocal tics, conduct and oppositional defiant disorders, depression and mania, episodes of violent aggression (especially against his mother), hyperactivity, poor concentration, learning disabilities, and sleep disorders. He has been in classes for the learning disabled and severely emotionally disturbed throughout his school career. He has been hospitalized in mental facilities multiple times for "unmanageable, oppositional, and aggressive behavior." Between hospitalizations he was often placed in group homes for his mother's protection. Despite treatment with a wide range of medications, most of the above behaviors have continued to paralyze his life.

At age 14 he was incarcerated for stealing. The evaluating psychologist stated:

> "In addition to the burglary charge, the defendant has been cited with two prior episodes of vandalism and battery. In regards to the last two mentioned charges, it is reported that the minor's mother called the police and indicated that their son was 'out of control.' She indicated that he was crushing their small dog.

When the animal yelped in pain and she demanded that he desist, he became even angrier and began chasing his mother. He then attempted to crush the dog's head between the bedroom door and the door jam. He was screaming at her, 'I'll kill you, you mother-fucker.' She reported that she locked herself in the bathroom with the telephone and he kicked a hole in the door in his efforts to reach her and the dog. He stated that he believed he behaved in the way that he did because his father 'used to be abusive to me.' He constantly minimized his responsibility for his behaviors."

The family history is shown in the following pedigree:

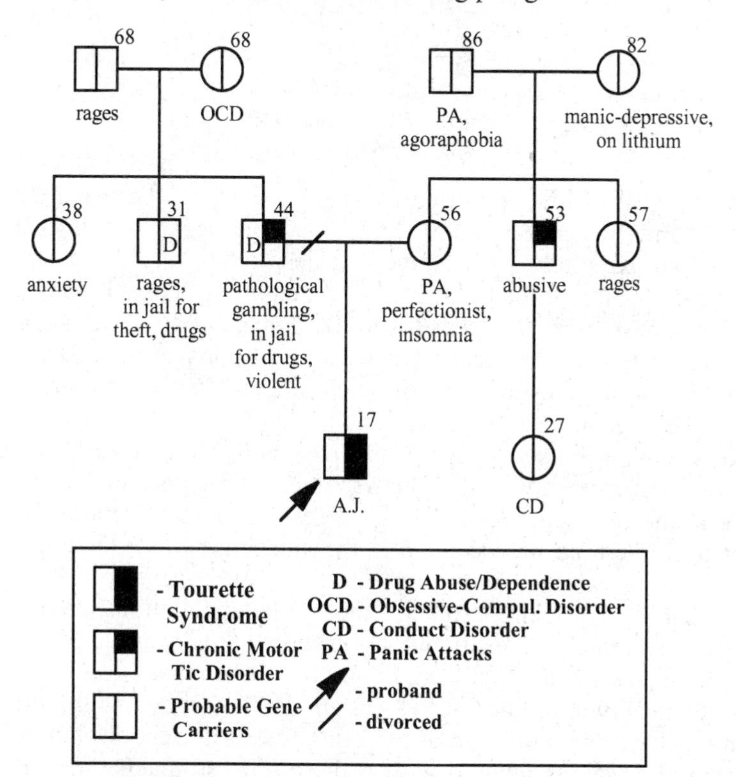

He was an only child. His mother has panic attacks and some perfectionistic behaviors. Her brother had chronic motor tics and an abusive personality. Her sister had problems with frequent rages. Her mother was a manic-depressive on lithium, and her father had panic attacks with agoraphobia. A.J.'s father has chronic motor tics, was a pathological gambler, violent, a drug addict and was in prison for selling drugs. His brother had similar problems. The father's father has rage attacks and his mother has obsessive-compulsive disorder. It is evident from this pedigree that a spectrum of impulsive, compulsive, addictive, anxiety, and mood disorders are present on both sides of the family and that A.J. is inheriting his *Gts* genes from both parents.

Family 2. This happened to be a pair of 6-year-old identical twin boys. As usual, the parents were at their wits' end struggling with these children and had taken them to many physicians and psychologists without help. To make the pedigree less overwhelming, I will present it in three parts – first the immediate family, then the mother's family, and finally the father's family. The names have been changed for confidentiality.

The immediate family was as follows: (ODD = Oppositional Defiant Disorder):

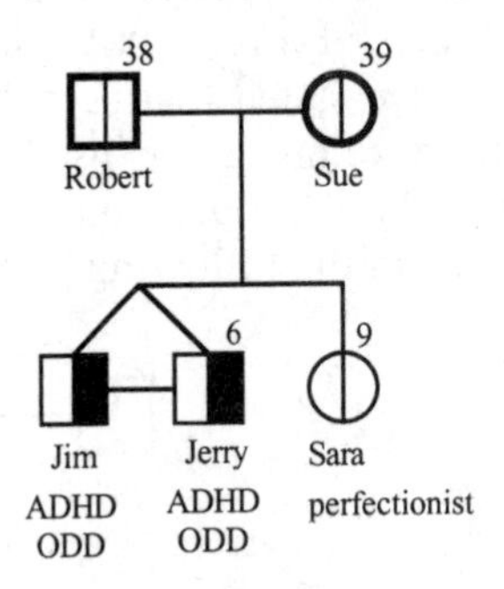

In his first year of life, Jim cried constantly, slept poorly, was diagnosed as having colic, and "was not a happy child." Once he began walking, he and his brother were so extremely hyperactive and destructive that the parents had to build a separate section of the living room, walled off with plastic rods, to protect the rest of the house from being destroyed. By 3 years of age, Jim had developed facial grimacing tics and a vocal tic consisting of constantly sniffing. He also seemed to have a heightened sense of smell and smelled everything, including other people. By age 4, he also had spitting tics and was constantly making "lip" noises like a braying horse. At age 6, in first grade, he was fearful of going to school. If his mother went with him, he was fine, but the instant she left he began to scream. In the first semester he missed fifty-four days of school. His parents relate that from the time he wakes up in the morning he is "non-stop" fighting, talking, yelling, screaming, running around, and swearing like a sailor – "that fucking teacher," "my fucking brother," and similar appellations – to everyone in range. He was extremely oppositional, never did what he was asked, and usually responded to requests with "fuck you, I don't have to do that."

Jim's identical twin brother, Jerry, was an identical copy of Jim, with virtually the same behaviors coming on at the same time. His school phobia began in kindergarten and was associated with bed-wetting, stomachaches, and bowel movements in his pants. This was so severe they eventually took him out of kindergarten. He barely managed to cope with school in first grade. His vocal tics consisted of throat clearing and spitting. He was also defiant and had a horrendous temper, accompanied by swearing, equal to that of his brother. Once, in a rage, he picked up a knife and threatened to kill his mother. Both children had learning problems in school. Both were sensitive to noise and stated that the school room was so noisy the sound hurt their ears.

The parents had taken the boys to a number of psychologists, only to be told either there was nothing wrong or their behavior was the parent's fault. Either there was some unrecognized conflict in the home, or the parents' had defective parenting skills. At age 5, one psychiatrist diagnosed both boys as having ADHD and possible TS, but after the failure of a single medication, no further medications were prescribed. The last professionals they had seen, two months before coming to the clinic, again stated the problem was poor parenting.

Those who labeled them poor parents seemed to miss the fact that the 9-year-old sister was the model of the almost perfect child. She got straight A's in school and was a darling of all her teachers; however, as much as her perfectionism was wonderful for her grades and behavior, it was taking its toll on her personally. Everything in her life had to be perfect and she would often cry if it was not.

The following is the mother's side of the family:

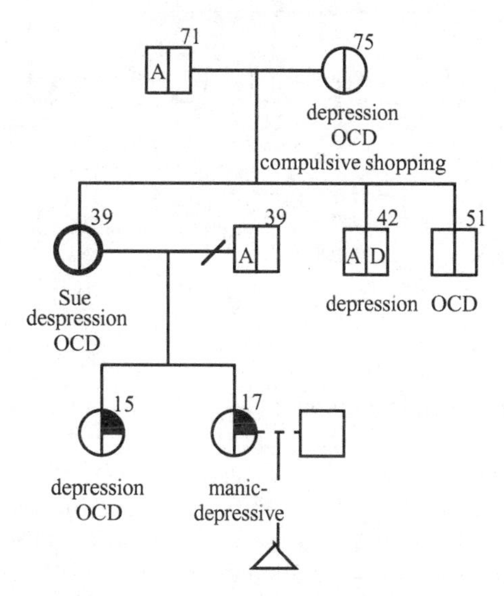

The mother, Sue, was also a perfectionist. Everyone called her a "clean freak," and the house had to be immaculate. She also had problems with chronic depression. A psychiatrist had recently began treating her depression with Prozac and "it has changed [her] life." Both her depression and her obsession with cleanliness disappeared, and she now felt normal. Sue had two children by a previous marriage.

The oldest daughter, 17, had some head jerking and was constantly touching her hair. She had also been diagnosed as manic-depressive and was on treatment for this. She had recently become pregnant but was not married; she was planning to keep the baby. Her younger sister was like her mother, depressed, and obsessive-compulsive. Their father was an alcoholic (A in the pedigree = alcoholism, D = drug addiction).

Sue's 42-year-old brother was an alcoholic and drug addict and had depression, and her 51-year-old brother was also obsessive-compulsive. Sue's father was an alcoholic, and her mother had depression and was obsessive-compulsive, including compulsive shopping. She had frequently put the family in debt with her shopping.

The father's side of the family is as follows (next page):

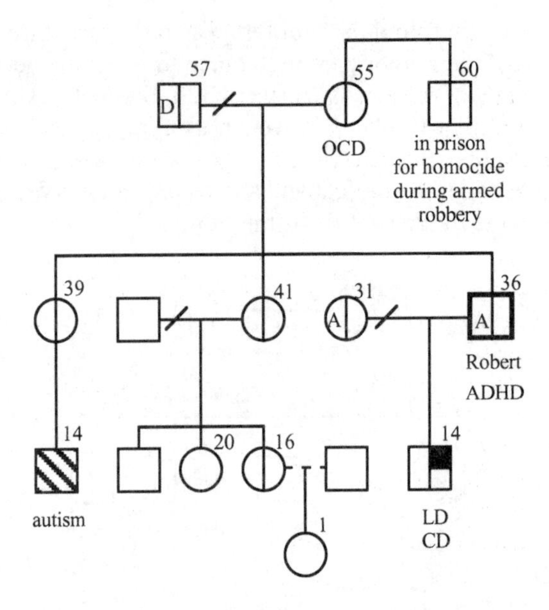

Like his two twin sons, Robert also had ADHD as a child. He also had a problem with alcohol abuse since his teen years, and a short temper. He had a son by a previous marriage who had some facial tics, learning disabilities, and was having behavior problems, including angry outbursts, lying, stealing, and other symptoms of conduct disorder (CD). The son's mother was an alcoholic. Robert had a 41-year-old sister with a 16-year-old daughter who had a 1-year-old daughter, out of wedlock. There was a 39-year-old sister with a son with autism. Robert's parents had been divorced for years because his father was a severe drug addict. His mother had obsessive-compulsive behaviors and had a brother who had spent his adult life in prison because he had killed someone in an armed robbery.

This family illustrates many of the important aspects of TS and ADHD. The presence of virtually identical severe behavioral problems in the two boys who have an identical set of genes provides a miniature twin study of the role of genes in behavior in general, and especially in TS and ADHD. While the consummate environmentalist might still say this was learned behavior, the sister, who grew up in the same family, had essentially the opposite behavior, yet her obsessions and compulsions are also part of the TS spectrum. The boys illustrate the spectrum of behaviors in TS – motor and vocal tics, ADHD, oppositional defiant behavior, separation anxiety, phobias, and learning disorders. The pedigree also shows the extent of the spectrum in relatives with chronic tics, ADHD, depression, obsessive-compulsive behaviors, manic-depressive disorder, alcoholism, drug abuse, and even autism. There is abundant evidence that the twin boys have received behavior disorder genes from both parents. Finally, and especially relevant to this book, there are two girls in this pedigree, who both undoubtedly carry different combinations of these genes, who had teenage, out-of-wedlock pregnancies – one at age 16, the other at age 17.

Molecular Genetics of TS and Related Disorders

***Dopamine D_2* receptor gene**. The genetics of a disorder is much clearer when the specific gene or genes involved are identified. While we initially thought TS might be inherited in a dominant fashion as a single gene disorder,[74] our continued studies led us to believe the genes were coming from both parents and that TS was a polygenic disorder – that is, it was caused by the coming together of a number of different genes.[56,57,60] Abnormalities in dopamine metabolism have often been implicated in TS. Dopamine is a neurotransmitter important in a wide range of behaviors, including movement, emotion, pleasure, and reward.[52] Neurotransmitters are effective only when there are specific proteins on nerve cells that recognize them. These are called receptors.[52] When Blum et al.[20] reported a significant increase in the prevalence of the *Taq* I A1 marker of the *dopamine D_2* receptor gene in alcoholism, we were interested in examining this marker in TS, since alcoholism was common in older TS patients and their relatives, and haloperidol – one of the commonly used medications in TS – specifically interacted with the dopamine D_2 receptor. Our results for TS and related disorders are illustrated as follows:

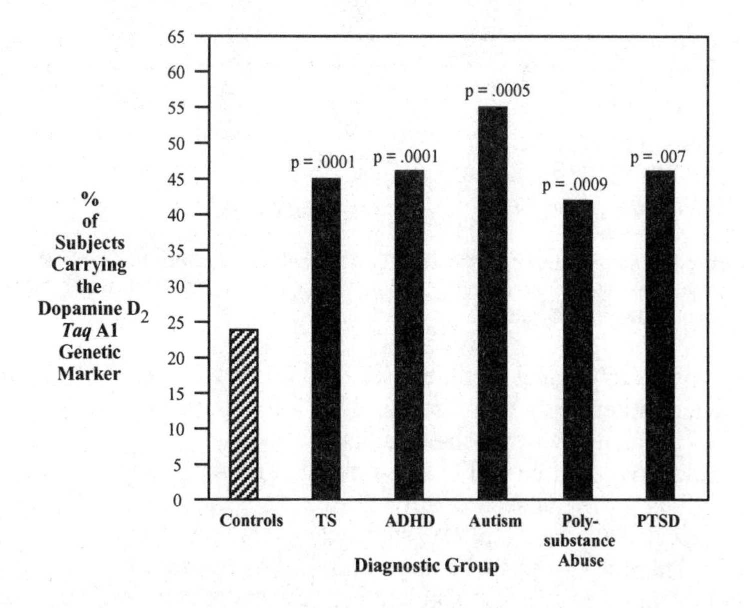

Figure 2. The prevalence of the *Taq* A1 dopamine D_2 receptor marker in controls and patients with TS, ADHD, autism, polysubstance abuse, and post-traumatic substance abuse (PTSD). From Comings et al.[75]

This showed that the prevalence of the *Taq* A1 marker of the *dopamine D_2* receptor gene was significantly increased in a wide range of impulsive, compulsive, addictive behaviors. These results led us to examine three additional addictive disorders – pathological gambling, smoking, and compulsive eating. Pathological gambling has been termed the "pure addiction," since nothing is being put into or injected into the body. It represents a unique combination of impulsive, compulsive and addictive behaviors. The results for 171 pathological gamblers are shown in Figure 3.

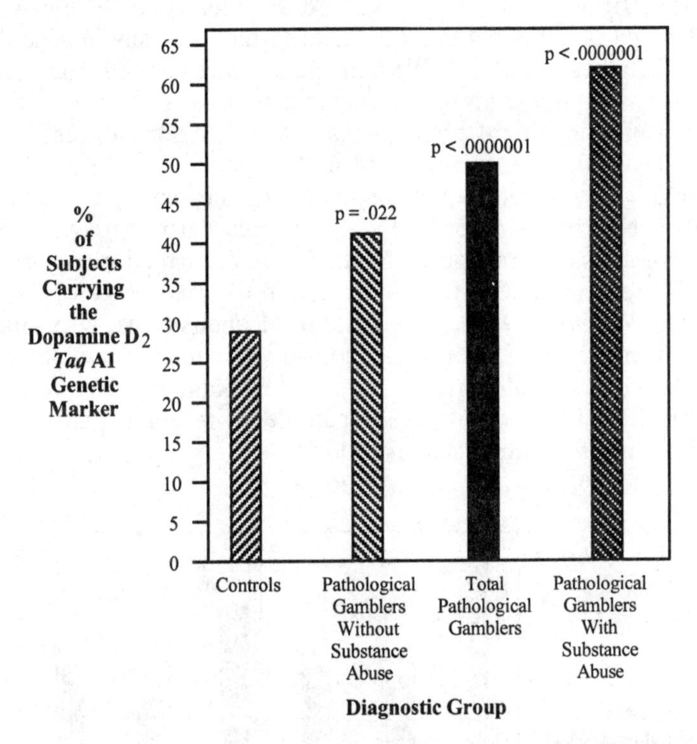

Figure 3. The prevalence of the *Taq* A1 dopamine D_2 receptor marker in controls and the total pathological gambling group, and in pathological gamblers without and with substance abuse. From Comings et al.[79]

Of the total pathological gamblers, 51.8% carried the *Taq* I A1 *dopamine* D_2 receptor gene marker. This increased to 63.8% for those pathological gamblers who also had problems with substance abuse.

The smokers we studied had to be smoking at least one pack per day and had to have previously tried unsuccessfully to stop smoking. Of the 322 such smokers, 48% carried the *Taq* A1 allele, significantly higher than in the controls ($p < .0000001$). There was a positive correlation between the prevalence of the A1 marker and the number of packs smoked per day, and a negative correlation with the age of initiation of smoking and the maximum duration of time that the subjects were able to stop smoking on their own. These results suggested that medications that increased the availability of dopamine in the brain might help them to stop smoking. In fact, a double-blind study using a placebo versus bupropion, a dopamine-acting antidepressant, showed a significantly greater degree of smoking cessation for those taking bupropion.[113]

Using a different set of genetic markers of the *dopamine* D_2 receptor gene, we found a significant correlation with obesity. These markers, called haplotypes, came in four varieties – I, II, III, and IV. The correlation between weight and the prevalence of the IV haplotype is shown in Figure 4.

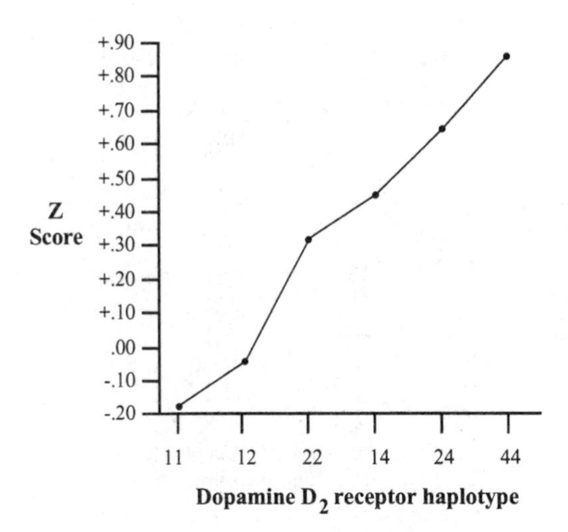

Figure 4. The Z score for weight by dopamine D_2 receptor haplotype. A Z score of +1.0 is one standard deviation greater than the average weight. From Comings et al.[81]

This showed a progressive, significant increase in the frequency of the 4 haplotype by body mass index. An independent study has also shown an association of markers of the *dopamine D_2* receptor gene and obesity.[200] This is consistent with the fact that appetite suppressants containing dopamine-like compounds (Dexamyl and others) are among the most widely used medications for weight control.

Several years after our initial study of the A1 allele in TS, we returned to the subject using additional cases and a sensitive technique for detecting the effect of a gene on the TS spectrum of disorders. This technique consisted of administering a detailed behavioral questionnaire to controls, TS patients, and their relatives. This allowed the evaluation of a wide range of different behavior problems. We then divided these subjects into three groups – controls who did not have the behavior problem in question (called *controls without*), TS patients or their relatives without the behavior problem in question (called *cases without*), and TS patients or their relatives with the behavior problem in question (called *cases with*).[81] If a gene contributed to any of the behaviors being studied, the frequency of specific markers of that gene should progressively and significantly increase in prevalence across these three groups. Figure 5 (next page) shows the results for the A1 marker of the *dopamine D_2* receptor gene in 319 controls and cases.

The four behaviors shown were significantly associated with the A1 marker. Other behaviors that were also significant were tics, schizoid behaviors, major depression, and conduct disorder.

***Dopamine β-hydroxylase (DβH)* Gene.** The DβH enzyme converts dopamine to norepinephrine, another important neurotransmitter. We examined the *Taq* B1 marker of the *DβH* gene to determine if it played a role in TS or its associated behaviors, using the same technique and same patients as for the dopamine D_2 receptor. Figure 6 shows the behaviors most significantly associated with this marker.

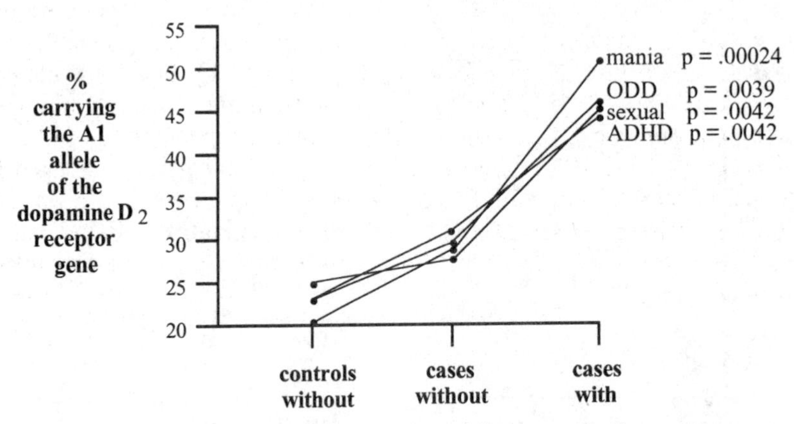

Figure 5. The prevalence of the A1 marker of the *dopamine D_2* receptor gene for various behaviors in controls, TS patients, and their relatives. From Comings et al.[81]

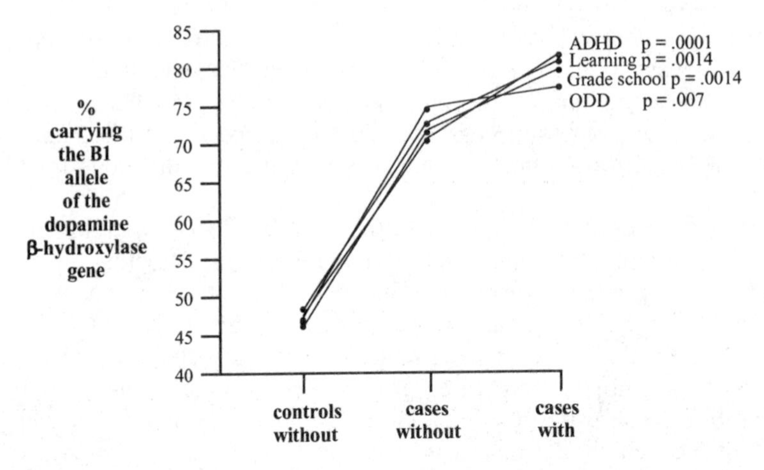

Figure 6. The prevalence of the B1 marker of the *dopamine β-hydroxylase* gene for various behaviors in controls, TS patients, and their relatives. From Comings et al.[81]

Here ADHD, learning disorders, poor grades in grade school, and oppositional defiant behaviors were all significant. Additional significant behaviors not shown were tics, mania, alcohol and drug abuse, reading and sleep problems, stuttering, and obsessive-compulsive behaviors.

***Dopamine Transporter* Gene.** When a nerve containing dopamine is stimulated, in order to pass that activity onto another nerve dopamine is released into the space between the nerves called the synapse. The next nerve is activated when the dopamine binds to the dopamine receptors such as the D_2 receptor. Once that has occurred, it is necessary to remove the rest of the dopamine from the synapse. This can be done by breaking it down by enzymes, such as monoamine oxidase, or by taking up the dopamine back into the nerve from which it was released. This is done by the dopamine transporter. The *dopamine* *transporter* gene has a section of DNA containing a series of repeated segments of DNA forty base pairs in length. In different people these repeats occur six to twelve times. When this segment of DNA is amplified by PCR (see glossary), and the resulting DNA is separated in an electric field, the different sizes can be easily seen. Chromosomes carrying ten copies of the repeat (the 10 allele) are most common, and the 10/10 genotype, where the 10 allele was inherited from both parents, is the most common. Using the same technique as for the *dopamine* D_2 receptor and the *DβH* genes, the frequency of carrying the 10/10 genotype was determined for controls, TS patients, and TS relatives. These results are shown in Figure 7.

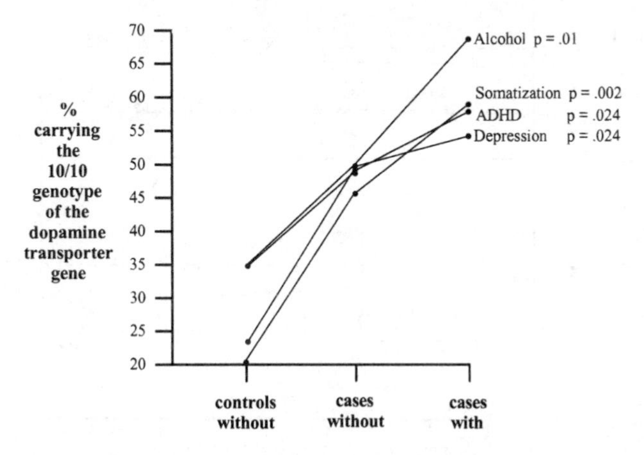

Figure 7. The prevalence of the 10/10 genotype for the forty bp repeat of the *dopamine transporter* gene for various behaviors in controls, TS patients, and their relatives. From Comings et al.[81]

The most significant correlation was with somatization. This refers to patients with many bodily (soma) complaints. These conditions used to be referred to as psychosomatic. Somatization has been linked to genetic factors and alcoholism.[48,240] The other significant behaviors shown were alcohol abuse, ADHD, and major depression. The other behaviors that were significant but not shown were panic attacks, obsessive-compulsive behaviors, general anxiety, and mania.

Polygenic Inheritance. The remarkable feature of all these associations is that the frequency of the alleles is high in the general population, but never 100% in the affected patients; thus, these cannot be examples of single gene disorders. The evidence that the genes come from both parents and that more than one gene is involved is consistent with a polygenic mode of inheritance. If that were the case, the above three genes should be additive in their effect. To test this we examined the magnitude of the different continuous behavioral variables in subjects that carried 3 of 3, 2 of 3, 1 of 3, and 0 of 3 of the above markers. If the magnitude of a continuous variable showed a progressive decrease across these four groups it would indicate an additive effect of the three genes. This progressive decrease was seen in seventeen of the twenty continuous variables and was significant in nine – ADHD, stuttering, oppositional defiant and conduct disorders, tics, obsessive-compulsive behaviors, mania, alcohol abuse, general anxiety disorder, and panic attacks. The results with the ADHD score are shown in Figure 8.

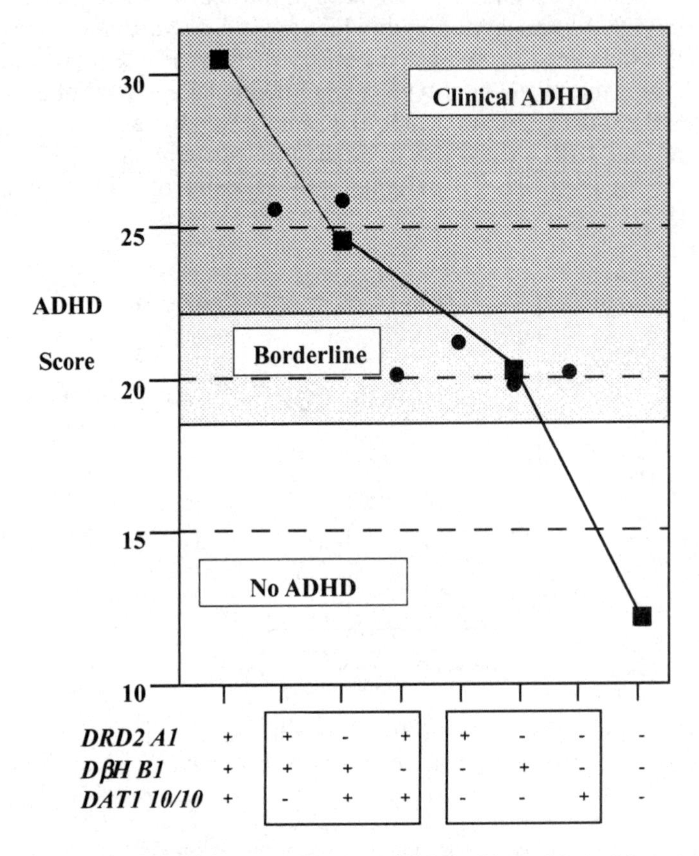

Figure 8. The ADHD score in individuals with 3 of 3, 2 of 3, 1 of 3, and 0 of 3 of the mutant alleles for the *dopamine D_2 receptor*, *dopamine β-hydroxylase* and *dopamine transporter* genes, showing the significant additive and subtractive effect of the combinations. From Comings et al.[81]

These results were important because they showed that these three genes were additive in their effect. Thus, individuals who had 3 of 3 had the highest score, 2 of 3 the next highest, then 1 of 3, and 0 of 3. The results were subtractive in that individuals with the fewest number of mutant alleles, especially 0 of 3, had the lowest scores. They were also consistent with our proposal that TS is a polygenic spectrum disorder in which the different associated behaviors are due to similar sets of genes, i.e. different psychiatric disorders have a number of genes in common.

As described before,[60] these results provide strong support at a molecular genetic level for the concept that TS, ADHD, stuttering, conduct, oppositional defiant, obsessive-compulsive, anxiety, substance abuse, and other disorders were:

- inherited in a polygenic fashion,
- part of a spectrum of related disorders,
- caused by shared genes,
- caused by alleles that are common in the population,
- caused by genes that are additive in their effect,
- caused by genes that upset the balance of dopamine, serotonin, and other neurotransmitters in the brain.

As reviewed in the companion book, *Search for the Tourette Syndrome and Human Behavior Genes,*[60] numerous other genes are also involved, including the *dopamine* D_1 and D_3 receptor genes and *tryptophan 2,3-dioxygenase*. The list will certainly continue to grow. The important point is that genes play a very important role in the impulsive, compulsive, addictive, depressive, anxiety, and learning disorders described in this book. As such, if individuals carrying these genes have children at an earlier age, have more children, and come from larger families than those who do not carry them, these genes will be selected for and thus increase in the general population.

Summary – Tourette syndrome is a complex hereditary neuropsychiatric disorder consisting of chronic motor and vocal tics and a wide range of impulsive, compulsive, addictive, mood, and anxiety disorders. As such, it forms a model for the study of genetic factors in these disorders. Three genes that regulate dopamine metabolism play a role in TS and its associated behaviors. Additional genes affecting serotonin and other neurotransmitters are also involved. The results of molecular genetic studies show that TS is a polygenic disorder, that the genes are common in the general population, and are additive in nature. Individuals who inherit a sufficient number of mutant genes are at risk to develop a wide range of interrelated impulsive, compulsive, and addictive behaviors.

Chapter 11

ADHD – A Hereditary Spectrum Disorder

Attention Deficit Hyperactivity Disorder is characterized by inattention, impulsivity, and hyperactivity.[52] It is the single most common behavioral disorder of childhood, estimated to be present in 4 to 8% of boys and 3 to 5% of girls. Like TS, it is associated with a wide range of comorbid behavioral disorders, including conduct disorder, oppositional defiant disorder, substance abuse, depression, anxiety, and others. While ADHD is clearly a genetic disorder,[33,75,108,109,248,250] its hereditary nature is less obvious than in TS, where the motor and vocal tics are so easily identified in children, parents, and other relatives. While once thought to be limited to childhood, it is now clear that in at least 50% of ADHD cases, significant symptoms persist into adulthood.

As with TS, one could claim that the high frequency of comorbid disorders is simply due to proband ascertainment bias. To avoid this problem, Biederman et al.[16] examined the lifetime frequency of a variety of behaviors in the ADHD fathers of children with ADHD. Since the fathers had not come to the doctor on their own, the problem of ascertainment bias was avoided. The results, compared to controls, are shown in Figure 1.

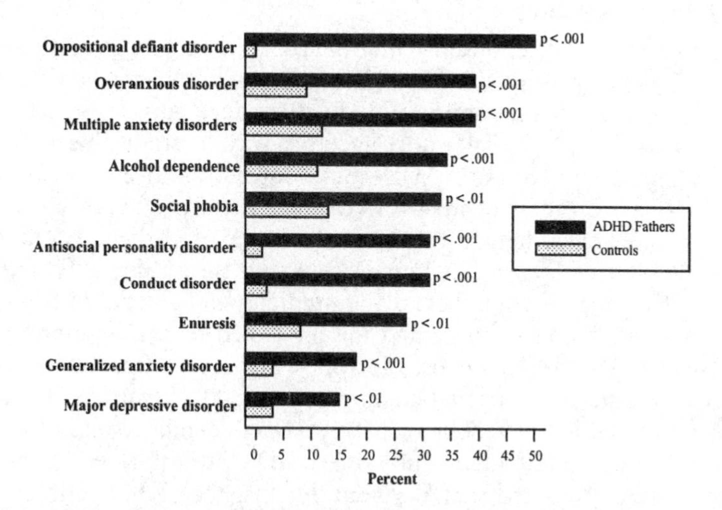

Figure 1. Frequency of comorbid behavioral disorders in the ADHD fathers of children with ADHD, compared to controls. From Biederman et al.[16]

There was a significantly increased frequency of a wide range of aggressive, conduct, impulsive, compulsive, addictive, anxiety, and depressive disorders in the non-proband fathers with ADHD. This indicated that like TS, ADHD was a hereditary neuropsychiatric spectrum disorder. The study of the relatives of TS patients shows how intimately related TS and ADHD are to each other. Figure 2 shows the frequency of ADHD in TS probands, relatives with and without TS, and controls.

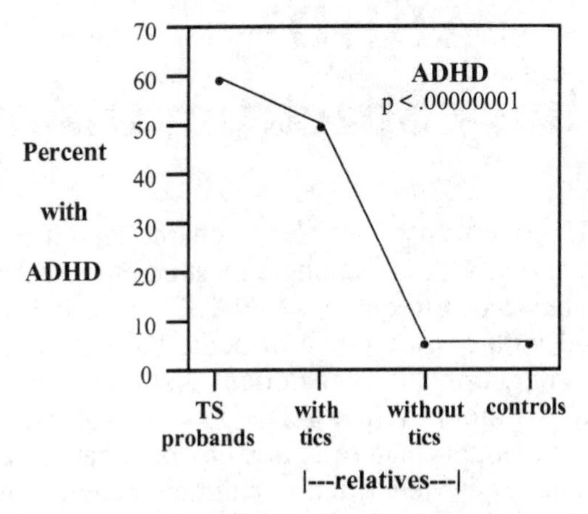

Figure 2. Frequency of ADHD in TS probands, relatives with and without tics, and controls. Comings.[59]

The frequency of ADHD in relatives with tics was significantly greater than in relatives without tics ($p < .00000001$), indicating ADHD is caused by the same genes causing TS. This was verified when the effect of individual genes was examined (see Chapter 10).

ADHD Children as Adults

It is well-known that children with ADHD are at increased risk when they grow older to develop problems with delinquency, conduct disorder, drug and alcohol abuse, antisocial personality disorder, and related difficulties.[14,52,111,174,225,231,249,250] The most accurate way to study this is to perform longitudinal studies where both ADHD and control children are followed for many years. These are called prospective studies and are much more accurate because they are begun before the individuals develop these problems, thus avoiding a number of biases inherent in selecting the subjects after they have problems. The results of a number of longitudinal studies of ADHD have been reviewed in a companion volume, and the interested reader is referred to those accounts.[52(p89-94),151-162] However, the following is an important longitudinal study not covered in those accounts, and is presented to illustrate the major life-long effect that a childhood genetic disorder, ADHD, can have on individuals.

This fifteen-year longitudinal study of ADHD was performed by Howell and colleagues[135] at the University of Vermont. It consisted of 369 children who had been examined in second, fourth, and fifth grade, and again in ninth and twelfth grades, and then finally re-interviewed three years after high school. A unique aspect of this study was that instead of simply comparing the outcome of those identified in grade school as having ADHD versus those who did not, it studied three groups based on their scores on a structured inventory of ADHD symptoms – those consistently scoring in the 80% or higher range (the ADHD group), those consistently scoring in the 20% or lower range (the low group), and the rest (the normal group). The addition of the low group allowed the evaluation of those who not only did not have ADHD, but who had virtually no symptoms of it.

The interviews in the ninth and twelfth grade indicated that those showing ADHD behaviors in early grade school performed much more poorly in high school and had poorer social adjustment. In early adulthood they continued to do more poorly than the normal group and much more poorly than the low group. The following are some diagrams that illustrate the differences between the ADHD, normal, and low groups for school performance.

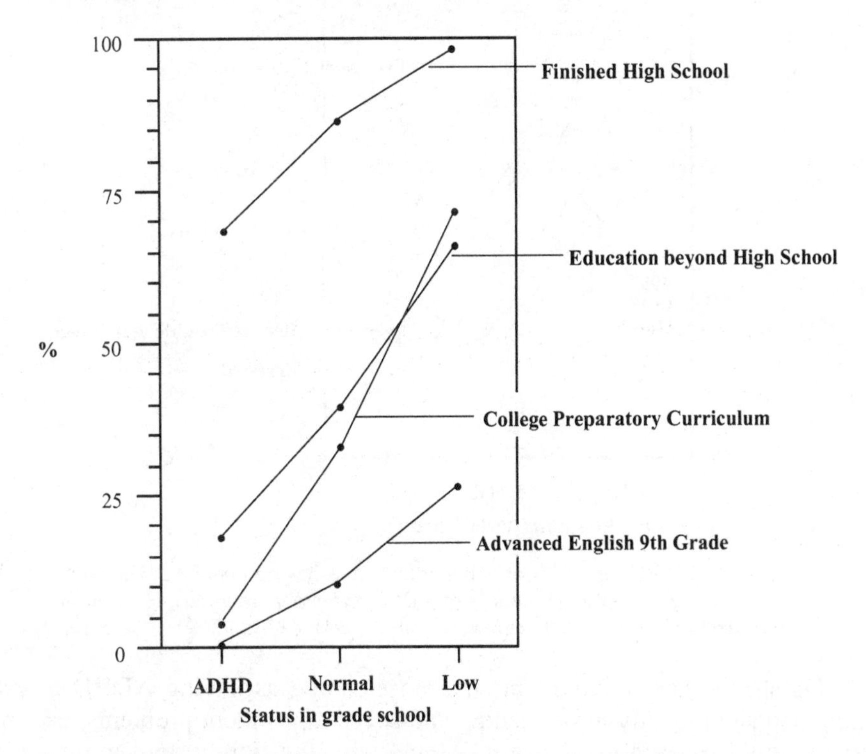

Figure 3. Educational outcome as adults of children ranked as having many ADHD symptoms (ADHD), an average number of ADHD symptoms (normal), or virtually no ADHD symptoms (low) in grade school. From Howell et al.[135]

The low group consistently did the best, the normal group was intermediate, and the ADHD group fared the worst for all measures of school performance. The difference in college preparatory curriculum was the most dramatic, with virtually none of the ADHD group taking college preparatory courses, while over 75% of those in the low group took such courses. Additional assessments relating to school, employment, and substance abuse are shown in the following figure.

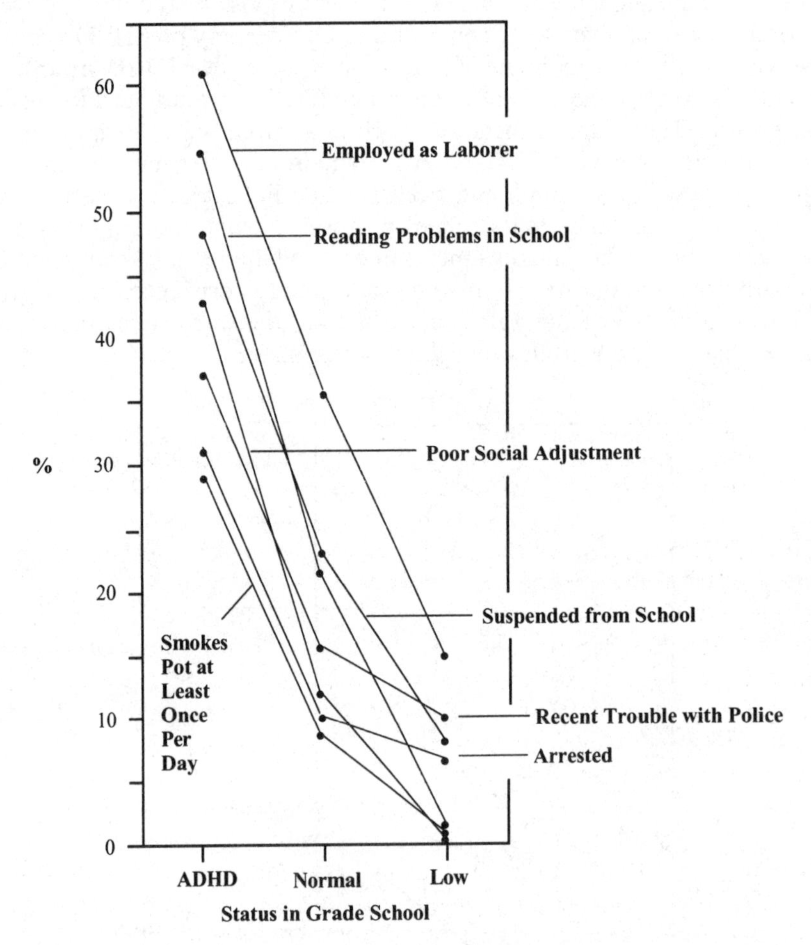

Figure 4. Outcome as adults of children ranked as having many ADHD symptoms (ADHD), an average number of ADHD symptoms (normal), or virtually no ADHD symptoms (low) in grade school. From Howell et al.[135]

This shows that, as young adults, those scored as having ADHD in grade school had significantly more reading and social adjustment problems, had more suspensions from school, more problems with the police, and more arrests. Almost 30% smoked pot daily, compared to none of those in the low group. Over 60% of the ADHD group were employed as laborers compared to 15% of the low group.

Not surprisingly, as shown in the following figure, the grade point average of the three groups was also significantly different.

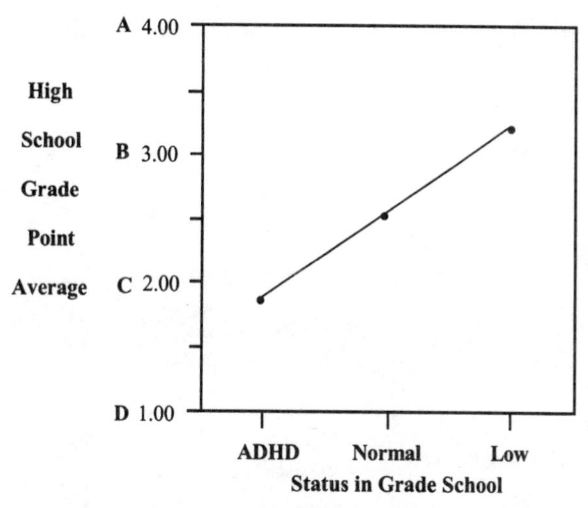

Figure 5. Grade point outcome in high school of children ranked as having many ADHD symptoms (ADHD), an average number of ADHD symptoms (normal), or virtually no ADHD symptoms (low) in grade school. From Howell et al.[135]

The ADHD group had a D+ average in high school, the normal group a C+ average, and the low ADHD symptom group, an A-B average.

These results illustrate the dramatic manner in which a high versus a low number of ADHD symptoms in early grade school can predict the academic, socioeconomic, psychiatric, and legal outcomes of these children as adults. The significant long-term stability of these behaviors is consistent with the genetic basis of ADHD. This suggests that both ADHD genes and IQ genes play a significant role in socioeconomic status.

Summary – ADHD is a hereditary neuropsychiatric spectrum disorder. The spectrum of associated conditions is the same as seen in TS and includes impulsive, compulsive, addictive, aggressive, depressive, anxious, and conduct disordered behaviors. As with TS, genetic variants of the dopamine D_2, dopamine β-hydroxylase, dopamine transporter, and other genes contribute to the clinical picture. The presence or absence of ADHD symptoms in childhood is strongly predictive of adult educational, occupational, psychiatric, and socioeconomic status. This long-term stability is consistent with the genetic basis of ADHD and related behaviors.

Chapter 12

Conduct Disorder

Conduct disorder (CD) is a childhood behavioral problem characterized by a continued disregard for the rights and property of others. Typical behaviors include lying, stealing, setting fires, playing hooky, breaking into others' property, destroying property, cruelty to pets, forcing sex, starting fights, using a weapon, and attacking others. Oppositional defiant disorder (ODD) is a related behavioral problem consisting of losing temper easily, constantly arguing with adults, defying adult rules, deliberately annoying others, failing to take responsibility, blaming others, and being angry, resentful, and spiteful.

While there have been few studies of genetics of ODD and CD, the involvement of genetic factors is strongly implied because of the high frequency of CD in children with ADHD,[17,109,249] which is a genetic disorder. This is supported by family studies of ADHD probands by Biederman and colleagues[17,109] where, of seventy-three ADHD probands, 45% had comorbid ODD and 33% had CD. Among relatives, the risk for any antisocial disorder (ODD, CD, antisocial personality disorder) was highest when the proband had ADHD and CD (34%) or ADHD and ODD (24%) versus 4% for controls and 7% for other psychiatric probands.

To examine the role of TS genes in CD and ODD, we examined the frequency of these two disorders in TS probands, their relatives, and controls. Figure 1 shows the results.[56]

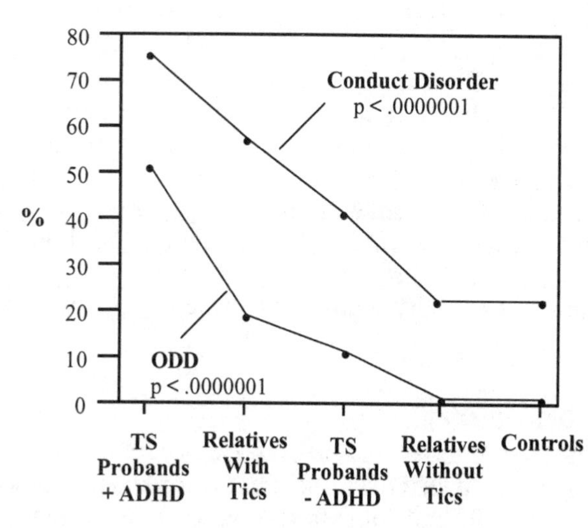

Figure 1. Frequency of conduct disorder and oppositional defiant disorder (ODD) in TS probands with ADHD, relatives with tics, TS probands without ADHD, relatives without tics, and controls. From Comings.[56]

To evaluate the combined effect of TS and ADHD, the TS probands were divided into those with and without TS. The frequency of both CD and ODD was highest in TS probands with ADHD, next highest in the relatives with TS (tics), next in the TS probands without ADHD, and lowest in controls and relatives without tics. In both cases the frequency of CD and ODD was significantly higher ($P < .0000001$) in the relatives with TS than the relatives without TS, indicating the TS genes were playing a significant role in the production of both CD and ODD. In similar studies of ADHD probands we found that 35% of the relatives with ADHD had been arrested, compared to 0% of relatives without ADHD.[56]

Antisocial Personality Disorder (ASP)

ASP is basically conduct disorder in adults. It is also characterized by a disregard for the feelings and rights of others, along with various types of criminal acts. In contrast to conduct disorder, numerous studies have described genetic factors in ASP and criminality. In one study of adult criminal behavior in twins,[44] if one identical twin had criminal behavior there was a high probability the other also had criminal behavior. This was much less likely to be the case for non-identical twins.

West and Farrington[271] found that criminality in parents was more strongly associated with delinquency in offspring after puberty than before puberty. The most powerful evidence for a genetic component to ASP comes from adoption studies which have shown a significantly increased rate of criminality in adoptees with criminal biological parents.[21,47,86,87,137,287]

Intriguing evidence for a combined role of genetic and environmental factors in violent criminal behavior comes from the studies of Brennan, Mednick, and Mednick.[24] They examined the interaction of two variables – parental history of psychiatric illness and low versus high frequency of complications at delivery. If there was no family history of mental illness, the frequency of violent offenses in adulthood was 5% or less. However, in the presence of a positive history of parental psychiatric illness, the frequency of violent offenses was 5% in those with a low frequency of delivery complications versus 33% in those with a high frequency of delivery complications. This suggests that in the presence of an appropriate genetic background, the addition of an environmental risk factor – in this case, complications at delivery – resulted in a higher frequency of violent offenses when those children became adults.

Finally, one of the most repeatedly documented findings in psychiatry is that aggressive behavior is a stable characteristic from early childhood to adulthood, implicating an important role of genetic factors.[110,111,125,169,173,174,176,177,225,226] The studies indicate that the more extreme the childhood antisocial behavior, the more stable it will remain over time.[169]

Frontal Lobe Dysfunction in Delinquency

As detailed elsewhere,[52] the frontal lobes are one of the major regions of the brain affected by the genes that cause ADHD, TS, conduct disorder, and other disruptive, addictive behaviors. The frontal lobes are important for attention span, concentration, long-term planning and goals, the critical evaluation of one's own behavior, the ability to shift activities, and inhibition of impulsive responses. These have been termed the executive functions of the brain. Defects in these frontal lobe functions result in short attention span, impassivity, poor judgment, inability to accept responsibility for one's behavior, poor self-control, and failure to put off immediate reward for future long-term goals. All of these behaviors are characteristic of individuals with ADHD, TS, conduct, and antisocial personality disorders. Defects in frontal lobe and executive functions have long been implicated in these disorders and especially in delinquent and antisocial behavior.[118,122,133,190]

There are a number of neuropsychiatric tests designed to evaluate the function of the executive functions of the frontal lobe. When abnormal, they suggest the defects are organic and neurological, rather than purely psychological in origin.

Between April 1, 1972 and March 31, 1973, all the children born at the Queen Mary Hospital in Dunedin, New Zealand were recruited to enter the Dunedin Multidisciplinary Health and Development Study. Since this was a prospective study, it eliminated the potential problems of ascertainment bias inherent in studies of delinquency based on arrested or incarcerated individuals.

At age 13, the 678 subjects in this study were given a battery of tests of executive function and a self-report test of delinquent behaviors. Of these, theft (especially shoplifting and burglary) accounted for 41.2%, minor assault for 24.7%, vandalism for 10.7%, and substance abuse for 9.9%. The results of the self-report were verified by parental interviews.

The subjects were divided into four groups – those without ADHD or delinquent behavior, those with delinquent behavior but not ADHD, those with ADHD but not delinquent behavior, and those with both ADHD and delinquent behavior. The results were reported in 1989 by Moffitt and Henry.[193] Figure 2 shows the findings for five of the tests that were significant.

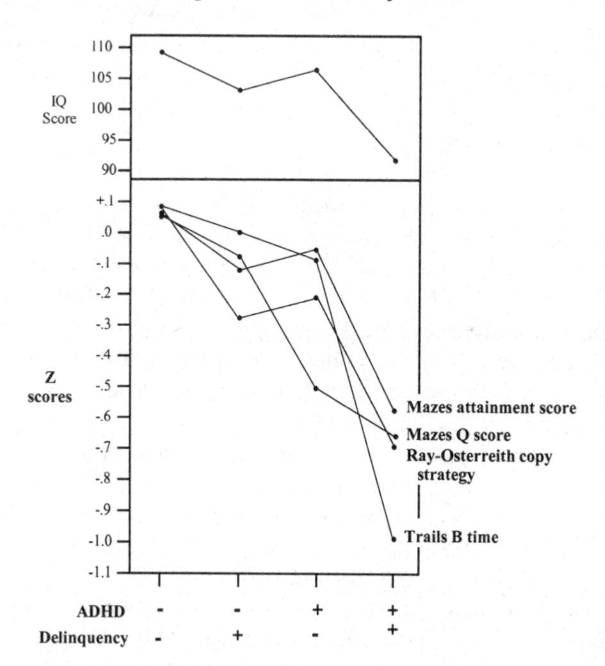

Figure 2. Results of five tests of frontal lobe executive function in subjects in the New Zealand Dunedin Multi-disciplinary Health Development Study. With Z scores -1 = -1 standard deviation from the mean. From Moffitt and Henry, *Development and Psychopathology* 1:105-118, 1989.[233]

What was of particular note was that the most dramatic defects were seen in those with both ADHD and delinquent behavior. These results confirm the important role of ADHD as a risk factor in conduct disorder and demonstrate a neurological rather than psychological cause. There was also a significantly lower IQ score in those with ADHD and delinquency (see also Chapter 21). Although the socioeconomic status of those with both ADHD and conduct disorder was lower than those with no disorder, significant differences among the four study groups remained after the effects of family adversity were partialled out.[180] Other studies have also shown the presence of significant cognitive defects in subjects with conduct disorder and a history of delinquency.[144,188]

Since both ADHD and conduct disorder have a strong genetic basis, the genetic aspect of the story comes full circle if we can identify the genes involved. Dopamine is one of the major neurotransmitters in the frontal lobe. It is thus not surprising that, as with ADHD in the previous chapter, the three dopamine genes examined were also found to play a role in conduct disorder and oppositional defiant disorder (ODD).

The Molecular Genetics of CD, ODD, and Aggression

In conjunction with our study of the role of the A1 allele of the *dopamine D_2* receptor gene in substance abuse,[78] we questioned the subjects for a history of ever having been arrested for a non-violent crime (mostly driving while intoxicated) and violent crime. The latter subjects were divided into those who had a childhood history of expulsion from school for fighting and non-expulsion for fighting. The results for Caucasians were as follows:[78]

	N	**%A1**
Jailed for nonviolent crime	111	28.8
Jailed for violent crime (total)	32	**53.1**
Violent crime, not expelled	19	**42.1**
Violent crime, expelled	13	**69.2**

Of those ever jailed for nonviolent crime, only 28.8% carried the D_2A1 allele. By contrast, of those arrested for violent crime and expelled from school for fighting, 69.2% carried the D_2A1 allele ($p < .001$). This correlation with conduct is supported by the studies of Noble et al. of the A1 allele in cocaine addiction.[198] They found a progressive increase in the prevalence of this allele from 16.6% in those with no risk factors to 87.5% in those with three risk factors. The risk factors were a history of childhood conduct disorder, the type of cocaine used, and family history of alcoholism.

The molecular genetic studies of the three dopaminergic genes separately and together[80] showed they played a significant role in CD and ODD (see Figure 3 on next page). As with the ADHD score, this shows that the three dopaminergic genes also play a role in conduct and oppositional defiant disorders. This is consistent with the higher frequency of these two disorders in individuals with TS and ADHD.

Serotonin. Serotonin is another major neurotransmitter involved in aggressive behavior. The correlation between low spinal fluid serotonin levels and a variety of aggressive behaviors has been reviewed in a companion book.[52] Since that time newer molecular genetic techniques have made it possible to grow mice that are missing specific genes. They are called *knockout mice*. In one study, knockout mice missing the serotonin 1B receptor were significantly more aggressive than normal mice of the same strain.[232] Our studies of the *tryptophan 2,3 dioxygenase* gene, which plays a major role in the control of blood serotonin and tryptophan levels, was described in a companion book.[60]

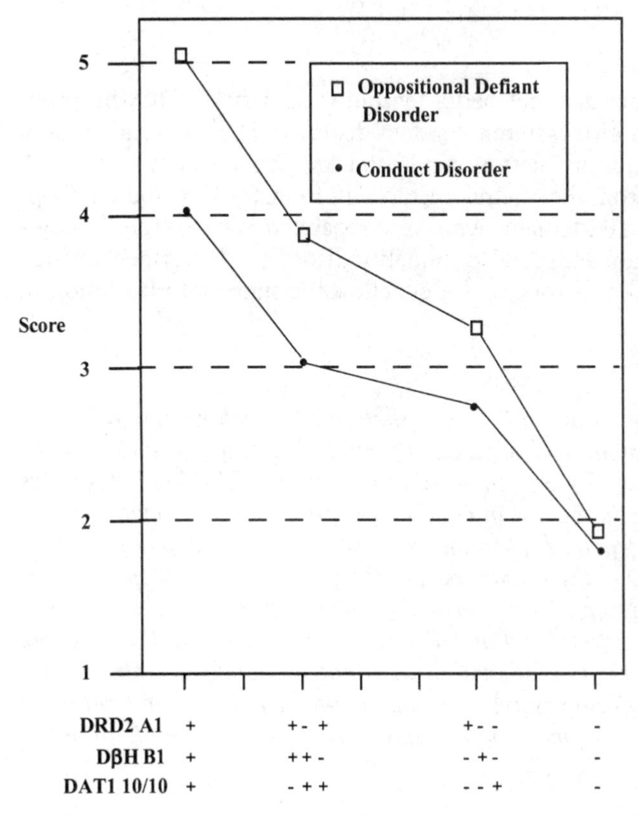

Figure 3. The conduct and oppositional defiant scores in individuals with 3 of 3, 2 of 3, 1 of 3, and 0 of 3 of the mutant alleles for the *dopamine* D_2 receptor, *dopamine β-hydroxylase,* and *dopamine transporter* genes, showing the significant additive and subtractive effect of the combinations. From Comings et al.[80]

Two different studies have shown that a specific allele of the *dopamine* D_4 receptor gene is associated with a novelty-seeking personality.[200a,15a] This personality variable is common in subjects with ADHD, Tourette syndrome, conduct disorder, and antisocial personality disorder. Gelernter and colleagues[122a] have shown an increase in the frequency of this same allele in subjects with Tourette syndrome.

Twin Studies of Conduct Disorder

Recently a large twin study of conduct disorder verified the important role of genetic factors and the concept that, like ADHD and Tourette syndrome, conduct disorder is a spectrum disorder associated with a wide range of other problem behaviors.[245a] For example, the study found that the heritability of conduct disorder was as high as .71, indicating that genetic factors contributed to 71% of the picture. In addition, the risk of other disorders was greatly increased. This is shown in the following table (next page):

Odds Ratio for Various Disorders in Subjects with Conduct Disorder[245a]

Disorder	**Males**	**Females**
Alcohol dependence	2.7	10.0
Depression	2.0	3.9
Panic attacks	2.1	2.7
Social phobias	1.7	3.6
Bulimia	1.9	5.4

Thus, girls with conduct disorder had a tenfold greater risk of having problems with alcoholism than girls without conduct disorder. They were also more likely to have problems with depression, panic attacks, phobias, and bulimia. A twin study of male twins from the Vietnam Era Twin Registry,[174a] and an adoption study of 95 males and 102 females who were separated at birth from biological parents with documented antisocial personality disorders,[30a] further verified the important role of genetic factors, and a genetic-environmental interaction in conduct disorder.

Summary – One of the major indications that genetic factors play a role in conduct and oppositional defiant disorder is the high frequency of these conditions in TS and ADHD patients and their relatives. Twin and adoption studies also support a role of genetic factors in conduct disorder and antisocial personality disorder. Molecular genetic studies indicate the same three dopamine genes involved in TS and ADHD – the dopamine D_2 receptor gene, the dopamine β-hydroxylase gene, and the dopamine transporter gene – also play a significant role in conduct and oppositional defiant disorder. The dopamine D_4 receptor gene also appears to be involved. Serotonin is another neurotransmitter that plays an important role in aggressive and antisocial behaviors, and relevant genetic defects in serotonin enzymes and receptors are rapidly being identified.

Chapter 13

Learning Disorders

The genetics of learning problems is complex because there are so many causes of learning disabilities. They can be broadly divided into those due to low IQ (exclusive of overt mental retardation) and those not due to low IQ. As discussed in Chapter 8, since genetic factors play a major role in IQ, they will also play a significant role in the cause of low IQ learning disorders. Thus, if both the mother and the father have an IQ in the 70 to 80 range, and birth trauma or childhood diseases are not a factor, it is highly likely that their children will also have an IQ in the 70 to 80 range.

It is the normal IQ type of learning disability that I will discuss in this chapters. These are often referred to as specific learning disabilities because they involve circumscribed cognitive difficulties, such as specific reading disorder (dyslexia), specific math disability (dyscalculia), auditory and visual perceptual disorders, and others. The detailed definition of specific learning disorders, as based on Public Law 94-142, is given in the glossary. The definition in practice is that there must be a significant discrepancy between a child's potential and actual achievement.

In many cases, such individuals with specific learning disabilities have an associated genetic disorder such as ADHD or TS. In fact, the definition in Public Law 94-142 includes "such conditions as...minimal brain dysfunction," the older term for ADHD.

ADHD and Learning Disorders. The association between ADHD and learning disabilities has long been recognized. On average, between 7 to 50% of children with ADHD are also diagnosed as having a learning disorder[17,236] and are either in or qualify for special education in the form of special day classes (all day), resource classes (one to three hours), or special schools. Studies have repeatedly shown that children with ADHD perform more poorly in school than controls and are more likely to repeat grades, require tutoring, or be placed in special classes, and do more poorly on measures of academic achievement.[31,97,163,242] Longitudinal studies show that this chronic school failure persists into adolescence.[125,270]

While it should come as no surprise that children who are inattentive, impulsive, and hyperactive would tend to perform poorly in academic subjects, this is not the whole explanation of the learning disorders in these children, as neuropsychiatric testing often shows various combinations of auditory and visual perceptual disorders, frontal and parietal lobe defects, short- and long-term memory problems, EEG abnormalities, changes in cerebral blood flow, and other objective measures of brain abnormalities.[52]

In one study,[236] the prevalence of reading disabilities in ADHD subjects was 15% versus 0% in controls when a rigid definition of learning disability was applied. When the definition was slightly less rigid the respective figures were 23% and 2%. In the same study, when defined fairly rigidly, the prevalence of math disabilities was 20% in ADHD subjects and 0% in the controls. When the definition was slightly less rigid, the prevalence did not increase in ADHD subjects but increased to 22% in the controls. Thus, the more rigid the criteria for learning disabilities, the greater the difference between children with ADHD and controls.

Tourette Syndrome and Learning Disorders. TS is very similar to ADHD, except that it is also associated with the presence of motor and vocal tics.[52] ADHD is present in 30 to 80% of cases of TS.[52,61,62,67] It is thus not surprising that the association with learning disorders is the same as that seen in ADHD. Specifically, in a study of 247 consecutive cases of TS, we found that 35% had already been placed in some type of special class predominantly for those with learning disorders. In those with severe TS, 60% had been placed in such classes.[62] In addition, 26% had been held back a grade, 13% had required a home teacher, and 41.5% had problems with reading comprehension, compared to 8.3% of the controls.[62]

In an epidemiologic study in a school district, we found that 28% of children in special education classes had TS.[77] Other studies have also documented the presence of learning disorders in subjects with TS.[105,106,127,162,166,203,244] In their own epidemiologic study of schools, Kurlan et al.[162] reported the presence of TS in 26% of children in special education classes, virtually identical to our findings.

The prevalence of learning disorders was also significantly increased in the relatives who also had TS compared to the relatives without TS,[59] indicating the learning problems are not due to ascertainment bias but are associated with the *Gts* genes. Since TS is so clearly a genetic disorder, this high degree of association of learning disorders with TS provides some of the strongest evidence for the important role of genetic factors in learning disabilities.

Conduct and Learning Disorders. Several studies have shown that children with ADHD and aggressive conduct disorder have more cognitive difficulties than children with just ADHD alone.[133,183,191,233] The significance of this for the present hypothesis is that if both are due in large part to genetic factors, because of the tendency for children with conduct disorder to drop out of school early as a result of learning disabilities, and then to start having children, there will be a simultaneous selection for the genes for aggression and conduct disorder. Further evidence for this is presented in Part III.

Learning Disorders and Behavioral Disorders. One might object that the learning disorders associated with behavioral disorders such as ADHD, TS, and conduct disorder might represent only a small percentage of the total and that most children with learning disabilities have no concomitant behavioral problems. While the finding of TS in approximately a third and of ADHD in 50 to 80% of children in learning disabled classes speaks against this, testing for the prevalence of behavioral disorders in unselected children with learning disorders would be more convincing.

Such a study was performed by Schachter and colleagues.[234] They examined[502] learning disordered children using the Achenbach Child Behavioral

Checklist described in Chapter 5. The prevalence of a range of behavioral problems was 43%, compared to 10% expected in the general population. They approached the possibility of teacher referral bias, consisting of mistaking behavioral problems for learning disorders, by comparing *children referred by teachers* to *children referred by parents*. This assumed the parents would be concerned about academic performance whether they were or were not associated with behavioral problems. There was no difference in the prevalence of behavior problems in the teacher versus the parent referred children.

The presence of learning disabilities can lead to frustration with school, and this may cause behavioral problems. However, the authors concluded that while this was a factor in the 1.7-fold greater frequency of behavioral problems in learning disordered adolescents, it was not a factor in the younger children. The results were consistent with numerous other studies showing a 24 to 52% prevalence of behavioral disorders in learning disordered children.[182,230,234]

Poor children, compared to middle class children, have a significantly higher frequency of ADHD.[205] In one study, 40.0% of male children of parents on welfare had a psychiatric disorder compared to 13.9% of male children of middle class parents.[204] In addition, 27.8% of female children of parents on welfare had poor school performance compared to 6.1% of female children of middle class parents. While one explanation is that the environment of poor children predisposes them to behavioral and learning disorders, an alternative explanation is that parents with genes for learning and disruptive behavioral disorders are more likely to require welfare assistance and they pass these genes on to their children. Most likely, both of these explanations provide part of the truth.

These results are consistent with the hypothesis that learning disorders are polygenically inherited, and the same genes contributing to learning problems also contribute to a range of behavioral problems.

Dyslexia. The simplest and most functional definition of dyslexia or specific reading disorder is that any child with otherwise normal intelligence whose reading skills and comprehension are two or more years behind same age norms without a clear cause (such as a chronic illness resulting in poor school attendance), has a specific reading disability. Many studies have documented the hereditary nature of dyslexia.[215] As with many neuropsychiatric disorders, there was a period when dyslexia was thought to be inherited in an autosomal dominant fashion as a single gene disorder. It is now clear that reading disabilities form a continuum from normal to severely impaired,[238] consistent with a polygenic mode of inheritance.

The Molecular Genetics of Learning Disorders

The reason for the high frequency of learning disabilities in children with ADHD or TS has often been debated. The molecular biology studies are beginning to shed some light on this question. For example, as shown in Figure 1, the B1 variant of the *dopamine β-hydroxylase* gene showed a significant association with learning and reading problems, and with performance in grade school, in subjects with TS.

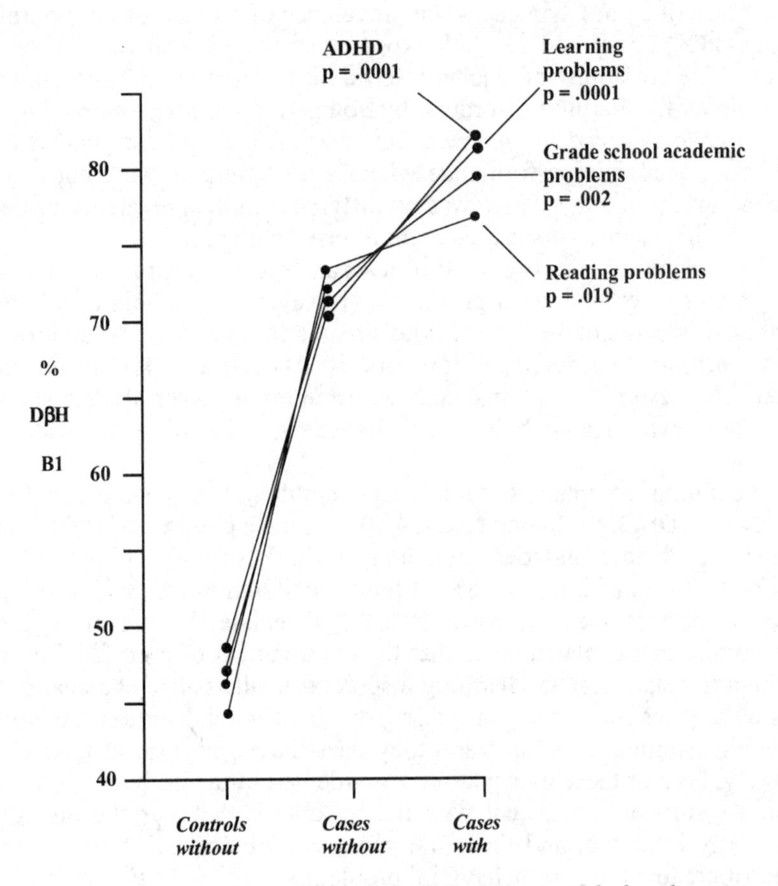

Figure 1. Prevalence of the *Taq* B1 allele of the *dopamine β-hydroxylase* gene in controls without the disorder (*controls without*), TS probands and relatives without the disorder (*cases without*), and TS probands and relatives with the disorder (*cases with*).

This is consistent with the assumption that ADHD, TS, learning, reading, and other disorders are the result of the polygenic inheritance of a number of genetic defects, and that these different disorders have many genes in common. The greater the number of adverse genes affecting neurotransmitter function that a child inherits, the greater the number and the more severe the disabilities, including learning disorders.

Summary – Learning problems can be due to a constitutionally low IQ or to specific learning disabilities defined as learning problems in the presence of a normal IQ. A high percentage of children with hereditary ADHD and TS have specific learning disabilities. In addition, dyslexia (specific reading disability) and dyscalculia (specific math disability) can occur as isolated, usually hereditary, entities. Molecular genetic studies have shown that dopamine genes, such as dopamine β-hydroxylase, play a role in learning disabilities.

These and other studies indicate the important role of genes in all types of learning disabilities.

[The genetic factors contributing to alcohol and drug addiction have been presented in detail in the two companion books *Tourette Syndrome and Human Behavior*[52] and *Search for the Tourette Syndrome and Human Behavior Genes.*[60]]

Part III

Evidence that People with Addictive-Disruptive Behaviors Have Children Earlier

Part II reviewed the evidence that genetic factors play an important role in many impulsive, compulsive, and addictive behaviors. This, by itself, provides no proof that selection for genes accounts for the increasing frequency of these disorders. To provide such proof requires that the individuals with these disorders are doing something that contributes to the selection of their genes. This "something" can be having more children, coming from larger families (having more siblings), or having children earlier. Part III reviews the evidence that any or all three of these may be occurring, with a special emphasis on having children earlier. I apologize for the number of graphs, charts, and statistics. However, a diagram can save many words and instantly portray the complex information that has been obtained from a number of different sources, and three different countries, to show that the trends described are real and not limited to a single study, state, or country.

Chapter 14

Gene Selection

Before presenting chapters showing a marked difference in age at the time of the birth of the first child for individuals with learning problems, drug and alcohol abuse, delinquent, and other behavior difficulties, compared to individuals without these conditions, I will first illustrate why these differences are of concern from a gene selection point of view. Let us suppose that a mutant gene called LD causes learning disabilities, and that individuals with these learning disabilities drop out of school earlier and start having children earlier than individuals who do not carry the mutant gene. Let us also assume that the average age at the birth of the first child of LD carriers is 20 years, while for those not carrying this mutation it is 25 years. As a result, the mutant form of the gene will reproduce faster, namely every 20 years, while the normal form of the gene will reproduce every 25 years. The ratio of 25/20 is 1.25. Figure 1 illustrates the results of using one of the equations[107] that calculates the rate at which the frequency of a gene with such a selective advantage will increase over succeeding generations.

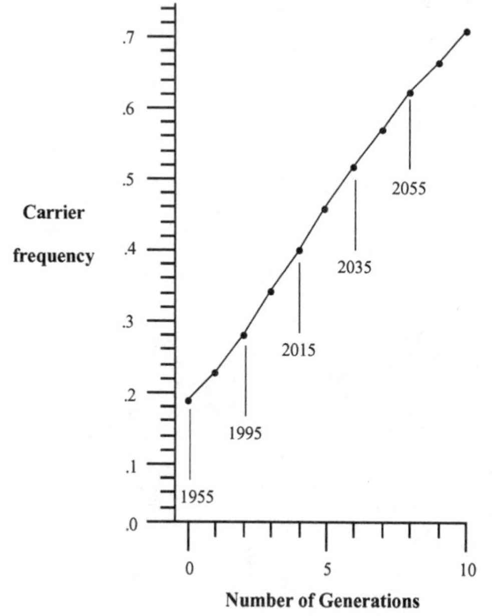

Figure 1. The rate at which a gene that has a 1.25-fold selective advantage will increase in frequency from generation to generation.

To provide a base for anchoring Figure 1 in the real world, I have proposed that the selection started picking up speed at around 1955. Subsequent times have been entered at 20-year intervals. This shows that under the above circumstances, the carrier frequency of such a gene could double from 1955 to 2015, and could be have increased 150% from 1955 to 1995.

There are many caveats about these figures that must be mentioned. First, in this book I am discussing polygenic inheritance rather than single gene inheritance. While the selection of many genes rather than one gene is clearly more complex, as discussed in Chapter 41, the end effect could be essentially the same. Second, many additional factors are involved, including the number of children, number of siblings, number of genes, different selective pressures on the different genes, and other factors. While I am not a population geneticist and would not purport to be able to derive the complicated formulas necessary to factor in all these variables, I also suspect the reader would not be interested in pages of dense formulas. The important point is that even if an army of mathematicians concluded these figures were not exactly accurate, I would contend that the precise rate of selection of these genes is irrelevant. If a blindfolded person is walking toward the edge of a thousand foot cliff, unless he stops and turns around, it makes little difference whether he is walking one mile an hour or five miles an hour. He will eventually reach the precipice and fall over.

The following chapters present evidence that the walk has begun.

Summary – A difference of four to five years in the age of mothers or fathers when they have their first children is sufficient to result in a significant and relatively rapid selection for genes carried by the group initiating childbearing at the earlier age.

Chapter 15

The Adolescent Problem Behavior Syndrome

A syndrome is a group of symptoms that cluster together more often than can be explained by chance. For many years various investigators have recognized that a number of problem behaviors in adolescents tend to cluster together – that is, any person who has one is more likely to have one or more of the others. These behaviors include alcohol abuse, illicit drug use, smoking, delinquent behavior and precocious sexual intercourse (see box).

Adolescent Problem Behavior Syndrome

- alcohol abuse
- drug abuse
- smoking
- delinquent behaviors
- precocious sexual intercourse
- teenage pregnacies

In 1977 Richard and Shirley Jessor from the University of Colorado in Boulder[140] reported their research on what they termed Problem Behavior Theory, which suggested that problem drinking, marijuana use, delinquent behaviors, and sexual intercourse may constitute a syndrome. Based on longitudinal studies of high school and college students, they found that these behaviors were positively associated with each other, i.e. if one occurred, the others were also more likely to be present. They also found that individuals with this syndrome were less likely to be involved in such accepted conforming and conventional behaviors as regular school attendance and going to church. They proposed that this syndrome was due to some unknown "latent variable of unconventionality in adolescence "[94]

The original studies were performed primarily on middle class White youths. In a further analysis of this data Donovan and Jessor[94] examined the variables *times drunk*, *frequency of marijuana use*, *frequency of sexual intercourse*, and *general deviant behavior* and found significant correlations between all four measures. They also found that "only a single common factor was needed to account for the correlations among the problem behaviors." This chapter will review additional evidence on the existence of this problem behavior syndrome independent of teen pregnancy. Since precocious sexual intercourse obviously increases the risk of teen pregnancies, subsequent chapters will address the issue of the co-occurrence of this syndrome with early pregnancy.

In a second part of their study, Donovan and Jessor[94] examined a more racially heterogeneous group of junior and senior high school students from a 1978 National Study of Adolescent Drinking. This time three additional variables were added – *number of cigarettes smoked per day in the last month, church attendance frequency in the past year*, and *school performance*. There was a positive correlation between each of the five variables on problem behavior (correlation coefficients of .32 to .59) and negative correlations between each of the problem behaviors and the two conventional behaviors (correlation coefficients of -.14 to -.28). Again, the results could be explained by a single unknown factor. This study indicated the results held true across widely differing socioeconomic and racial groups. They were able to rule out the "counter-culture" of the late 60s and early 70s as the common factor. They re-studied the sample examined in the 1977 when the subjects were young adults in their middle and late 20s. They found that although the majority of the problem drinkers no longer abused alcohol,[95] the single factor model still held; thus the problem behavior syndrome was a stable rather than a transient phenomena.

In attempting to identify the nature of this common factor, Donovan and Jessor[94] concluded it reflects a general dimension of *unconventionality* – in both personality and social environment. For example, in the 1977 study, these problem behaviors were also associated with the following psychosocial attributes: lower value on academic achievement or recognition, greater value on independence rather than achievement, lower involvement in religion, greater tolerance for socially disapproved behaviors, and greater weight placed on positive reasons for drinking, drug use, and sex. The suggested explanations of the co-occurrence of these problem behaviors were based on social learning theory and concluded that possible reasons for this clustering might be that behaviors that are learned together tend to continue to be performed together, or that there was peer pressure for the involvement in multiple problem behaviors.

Kaplan[45-147] has also proposed a common underlying factor for what he termed adolescent "deviant" behavior consisting of alcohol and drug use, delinquency, aggression, and mental illness. He proposed the common factor was psychological – a self-rejecting attitude. In a further study of the general deviance model, Osgood and colleagues[212] also concluded that a general deviance factor was involved, but that an explanation of the total picture also required some behavior specific factors.

Other studies have shown correlations between different parts of the adolescent problem behavior syndrome. These include alcohol and drug abuse,[136,143] alcohol, drug, and tobacco use,[12] illicit drug use and criminality,[100,141] alcohol, drug use, and sexual activity,[189,196,278] and sexual activity and problem behaviors.[104]

A study by Bachman and coworkers[12] of smoking, drinking, and drug use among American high school seniors showed that above average drug use occurred among students less successful in adapting to the educational environment. Some of these correlations are shown in Figure 1.

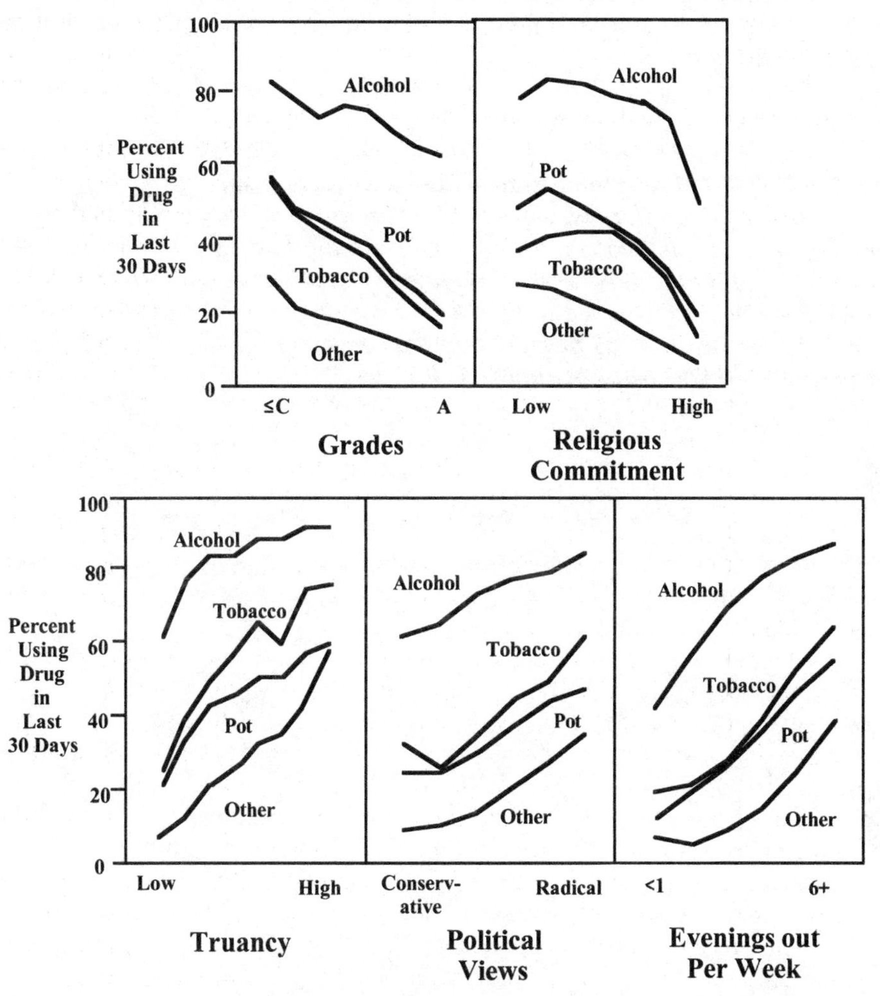

Figure 1. Negative correlations between grades and religious commitment and positive correlations between truancy, political views, and evenings out per week with drug use in high school seniors. Redrawn from Bachman et al. *Am. J. Public Health* 71:59-69, 1981.

The highest correlations with drug use were with truancy, grades, religious commitment, political views, and evenings out per week. These correlations relate to the degree to which young persons were under the direct influence or supervision of adult-run institutions – the school, home, and church. As in the earlier studies, the authors also found that some individuals were especially predisposed toward problem behaviors.[141] Although the specific drug used, such as a preference for tobacco, alcohol, marijuana, or cocaine, changed over the years, the above risk factors remained the same, *indicating the common factor is with a tendency for general deviant behavior rather than with any given substance.*

None of these studies clearly identified the nature of the unconventionality factor. As detailed in this book, I believe the common factor is the inheritance of a group of genes leading to a predisposition to addictive, aggressive, and other problem behaviors.

Summary – The adolescent problem behavior syndrome consists of a clustering of alcohol and drug abuse, smoking, delinquent behaviors, and precocious sexual intercourse often associated with teenage pregnancy. On average, the greater the drug and alcohol use, the poorer the grades, the lower the commitment to religion, the greater the truancy and evenings spent away from home. This clustering was assumed to be due to some unknown common factor related to all of these behaviors. I propose that the common factor is a set of genes predisposing to addictive and disruptive behaviors.

Chapter 16

Conduct Disorder and Teenage Pregnancy

The second half of the twentieth century has witnessed a dramatic increase in teenage pregnancies, many to single mothers. The percentage of teen births out of wedlock increased from 15% in 1960 to 48% in 1980.[13] Having a child out of wedlock has passed from being a shameful event to being widely accepted. Is being a teenage single mother a random event that equally affects all teenagers, or is it concentrated in certain groups?

The previous chapter described a syndrome of adolescent problem behaviors consisting of substance abuse, delinquency, and early sexual intercourse. Since the early initiation of sexual intercourse clearly increases the risk of teenage pregnancy, this suggests that teenage mothers and fathers would show an increased frequency of these problem behaviors. Do they? A number of studies indicate they do.

Using data on 706 young women from a longitudinal survey of high school youth, Yamaguchi and Kandel[277] found that such variables as *cohabitation, having had poor grades, a lot of social involvement with peers, use of illicit drugs other than marijuana*, and *having dropped out of high school* were associated with a two- to threefold increase in the risk of a premarital pregnancy. Their results supported the statement of Teachman and Polonko[259] that "education operates to defer premarital pregnancies that end in live births."

Elster and colleagues[101] at the University of Utah examined a national sample of urban and rural youth and divided women into three groups a) school age mothers, b) mothers who had a child between age 19 and 21, and c) women who had not had a child by age 21. The school age mothers had been engaged in the most problem behaviors, followed by the 19- to 21-year-old mothers, followed by women who had not children by age 21. The school age mothers were more likely than the other groups to have been involved in school suspensions, truancy, running away, smoking marijuana, and fighting. Although the results were similar for urban and rural mothers, the effects were more consistent for the urban mothers.

While the possible association between teenage fathers and problem behaviors has been investigated less than problem behaviors in teenage mothers, a number of studies have been reported. Elster and colleagues,[103] found that teenage fatherhood was associated with school problems, excessive use of marijuana, and delinquency, regardless of the family income. The association was more prominent in White and Hispanic youth than Black youth. They also found[102] that in a clinic sample of mostly White youths from lower middle class families, youthful fathers had a higher cumulative arrest rate than males in the general population. For example, for fathers younger than 18 years of age, 61% had an arrest record, compared to 12 to 29% for general youth from a comparable social background. For fathers 21 years of age or less, the comparable rates were 51% versus 13 to 30% for non-fathers.

While the studies described previously showed that conduct disorder, substance abuse, sexuality, and teenage pregnancy formed a unitary cluster of behaviors, the design of the studies did not demonstrate to what degree, if any, the conduct disorder problems occurred before the pregnancies. One way to approach this would be to carry out observations both before and after a subject's pregnancy, making it possible to more clearly identify which behaviors came first.

Maria Kovacs and her associates at the University of Pittsburgh School of Medicine,[161] were able to examine this because they had already begun a longitudinal study of depression in 183 children. Since the study was initiated when the subjects were 8 to 13 years old, it predated the onset of possible teenage pregnancy. Of these, 83 were girls who formed the basis of the investigation. These were originally students who had been referred to the psychiatric outpatient services for a variety of reasons. Of the group, 59% were White and 41% were Black, 72% were from low socioeconomic status, 28% from high. The girls averaged 11.5 years of age at the beginning of the study and they were followed for up to 12 years. By the end of the study 29 of the girls had become pregnant at ages ranging from 13 to 23 years. They had a total of 65 pregnancies, with some girls having up to 6.

The authors had assumed that depression might be a risk factor for the development of teenage pregnancy. They found it was not. Only 28% of the pregnant teens had an early onset of depression as compared to 66% of the nonpregnant teens. What they did find was that conduct disorder was a major risk factor. Thus 76% of the pregnant teens had had conduct disorder by the time they were 18 years old, as compared to 24% of the nonpregnant girls. The most dramatic finding of this study was the degree to which the girls with conduct disorder became pregnant at an earlier age than those without conduct disorder. This is shown in Figure 1 (next page).

Socioeconomic status was not a significant factor in teenage pregnancy, except that three-fourths of all the girls were of low socioeconomic status. Living in a non-intact family was also not a factor. The authors stated that "...the presence of conduct disorder is the only psychological variable that predicts teenage pregnancy in this sample." "Of the 31 girls with conduct disorders, 54.8% eventually became pregnant as teenagers [13-18 years], whereas of the 50 girls who did not have conduct disorder (but had other psychiatric diagnoses), only 12% became pregnant as teenagers."

The mean age at which the conduct disordered girls became pregnant was 17.3 years, compared to 18.7 years for these without conduct disorder. From a gene selection point of view, this would be trivial; however, this is only the mean age of first pregnancy of those who became pregnant. As shown in Figure 1, approximately 72% of the girls with conduct disorder had become pregnant by age 20, while only 35% of those without conduct disorder had become pregnant. Figure 2

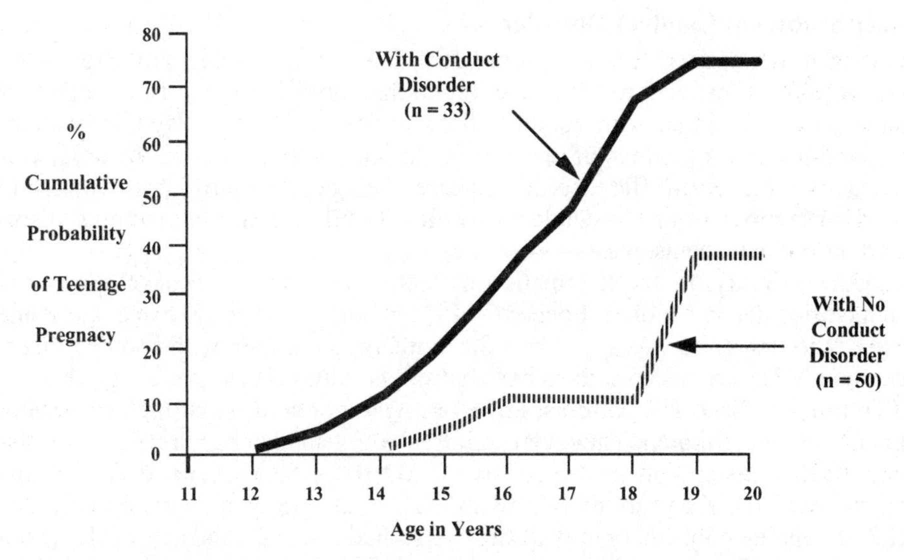

Figure 1. Comparison of the cumulative frequency of pregnancy by age, in girls with conduct disorder versus those without conduct disorder. Redrawn from Kovacs M. et al. *J. Am. Acad. Child and Adolescent Psychiatry* 33:106-113, 1994.

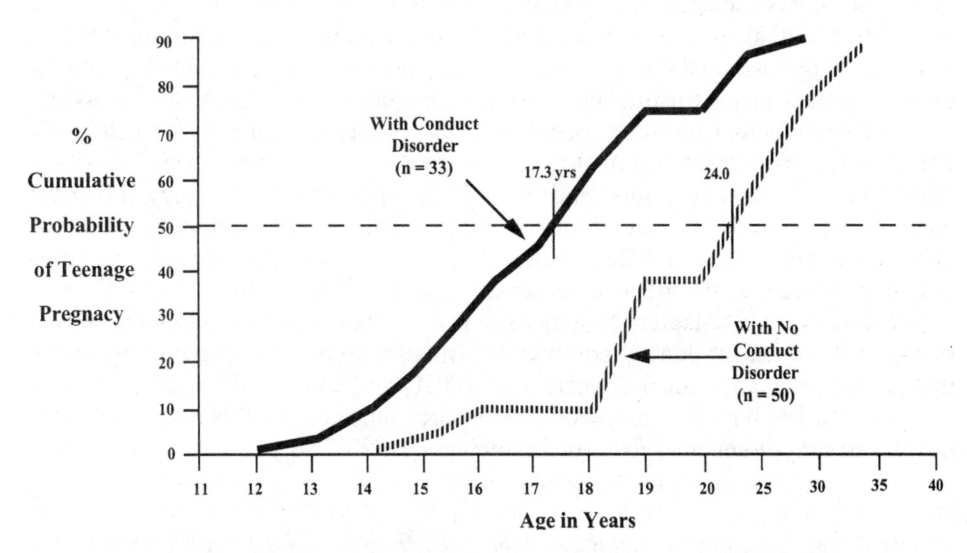

Figure 2. Extrapolation of data from Figure 1 to lifetime cumulative probability of pregnancy by age of first pregnancy.

illustrates what happens if these pregnancy rates are extrapolated to age 40.

Since almost all of those girls with conduct disorder were pregnant by age 20, the mean age of first pregnancy was kept at 17.3 years as reported by Kovacs. The extrapolated rate for the girls without conduct disorder produced an approximate mean age of first pregnancy of 24 years; this is a difference of 6.7 years.

Other Studies of Conduct Disorder

In a much earlier, totally independent 1982 study, Behar and Stewart[251] reported almost identical results. They found that the mean age at first pregnancy of mothers of children with conduct disorder was 18.7 years. By contrast, the comparable age for mothers of the control children without conduct disorder was 23.0 years, a 4.3-year difference. In this case, the age at the birth of the first child was for the mothers of the conduct disordered children, not the conduct disordered individuals themselves.

More recently, in an examination of factors leading to the development of conduct disorder in children, Loeber et al.[170] reported that the average age at the birth of the first child was 20.7 years for mothers of children with conduct disorder versus 23.5 years for mothers of "control" children. These were significantly different ($p \leq .001$). The controls, however, were not controls in the usual sense of being normal children. They were children who had also been referred for disruptive behaviors and diagnosed as having ADHD or oppositional defiant disorder, but over the six years of follow-up did not develop conduct disorder. The children who developed conduct disorder also had a significantly lower IQ (both verbal and performance), more symptoms of concomitant oppositional defiant behavior, more fighting, more resistance to discipline, more paternal and maternal substance abuse, poorer supervision, more inconsistent or ineffective discipline, and a lower socioeconomic status.

Further statistical tests showed that of these variables, socioeconomic status, presence of oppositional defiant behaviors, and parental substance abuse were the most important factor for predicting the development of conduct disorder. While many of these factors are often considered to be purely environmental causes, it is important to realize that one of the reasons the parents abuse drugs and are ineffective providers, supervisors, and disciplinarians, is that they carry the same genes that have been passed on to their children. This is supported by the observation[172] that children with ADHD at the time of the first interview were 5.6 times more likely to develop conduct disorder than those children without ADHD.

As discussed in Chapters 11 and 12, it is clear that ADHD and conduct disorder have a significant genetic component. As discussed in Chapter 14, when the age of first pregnancy for those carrying ADHD and conduct disorder genes is 4 to 7 years earlier than for those who do not carry those genes, this results in a significant degree of genetic selection for such genes.

Summary – Studies of teenage pregnancies show a high correlation with conduct disorder. In one study the mean age of first pregnancy for girls with conduct disorder was 17.3 years, versus approximately 24 years for those without conduct disorder – a 6.7-year difference. To the extent that conduct disorder is a genetically influenced condition, this earlier age of initiation of childbearing would strongly select for the genes involved.

Chapter 17

Teen Attitudes About Getting Pregnant

In addition to examining the characteristics of teenagers who get pregnant, it is also informative to ask nonpregnant, unmarried teenagers about their attitudes toward pregnancy, especially if it is followed up later by determining how they acted on those attitudes. This type of study was done by Allan Abramamse, Peter Morrison, and Linda Waite from the Rand Corporation in Santa Monica, California.[2,3] They obtained their data from the *High School and Beyond* study, a large, nationally representative investigation begun in the spring of 1980, with follow-up surveys at two-year intervals. The initial survey involved men and women in the sophomore and senior classes of 1,015 schools across the U.S. In 1982, 88 to 96% of these students were found and re-examined as to whether they were still in school or not. The authors reported on the follow-up of 13,061 of the female sophomores of the initial study. Since this was a study about attitudes of unmarried women toward childbearing, it excluded those who were married at the time of the initial survey. The 3,293 women who claimed in 1980 that they would consider having a child out of wedlock were compared to their peers who rejected this idea.

In the initial survey, each sophomore was asked, "Would you consider having a child if you weren't married?" They had one of three choices – "yes," "maybe," or "no." The first two were considered as willing, the no's as unwilling. Eliminating non-respondents, 48% of Blacks, 32% of Hispanics, and 24% of Whites were willing to have a child out of wedlock.

To examine this in more detail, the women were stratified into groups based on whether they were considered at high or low risk for pregnancy. This assessment was based on such factors as academic ability, whether they were from a single parent family, and socioeconomic status.[2] Among the Black women, of those in the highest risk group, 69% were willing to become single mothers, versus 47% willing in the lowest risk group. For Hispanics the figures were 52 and 30%, and for Whites they were 60 and 22%.

The important questions are "How did the women act on these attitudes?" and "Were the women in the high risk groups actually more likely to become single mothers two years later than those in the low risk groups?" To determine this, the authors studied the 342 women who went on to become single mothers by the time of the re-survey in 1982. The results are shown in Figure 1 (next page).

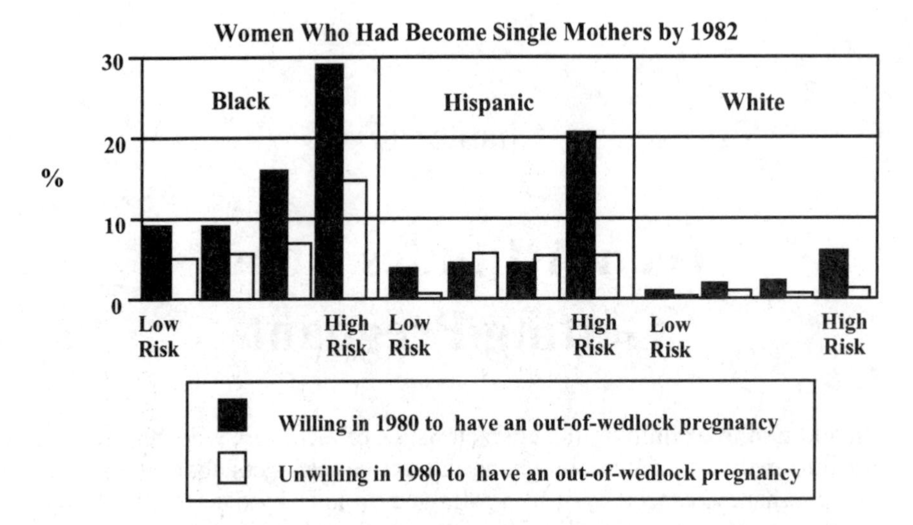

Figure 1. Percentage of female high school sophomores surveyed in 1980 who were single mothers by 1982, by background risk of non-marital childbearing and previous willingness to consider a pregnancy. Redrawn from Abrahamse et al. *Family Planning Perspectives* 20:13-18, 1988.

Except for Hispanics in the mid-risk range, in all cases women who expressed a willingness in 1980 to become unwed mothers were much more likely to become unwed mothers in the following two years than those who were unwilling. For example, in the highest risk groups among the Blacks willing to become unwed mothers in 1980, by 1982 29% had become unwed mothers, versus 14% of those unwilling. For the Hispanics the comparable figures were 20 and 5%, and for Whites, 6 and 2%. Except for the Hispanics, this ratio was similar for all risk groups.

These results indicate that teenage unwed pregnancy is not simply an unavoidable "accident." *Regardless of risk category, those women who were unwilling to be single mothers were much less likely to become single mothers than those who were willing.* While the way in which they avoided pregnancy was not identified, the authors state that, "They may choose to be less sexually active or to initiate sexual activity later, they may be more diligent in their contraceptive use, or they may have more favorable attitudes toward marriage as a means of resolving a premarital pregnancy."

Relevant to the thesis of this book, the authors then went on to determine if there was a relationship between problem behavior and unwed motherhood. To determine problem behavior the authors examined three variables: *disciplinary problems in school, class-cutting,* and *truancy*. While substance use was not examined, as shown in the previous chapters, the variables the authors chose are highly correlated with substance use and abuse. Based on these three variables, subjects were ranked as high, moderate, and low in problem behaviors. Figure 2 (next page) shows the results.

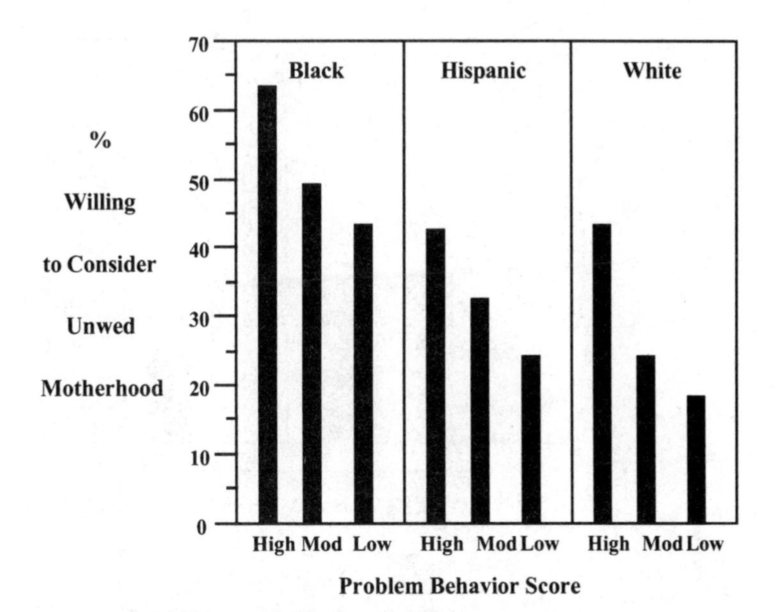

Figure 2. Percentage of female high school sophomores who were willing to consider non-marital childbearing, by their ranking on a problem behavior score. Based on Abrahamse et al. *Family Planning Perspectives* 20:13-18, 1988.

Women with problem behaviors were more willing to consider unwed pregnancies than women without problem behaviors. Since, as shown above, this translates into actual unwed pregnancies, to the extent that these problem behaviors have a genetic component, there would be selection for those genes. The degree of the selection is related to the ratio of early pregnancies occurring in those ranking high on problem behaviors versus those ranking low. It can be seen that despite the higher willingness of Black teenage girls to consider an unwed pregnancy, this ratio is actually higher in Whites (2.36) than for Blacks (1.58). Thus, these figures suggest that, at the present time, this gene selection may be greater in Whites than in Blacks.

As dramatic as this figure is, when presented in a slightly different manner – that is, showing the percent exhibiting problem behaviors for non-mothers compared to those who became mothers – the results are even more striking (Figure 3, next page). Viewing the data in this fashion illustrates the much higher frequency of problem behaviors in those individuals, especially Whites, who go on to become single mothers.

The authors also examined a second variable relevant to the thesis of this book – the relationship between plans for higher education and involvement in unwed motherhood. This was cast in terms of opportunity costs to the women if they became pregnant. Here, the concept was that women who feel they have much to lose if they get pregnant may be less likely to get pregnant than women who feel they have little to lose. This opportunity cost was based on whether the women expected to pursue two or more years of college education, or whether they planned "to go to college at sometime in the future." The scale ranged from low for no plans for college, to moderate for future plans, to high for immediate plans. These results are shown in Figure 4 (next page).

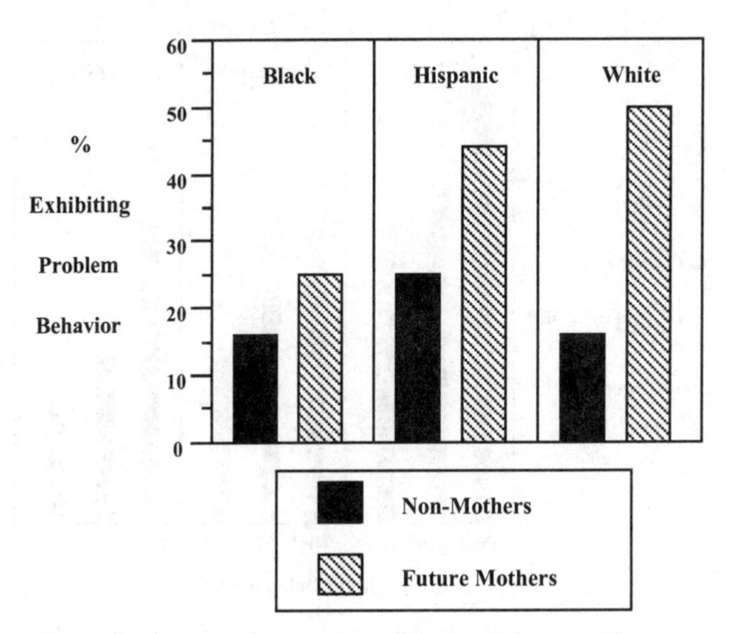

Figure 3. Frequency of problem behavior in non-mothers versus future mothers. Redrawn from Abrahamse et al. *Beyond Stereotypes. Who Becomes a Single Teenage Mother?* 1988. Rand Corporation, Santa Monica, CA.

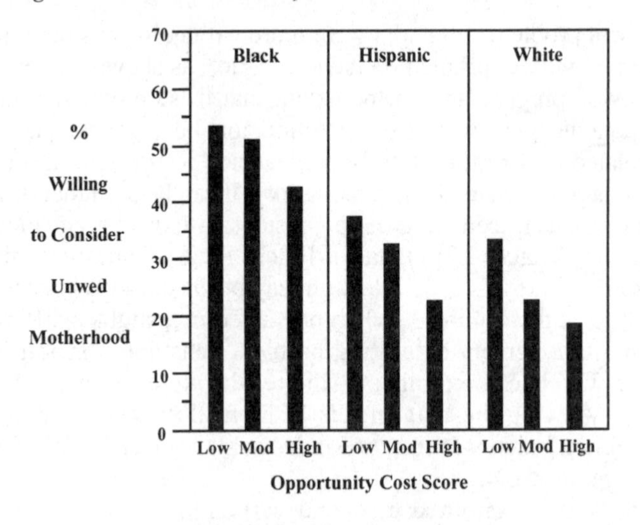

Figure 4. Percentage of female high school sophomores who were willing to consider non-marital childbearing, by their ranking on an opportunity cost score. Based on Abrahamse et al. *Family Planning Perspectives* 20:13-18, 1988.

These results show that *the young women who felt they had the most to lose because they were planning to go on to college, were the least willing to have an unwed pregnancy.* Again, as shown above, this translates into actually being less likely to become a single mother. As with the problem behavior score, the ratio for those with the highest versus the lowest perceived opportunity cost was greater for Whites (1.85) than for Blacks (1.24).

The above figures clearly show that those assessed as being at increased risk for an unmarried teen pregnancy do in fact have the highest pregnancy rates. Further evidence for the critical role of academic performance comes from examining what goes into this risk assessment. The authors state that, "Our preliminary analyses also showed that the overall measure of academic aptitude (consisting of standardized subtests on vocabulary, reading, mathematics, civics, writing, and science) distinguished future single mothers within the entire sample better than any of its individual components."[2] Thus, *high academic ability is one of the major deterrents to teenage unwed pregnancy.*

The authors also asked, "During the past month, have you felt so sad, or had so many problems, that you wondered if anything was worthwhile?" The answer could be "yes, more than once," "yes, once," and "no." For the Whites, of those who answered "Yes, more than once" on this depression score, 33.6% were willing to consider an unwed pregnancy, versus 18.1% of those who said "no." For Hispanics the comparable responses were 37.7% versus 23.6%. This was not a factor in Blacks. Thus, to the extent that this response reflects depression, or a related disorder, and to the extent that it is, in part, genetic, those genes would be selected for.

Summary – Teenage unwed pregnancy is not an accident. Teens who expressed a willingness to become unwed mothers were much more likely to actually become unwed mothers than those who were unwilling. This willingness was significantly greater for women with problem behaviors than those without problem behaviors, less for women with higher academic ability, and much less for women who wanted to go onto college, than those who did not want to go to college. Finally, Whites and Hispanics who had feelings of depression were much more willing to become unwed mothers than those without such feelings. To the extent that genetic factors are involved in each of these variables: problem behaviors, school problems, learning disabilities with resultant dropping out of school, and depression, the association with an earlier age of first pregnancy will select for the genes involved.

Chapter 18

The Berkeley Study

In 1959 the School of Public Health at Berkeley initiated a large longitudinal project called the Child Health and Development Studies. This involved over 20,000 pregnant women in the Kaiser health system and many of their children. The latter were examined multiple times, including in adolescence. This made it possible to relate various behaviors in the teenagers to their mother's age at the birth of her first child (MAFB). The results for drinking behavior were as follows:

How often have you gotten tight (drunk) from drinking?

	MAFB	N
Never, never last year	25.03	356
About once a year	23.60	85
A few times a year	24.45	156
About twice a month	23.11	114
A few times a month	22.89	98
About once a week	22.57	77
A few times a week	23.61	29
Every day	20.00	4

This showed that for adolescents who never got drunk, the average age at the birth of their mother's first child was 25.03 years. By contrast, the MAFB for those adolescents who got drunk once a week or more ranged from 22.57 to 20.00 years of age. These differences were highly significant ($p < .0001$). Figure 1 diagrams the results.

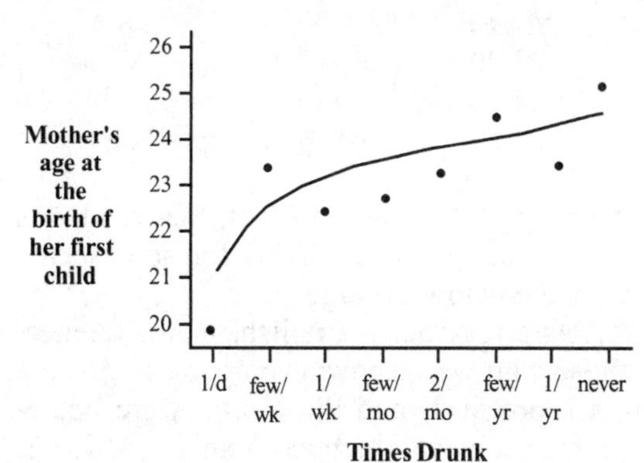

Figure 1. Mother's age at the birth of her first child versus the frequency that her adolescent child got drunk. From the Berkeley Child Health and Development Studies.

For smoking the results were:

Do you smoke cigarettes now?	MAFB	N	
No	24.28	794	
Yes, less than 1 a day	22.64	25	
Yes, regularly	22.70	116	p = .003

If the adolescents did not smoke, the average MAFB was 24.28 years. If they smoked regularly, it was 22.7 years.

For examining school performance, one of the questions concerned school work in general.

How good a student are you in general?	MAFB	N	
Very good student	25.82	196	
Good student	23.74	492	
Fair student	23.52	192	
Poor student	22.02	15	p < .0001

Here the average MAFB was 25.82 years for very good students versus 22.02 years for very poor students, with the fair and good students intermediate.

The results for science classes were:

How good at science?	MAFB	N	
Above average	25.31	263	
Average	23.71	508	
Below average	23.31	124	p < .0001

Even though they may do poorly in academic subjects, individuals with ADHD, TS, or learning disorders often do well in sports. Here the results were:

How good at competitive sports?	MAFB	N	
Above average	23.70	315	
Average	24.33	430	
Below average	24.53	146	p = N.S.

This time the results were no longer significant, and, in fact, the trend was slightly in the opposite direction, with a lower average MAFB for adolescents who did well in sports versus those who were below average.

To compare all the academic subjects altogether – English, math, science, and social studies – for each subject, those who were above average were given a score of 2, those who were average, a score of 1, and those who were below average, a score of 0. Since there were four subjects the maximum score was 8. The results for adolescents with scores from 1 to 8 were:

Academic Score	N	MAFB	
8	77	25.85	
7	104	25.30	
6	163	24.84	
5	174	23.93	
4	208	23.73	
3	109	22.66	
2	39	22.72	
1	12	20.77	$p < .001$

Except for a score of 0, where the number of adolescents was too small to be accurate, there was a progressive decrease in MAFB from 25.85 years for the students who were above average in all academic subjects, to 20.77 years for those with a score of 1. This was significant ($p < .001$). Figure 2 shows the results.

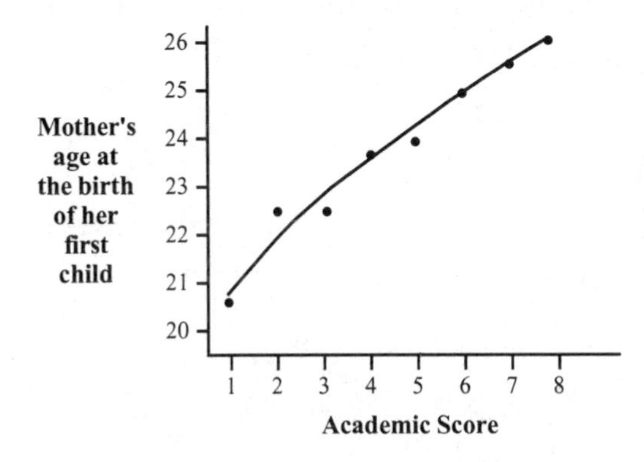

Figure 2. Mother's age at the birth of her first child versus the academic score of her adolescent child. From the Berkeley Child Health and Development Studies.

This showed that when adolescents who did poorly in school, smoked, or drank, their mothers showed a significant tendency to start having children earlier than for those adolescents who did well in school and didn't smoke or drink. While environmental factors could be playing a role, the socioeconomic status of these subjects tended to be fairly uniform. The most reasonable explanation is that the mothers who started having children at an early age were more likely to have genetically caused learning disorders and impulsive, compulsive, and addictive behaviors. These genes were passed to their offspring, who then tended to have these problems themselves. The earlier age at the birth of the mother's first child would select for the genes involved.

Summary – The Berkeley Child Health and Development database was used to examine the relationship between poor academic performance and addictive behaviors in teenagers and the age of the mothers at the birth of their first child. This age was significantly lower for teenagers who more often drank alcohol, smoked cigarettes, and did poorly in academic subjects in school. To the extent that genetic factors play a role in these conditions, the earlier age at first birth of their mothers would select for those genes.

Chapter 19

The NLSY

The Berkeley Study was a longitudinal study that started at birth. Another important large longitudinal study carried out in the United States included 12,686 youthful subjects first examined when they were 14 to 22 years of age. This has been called the National Longitudinal Surveys of Youth, or NLSY. It consisted of a nationally representative sample of youth, including Whites, Blacks, and Latinos and was begun in 1979. The extensive follow-up studies were performed by the Center for Human Resources at Ohio State University. The study gathered a rich and wide range of information concerning socioeconomic status, race, education, IQ and other psychometric studies, occupation and work history, drug use, legal problems, and reproductive history.

Because of the extensive battery of tests used to assess IQ and other cognitive skills, the NLSY was extensively used in the popular book, *The Bell Curve,* by Richard Herrnstein and Charles Murray. In addition to its controversial conclusion about race and IQ, the book presented large amounts of data about the correlation between IQ and a wide range of other behaviors. The portions of this data that are relevant to the theme of this book will be reviewed later in this chapter.

Before the present book was written, I had pointed out that having children early can provide an even stronger force for gene selection than having more children.[52,53p278] The authors of *The Bell Curve* also recognized that if a behavior such as IQ was controlled in part by genes, and if individuals with a lower IQ began having children earlier than those with a higher IQ, this faster generational cycle would select for the genes contributing to a lower IQ. They and others have used the term dysgenic to refer to any force that tends to have a negative influence on the human gene pool. Using data from the NLSY study, they showed the following relationship between cognitive class (IQ) of the mother and the age she first began having children (see Figure 1, next page).

This showed a dramatic difference of 7.4 years between the mean age of first birth for those in the lowest versus the highest cognitive class. Since having a higher number of children also provides a force for gene selection, Herrnstein and Murray examined the relationship between education and the average number of children born to women ages 35 to 44 in 1992. These results are shown in Figure 2 (next page).

Figure 1. Cognitive class (IQ) of women and their average age when they had their first child, from subjects in the NLSY. Redrawn from data in *The Bell Curve*, Richard Herrnstein and Charles Murray, The Free Press, 1994, p352.

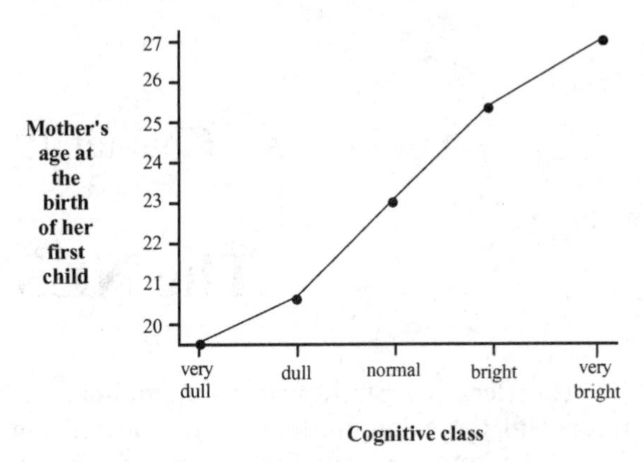

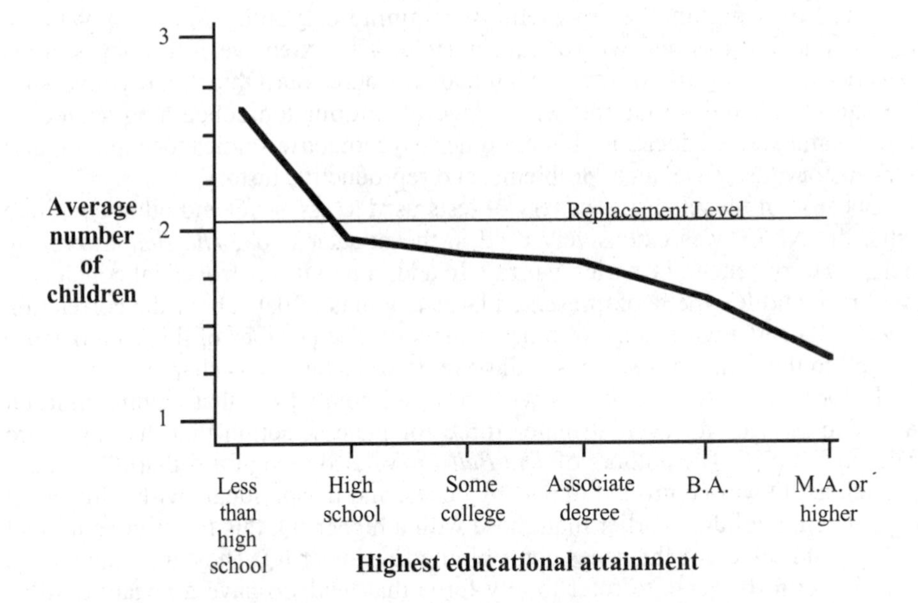

Figure 2. Educational achievement and average number of children born to 35 - to 44-year-old women in 1992, from the NLSY. Redrawn from *The Bell Curve*, Richard Herrnstein and Charles Murray, The Free Press, 1994, p349.

This indicates there was a progressive decrease in the number of children born to women with progressively greater levels of education. This in itself is of little import from a gene selection perspective unless it can be shown that genetic factors play a role in the average level of education. Herrnstein and Murray also examined the correlation between cognitive class and education. These results were as follows (next page):

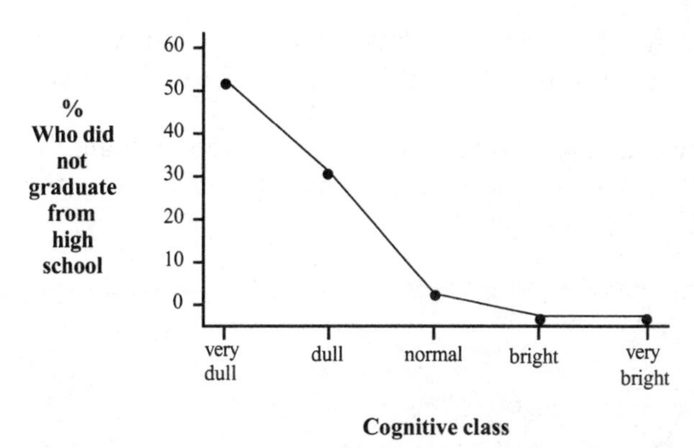

Figure 3. Percent of White subjects from the NLSY who did not graduate from high school or pass the high school equivalency exam, by cognitive class. Redrawn from data in *The Bell Curve*, Richard Herrnstein and Charles Murray, The Free Press, 1994, p.146.

Thus, there was a high correlation between low cognitive class and failure to complete high school. To the degree that cognitive class and IQ are influenced by genetic factors (Chapter 8), those genes for low cognitive class will be selected for because such women both have more children and have children earlier than women in higher cognitive classes.

Other Correlations with IQ

Several other aspects of behavior relevant to the subject of this book, such as the relationship between IQ and work, poverty, welfare, and crime, were reviewed by Herrnstein and Murray and are summarized below.

Work and IQ. The relationship between IQ and being out of work for month or more is shown in Figure 4.

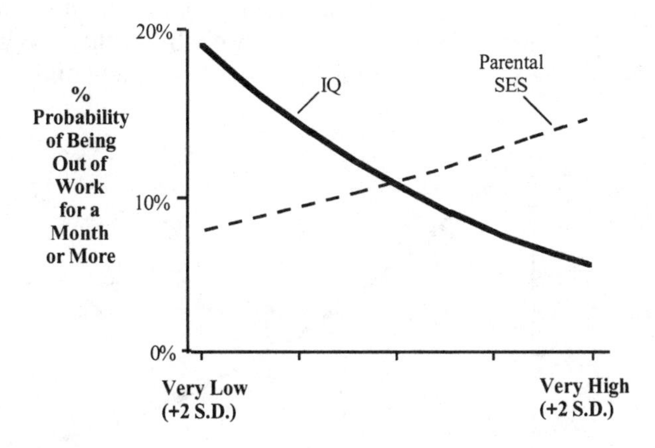

Figure 4. Percent of White men from the NLSY who were out of work for a month or more versus IQ. Redrawn from *The Bell Curve*, Richard Herrnstein and Charles Murray, The Free Press, 1994, p. 159.

While such a strong correlation between IQ and unemployment is not inherently obvious, it is not unreasonable. Individuals with a higher IQ tend to have more professional and more career-oriented jobs and such jobs tend to be more stable. While parental socioeconomic status (SES) played some role, it was less important than IQ. In addition, IQ is strongly hereditary and SES itself is highly correlated with IQ.

Poverty and IQ. There was an even higher correlation between IQ and the probability that young adults in the NLSY would have an income below the poverty line. This relationship is shown in Figure 5.

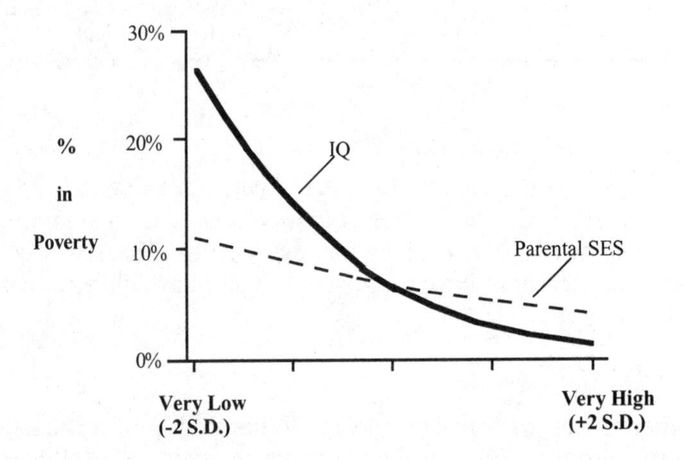

Figure 5. Percent probability of young White adults from the NLSY being in poverty versus IQ. Redrawn from *The Bell Curve*, Richard Herrnstein and Charles Murray, The Free Press, 1994, p 134.

By comparison to IQ, especially for those between -1 and -2 standard deviations (S.D.), there was little correlation with parental socioeconomic status.

Welfare and IQ. Figure 6 shows the percent probability that a White mother will have to go on welfare within a year of the birth of her first child.

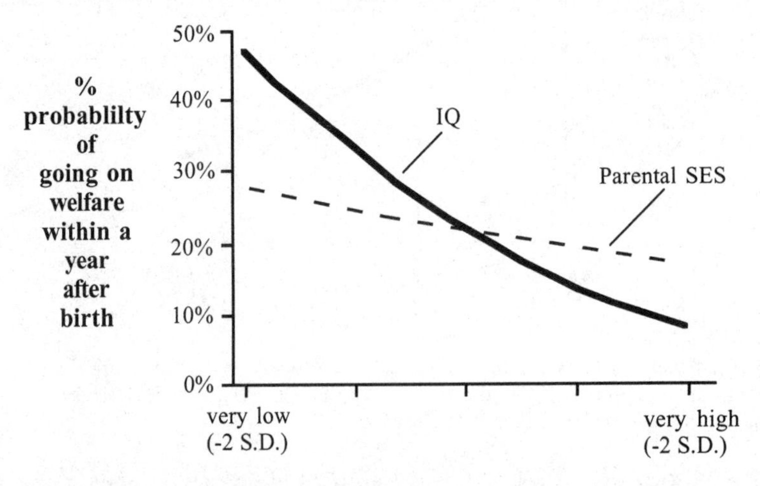

Figure 6. Percent probability of White women from the NLSY going on welfare within a year of the birth of their first child. Redrawn from data in *The Bell Curve*, Richard Herrnstein and Charles Murray, The Free Press, 1994, p 195.

This shows that the lower the IQ of the mother, the greater the probability that she will be on welfare after her first child is born. A reasonable explanation could be that those with a higher IQ simply come from more affluent families which are able to lend sufficient support to keep from requiring welfare. While this played some role, the line showing the parental SES indicates it was not nearly as important as the mother's IQ.

Crime and IQ. Crime can be assessed by a wide variety of means from self reports of various criminal activities, to official arrest records, to actually being arrested and incarcerated. The IQs of White subjects who were incarcerated at the time they participated in the NLSY study are shown in Figure 7.

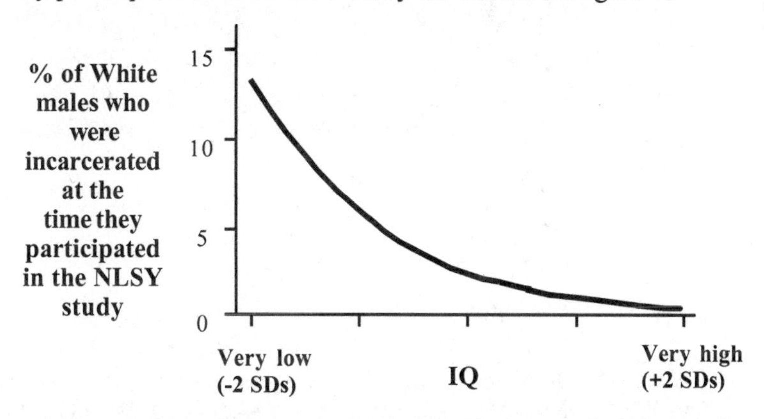

Figure 7. IQ of White subjects from the NLSY who participated in the study while incarcerated for a crime. Redrawn from *The Bell Curve*, Richard Herrnstein and Charles Murray, The Free Press, 1994, p 249.

The lower the IQ, the higher the percentage of subjects who were incarcerated at the time of participation in the study. This was not simply due to a lower socioeconomic status of these with a lower IQ, since the role of SES was close to nil after controlling for IQ.

It bears emphasizing that despite the significant association between these different behaviors and IQ, IQ itself is only one of a number of factors responsible for the behaviors. In statistical terms the percentage of the total accounted for by IQ ranges from 5 to 20% (percent of a variance).[131]

Summary – Based on data from the National Longitudinal Surveys of Youth (NLSY) women in the lowest cognitive class (very dull) had their first child at an average age of 19.8 years while those in the highest cognitive class (very bright) had their first child at an average age of 27.2 years. In addition, women who failed to finish high school, a group in which 55% were in the lowest cognitive class, had an average of 2.7 children, well above replacement levels, while women with a college degree or higher had 1.8 to 1.4 children, well below replacement levels. The factors of a greater number of children and having children earlier combine to form a strong force in the selection for genes associated with a lower than average IQ.

Chapter 20

NLSY – IQ

The NLSY was such a rich lode of information relevant to the subject of the potential selection for genes for disruptive, addictive, and cognitive disorders, that I obtained the data and its documentation for my own analysis. By this time the 1993 follow-up survey had been completed. Since the original set of 12,686 subjects interviewed in 1979 were between 14 and 22 years of age, by 1993 they were 28 to 36 years of age. Thus, a sufficient period of time had passed to allow a reasonably complete study of many of the behaviors documented in the NLSY, including childbirth. In this chapter I will examine the relationship between IQ, the age the subjects were when they had their first child, and the number of children they had. For the measurement of IQ, the Armed Services Vocational Aptitude Battery (ASVAB) was used. This consisted of a battery of ten tests that measured knowledge and skill in the following areas: 1.) general science, 2.) arithmetic reasoning, 3.) word knowledge, 4.) paragraph comprehension, 5.) numerical operations, 6.) coding speed, 7.) auto and shop information, 8.) mathematics, 9.) mechanical comprehension, and 10.) electronics information. A composite score derived from selected sections of the battery were used to construct an Armed Forces Qualification Test score (AFQT) for each subject. This correlated well with general intelligence.[131] The scoring was revised in 1989, and these revised scores were used. The norms for the AFQT were based on persons who were at least 17 years old.

Since the NLSY database will be used not only for this chapter, but a number of subsequent chapters, before progressing a few caveats are important. For all of these chapters I have attempted to keep any manipulation of the variables to a minimum. For example, in the NLSY there was an oversampling of Blacks and Hispanics to obtain adequate numbers for statistical validity. If I was interested in using the NLSY data to obtain estimates for different variables that held for the United States population as a whole, it would be necessary to use the weighting scores assigned to each subject to compensate for this oversampling.

I have not done that for several reasons. First, my primary interest is in the relative rather than absolute values. Thus, for the question, "How does the average age at the birth of the first child compare for subjects that have only completed high school compared to those that have completed college?" the important result is the difference between the two, not the absolute U.S. averages. Second, the more the data is manipulated, the more the results, and reasons for the manipulation, can come into question. Third, by manipulating the data as little as possible, others could, if they wished, obtain the same database and easily verify the results for themselves. This becomes more difficult when a series of complex manipulations of the data are involved. Finally, for all the following results I have examined both the total set of subjects, Hispanics, Blacks, and Others (mostly Caucasian) as well as the Others alone, and rarely have the results been any different. When they are I will point that out.

The first question I examined was whether those subjects who had their first child at earlier ages tended to have lower IQs than subjects having their first child at a later age. Because the range of AFQT scores was not a normal distribution, the scores were simply divided onto ten percentile groups (based on a prior standardization in predominantly Caucasian subjects), with the lowest group representing those in the 1st to 10th IQ percentile, and the tenth group, those in the 90th to 99th percentile. The results are shown in Figure 1.

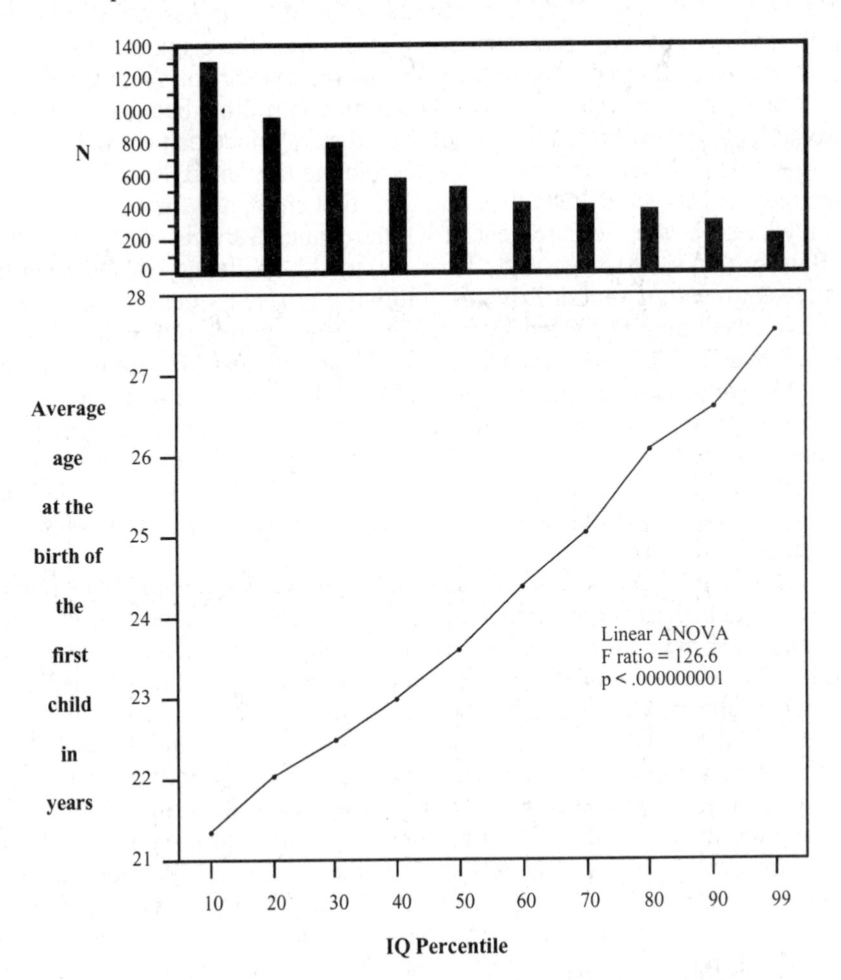

Figure 1. Bottom: Relationship between IQ percentile as measured by the AFQT (revised) and average age at the birth of their first child. Top: The number of subjects in each percentile group. From NLSY database.

This showed a dramatic decrease in IQ percentile, with decreasing age of subjects at the time they had their first child. To test whether this progressive series of averages was significant, a technique called linear ANOVA was used. This showed the results were highly significant ($p < .000000001$). For example, the average age at the birth of their first child for those in the 1st to 10th percentile group was 21.3 years versus 27.6 years for those in the 90th to 99th percentile group. The results concur with those presented in the previous chapter by Herrnstein and Murray. They obtained an even lower average at having the first child of 19.8 years by splitting up the 1st to 10th percentile group into a "very dull" group.

A further contribution to this dysgenic effect is evidenced by the fact that for this total set, greater numbers of subjects were in the lower IQ percentile groups. When different racial groups were examined separately, the curves for the IQ percentile versus age at birth of the first child were similar, but the distribution of cases in the different IQ percentile groups were different. This is shown in Figure 2.

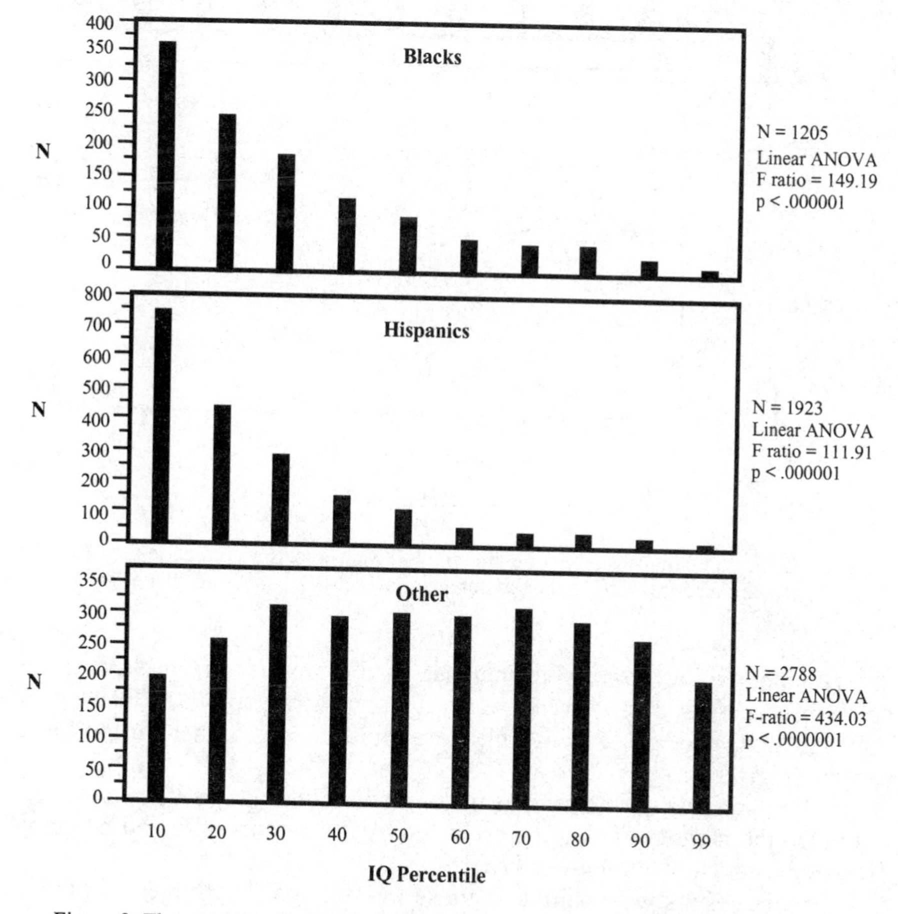

Figure 2. The number of subjects in the different IQ percentile ranges subdivided by racial group. From NLSY database. The linear ANOVA results on the right refer to the significance of the IQ percentile group versus average age at the birth of the first child, as shown in the bottom of Figure 1.

This showed a fairly uniform distribution of subjects by IQ percentile group in the Other set, consisting of Caucasians, a minor group of Orientals, and an unequal distribution for Blacks and Hispanics.

A second potential dysgenic effect on IQ would occur if subjects in the lower percentile groups had significantly more children than subjects in the higher percentile groups. These results are shown in Figure 3.

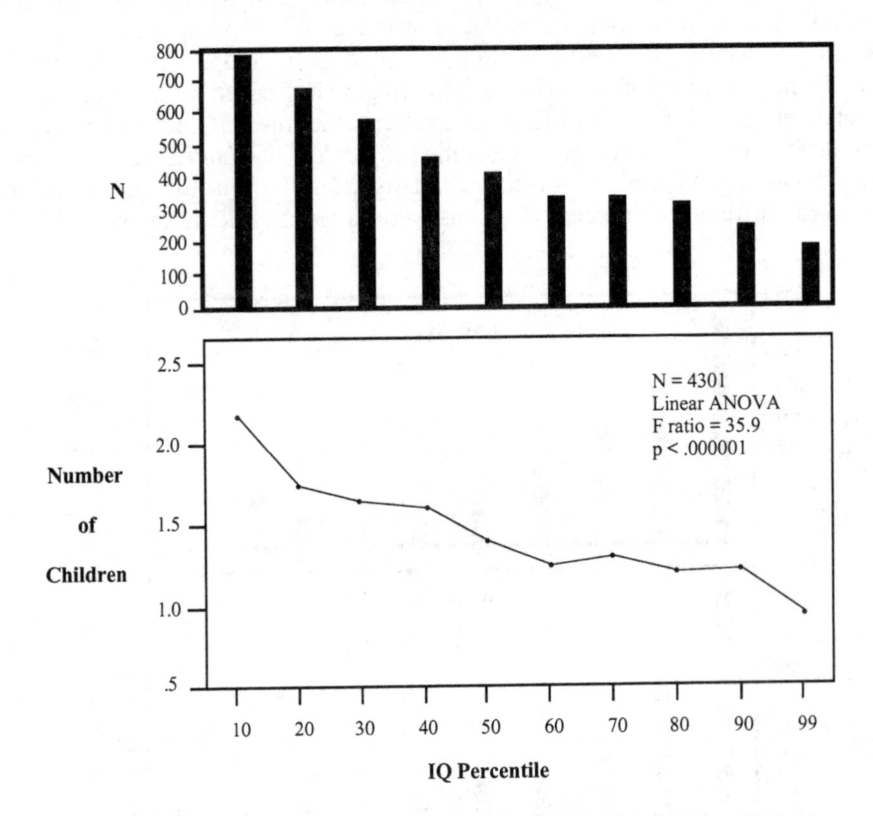

Figure 3. Bottom: Number of children of NLSY subjects according to their IQ percentile. Top: The number of subjects in each IQ percentile group. From NLSY database.

This showed a progressive increase in the number of children as the IQ percentile decreased. The trend was very significant ($p < .000001$). In addition, the number of subjects having those children was higher for the lower IQ percentile groups.

A third variable of interest was the number of siblings by IQ percentile. This examined the number of children produced by the parents of the NLSY subjects. The results are shown in Figure 4 (next page).

Here the results were similar to those for siblings. The subjects in the lower percentile groups came from significantly larger families than those in the higher IQ percentile groups.

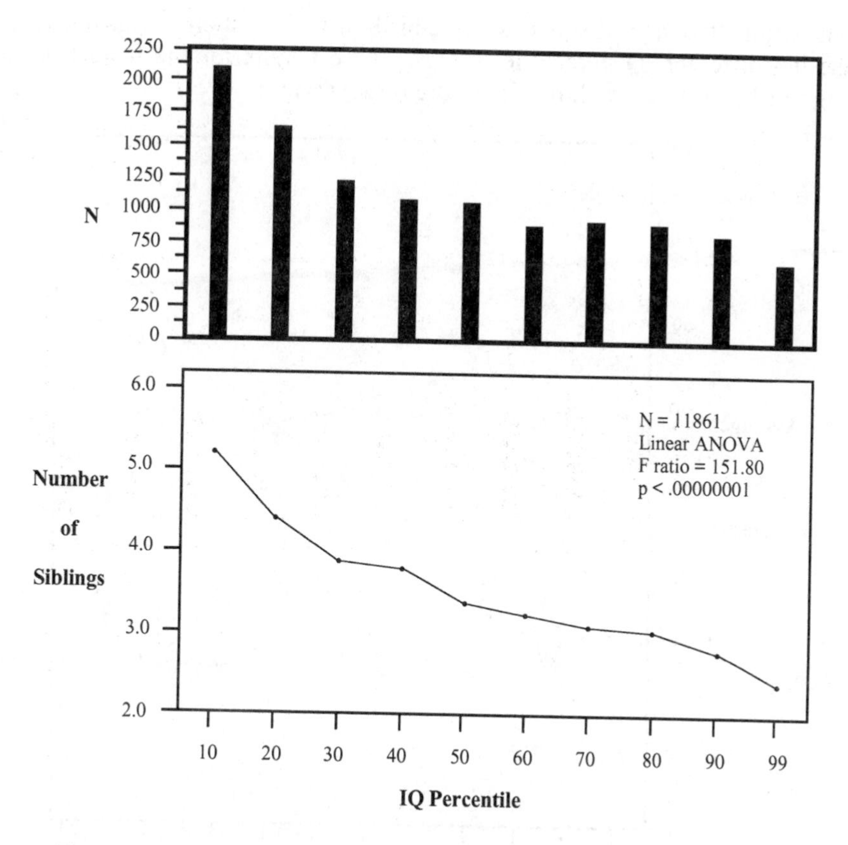

Figure 4. Bottom: Number of siblings for subjects in various IQ percentile groups. Top: Number of subjects in each group. From the NLSY database.

Figure 1 examined the relationship between IQ percentile and average age at the birth of the first child. Figure 5 (next page) examines this relationship in a slightly different way by dividing the subjects into groups on the basis of their age when they had their first child and asks if these groups differ by average IQ percentile.

Here the results were equally dramatic. There was a highly significant trend toward a lower IQ percentile as the age at which the subjects had their first child decreased. For example, the mean IQ percentile of the subjects who had their first child at age 15 was 16.1. There was a progressive increase such that the average IQ percentile of the subjects who had their first child at age 35 was 73.0.

A unique aspect of the NLSY database was that by 1993 the majority of subjects had passed through the peak of childbearing, especially of having their first child. From the point of view of gene selection, in addition to the age of the subjects when they had their first child, the number of children they had, and the number of subjects in each IQ percentile group, an additional important variable was the percentage of subjects in the different IQ groups that had children. It is obvious that if many more of the higher IQ subjects than lower IQ subjects opted to have no children, this would have a significant dysgenic effect. To examine this, the cumulative age of subjects at the birth of their first child was determined for the ten different IQ percentile groups. These results for the lowest, middle and highest IQ groups are shown in Figure 6 (next page).

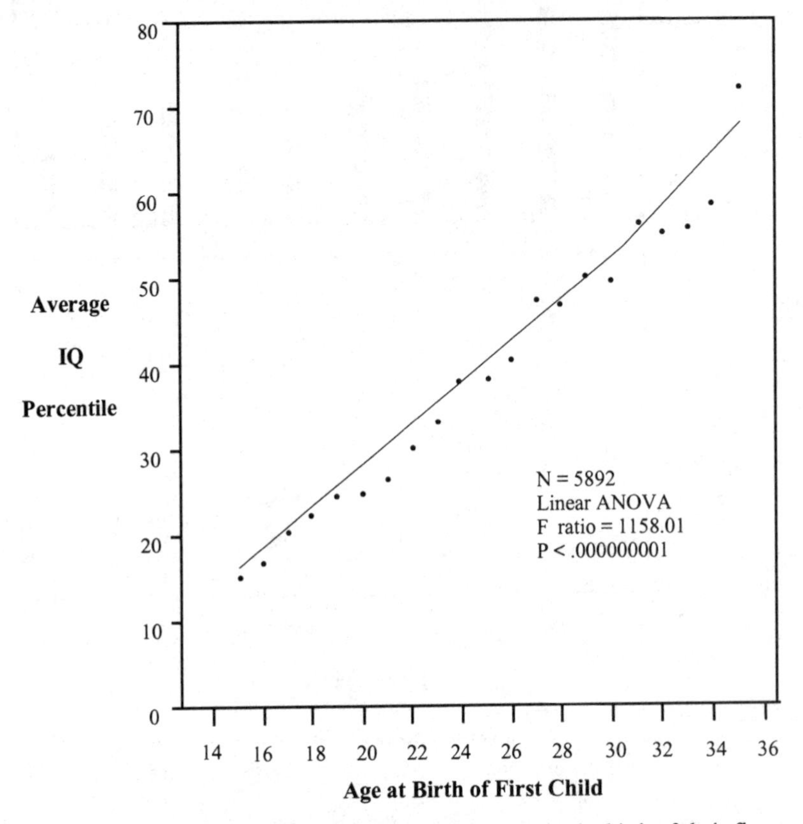

Figure 5. Average IQ percentile by the age of subjects at the birth of their first child. Form NLSY database.

Those subjects in the 1st to 10th IQ percentile group began having children at age 14. By age 24, 50% had given birth to their first child. By age 32 no new subjects in this group were having their first child, and a total of 62% had one or more children. By contrast, those subjects in the 90th to 99th IQ percentile group began having children at age 18, were having few new first children by age 34, and only 33% had any children at all. All other IQ percentile groups were intermediate. Since in most of the groups less than 50% of the subjects had children, it was not possible to compare the groups by examining their 50% points. Thus, the 25% points were chosen. This showed that in the 1st to 10th IQ percentile group, by age 18.5 years, 25% had given birth to their first child. By contrast, for the 90th to 99th IQ percentile group, by age 30 years 25% had given birth to their first child. There was a 11.5-year difference in the two groups.

These results also raised the question, was the average IQ percentile greater for those who opted to have no children than those who had children? For the total group, by 1993 50% had children and 50% did not have children. The mean

IQ percentile of those who had children was 35.9 versus 45.9 for those who had no children. This was a highly significant difference ($p < .000000001$). As discussed in Chapter 8, it had been proposed by Higgins, Reed, and Reed[132] that the tendency for those with a lower average IQ to have larger families did not actually lead to a progressive decrease in IQ because those who had no children had a lower IQ than those who had children. These results clearly show this is not the case.

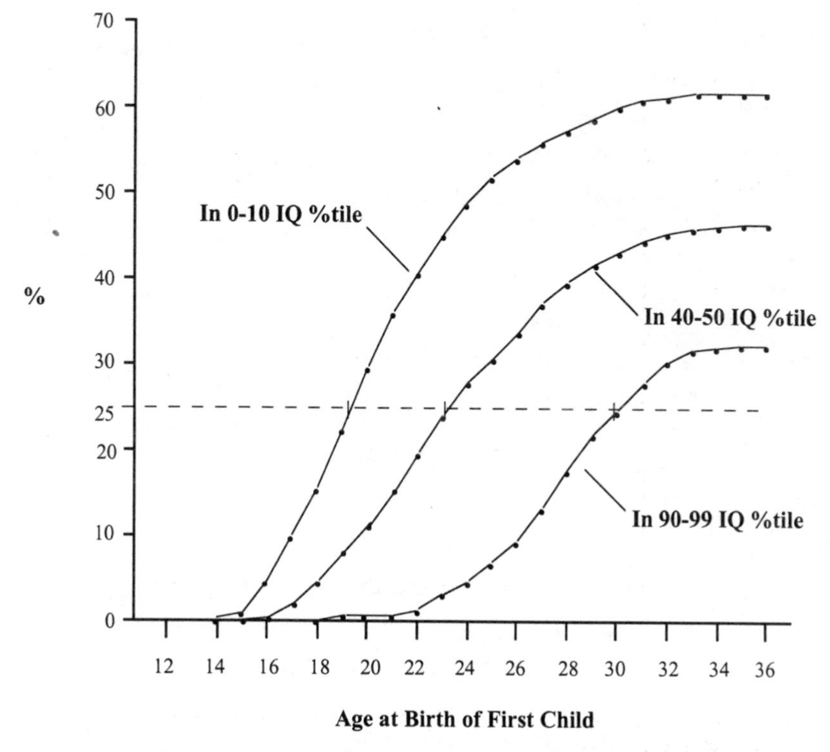

Figure 6. Cumulative percent of subjects having their first child at different ages for those in different IQ percentile groups. From NLSY database.

Summary – Since there is little dispute that IQ is at least 50% genetically controlled, the genes contributing to a lower-than-average IQ could be selected for by any of four different mechanisms: 1.) If subjects with lower average IQs had their first child earlier than those with higher IQs, 2.) If those with lower IQs had a greater number of children, 3.) If those with lower IQs came from larger families, i.e. had more siblings, 4.) If a higher percentage of subjects with lower IQs had children. The above results show that all four of these mechanisms are occurring and to a very highly significant degree. These observations suggest that by the latter part of the twentieth century, when these studies were performed, strong selective forces were in place to result in a progressive lowering of the IQ in successive generations.

Chapter 21

NLSY – Crime

Many studies have suggested that genetic factors can play a role in the propensity to commit various criminal acts.[21,45,47,137,186,187,241] The role of genetic factors in conduct disorder, a frequent precursor of antisocial behavior in adults, was reviewed in Chapter 12. Given that both environmental and genetic factors play a role in criminal behaviors, the present chapter examines the possibility that there there may be some selection for these genes.

To explore this possibility, the variable of *Number of times sent to an adult correctional institution* was particularly useful, because it eliminated minor crimes and required both a conviction and actual incarceration, rather than just probation. Since the vast majority of responses were "1" or "2 times," those who had not been sent to an adult correctional institution were simply compared to those who had. The results for the five variables related to gene selection are shown in Table 1.

Table 1. Ever sent to an adult institution?

	No		Yes		
	N	**Mean**	**N**	**Mean**	**p**
IQ Percentile	11,231	41.57	294	24.96	<.000001
Age at birth of first child	5,755	23.39	146	21.99	.0002
No. siblings	11,660	3.82	307	4.59	<.000001
No. children	4,300	1.62	25	2.16	.044
% with children	11,675	49.3	308	47.4	N.S.

The most striking difference was in IQ percentile. The mean was 41.57 for those never sent to an institution, versus 24.96 for those who had been incarcerated. This is consistent with the previous data on the relationship between IQ and incarceration. The difference in age at the time of the birth of their first child was less dramatic, but still significant. The number of siblings was higher for those answering yes, indicating they came from significantly larger families. The incarcerated subjects also had more children than those not incarcerated, but this was barely significant, and the percent of those incarcerated who had children was not significantly different from those not incarcerated. The magnitude of the latter two variables may have been lower for the incarcerated subjects simply because they *were* incarcerated.

The lower IQ for subjects with criminal records or delinquency has been frequently reported in the literature. While some have suggested this is entirely due to a lower socioeconomic status for those with problems with delinquency, several studies have shown this is not the explanation.[134,192,271] For example, Moffitt et al.[192] studied the correlation between IQ and number of offenses in two Danish samples. In the first the correlation was -.27, p = < .001, i.e. the lower the IQ the greater the number of offenses. When socioeconomic status was factored out, the correlation coefficient was unchanged at -.28, p < .01. In the second group the correlation coefficient was -.19, p < .0001, and after socioeconomic status was factored out, it was -.17, p < .0001.

Wolfgang and coworkers[273] obtained IQ scores on 8,700 boys in a Philadelphia cohort. These results in relation to criminal behavior are shown in Figure 1.

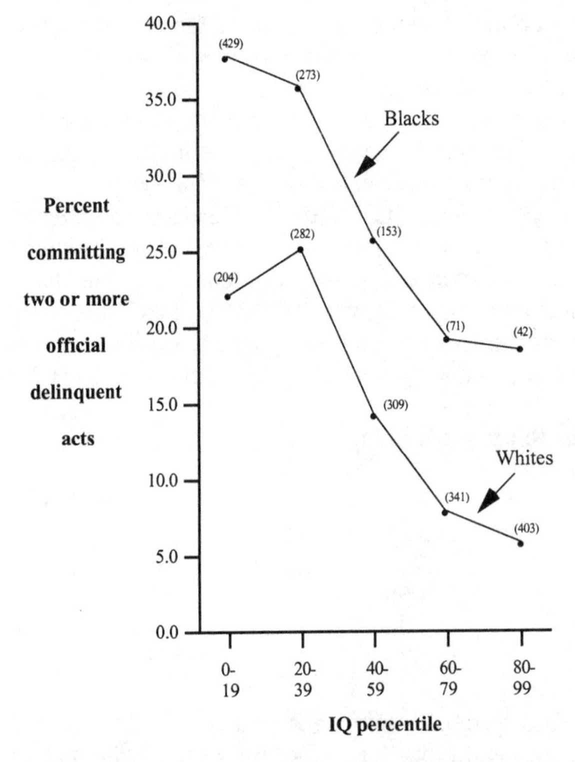

Figure 1. Correlation between the percent of boys committing two or more official delinquent acts by IQ versus IQ percentile for Whites and Blacks. From Wolfgang et al. *Delinquency in a Birth Cohort*, 1972, University of Chicago.

There was again a significant negative correlation between IQ and the percent of boys having committed two or more officially registered delinquent acts. The relationship was present in both Whites and Blacks and was independent of socioeconomic class.

Summary – Both genetic and environmental factors play a role in criminal behavior. Those subjects in the NLSY who had been convicted of and incarcerated for a criminal act had a significantly lower IQ (24.96 versus 41.57 percentile), more children, more siblings, and started having children earlier than those without such a record. All of these factors can lead to a selection of the genes involved in a predisposition to some criminal behaviors.

Chapter 22

NLSY – Drug Abuse

Genetic factors play a significant role as risk factors in drug abuse.[60] Generally, the more severe the abuse or addiction, the more likely it is that genetic factors will be involved. This chapter utilizes the NLSY data to ask if there is any evidence for dysgenic factors leading to a selection for the genes associated with drug abuse. Specifically, it examines whether individuals with more severe drug abuse were more likely to have children earlier, were more likely to have a lower IQ, or were more likely to have more children than individuals with a minimal or no history of drug abuse.

While cigarette smoking is not generally considered to be in the same class as "street" drugs, twin studies show that genetic factors play a role in smoking,[35,257] and the *dopamine* D_2 receptor is one of the genes involved [76,199]. Thus, I will start with cigarette smoking. Figure 1 shows the relationship between the number of cigarettes smoked per day in the month prior to the interview, the age of the subjects when they had their first child, and their IQ.

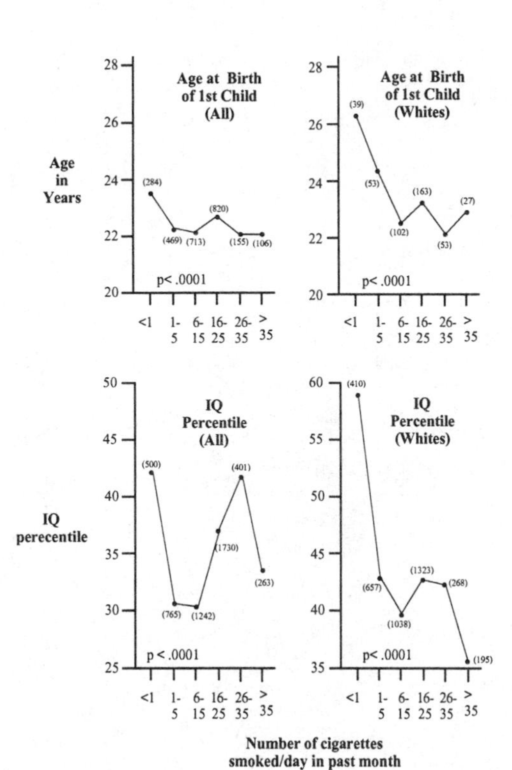

Figure 1. Age at birth of their first child and IQ percentile for subjects smoking less than 1 to more than 35 cigarettes per day. Figures in ( ) = number of subjects. P values are based on linear ANOVA. All = All races. Whites = non-Blacks and non-Hispanics. From the NLSY database.

For all races together, there was only a modest decrease in the age at birth of the first child with increasing numbers of cigarettes smoked. However, when limited to Whites (the designation Whites includes a small number of Orientals and other non-Black, non-Hispanic subjects), the trend was more noticeable. Thus, the age at birth of the first child for nonsmokers (<1 cigarette per month) was 26.5 years, while for those smoking more than a 1/2 pack of cigarettes per day (6-15 or more cigarettes) this age ranged from 23.4 to 22.3 years. For the total group, the IQ percentiles bounced all around. However, for the White group the nonsmokers averaged at the 59th IQ percentile, with a dramatic drop to the 44th percentile or lower for smokers, and a low of the 35th IQ percentile for those smoking more than a pack and a half per day.

Pot, or marijuana, could be considered to be intermediate between cigarette smoking and the use of more potent street drugs. Figure 2 shows the results for the age at the birth of their first child, and IQ percentile, in relation to the age of onset of smoking marijuana.

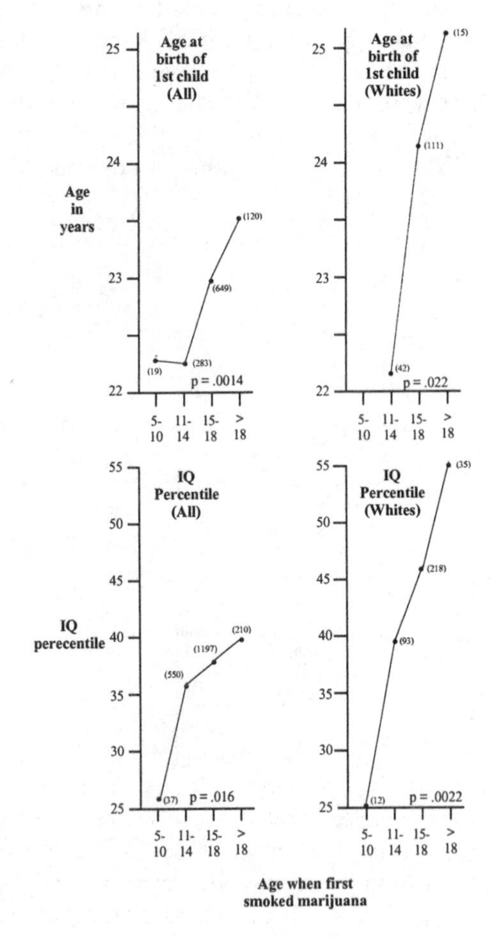

Figure 2. Age at birth of first child and IQ percentile for subjects who began smoking marijuana between ages 5 to greater than 18 years. Figures in ( ) = number of subjects. P values are based on linear ANOVA. All = All races. Whites = non-Blacks and non-Hispanics. From the NLSY database.

In general, the earlier an individual became involved in using a given drug the greater the genetic loading for genes contributing to their addictive potential. In contrast to cigarette smoking, for the total group there was a more dramatic decrease in the age at the birth of their first child for subjects who began to smoke marijuana at an early age (5 to 14 years) than those who began smoking at a later age (15 to > 18). This was even more dramatic for Whites. There were too few White subjects who began smoking at 5 to 10 years of age for analysis. The age at the birth of their first child increased from 22.3 years for those who began smoking at 11-14 years of age, to 25.2 years for those starting later than age 18.

There was a comparable significant correlation with IQ percentile. For the total group, those who began smoking pot at 5 to 11 years of age averaged in the 26th IQ percentile. This increased to the 40th IQ percentile for those starting after age 18. Again, the results were more striking for the Whites, increasing from the 25th IQ percentile for those starting to use pot at 5 to 11 years of age, to the 55th IQ percentile for those starting later than 18 years of age.

Examining the number of times subjects smoked pot gave somewhat similar results (Figure 3).

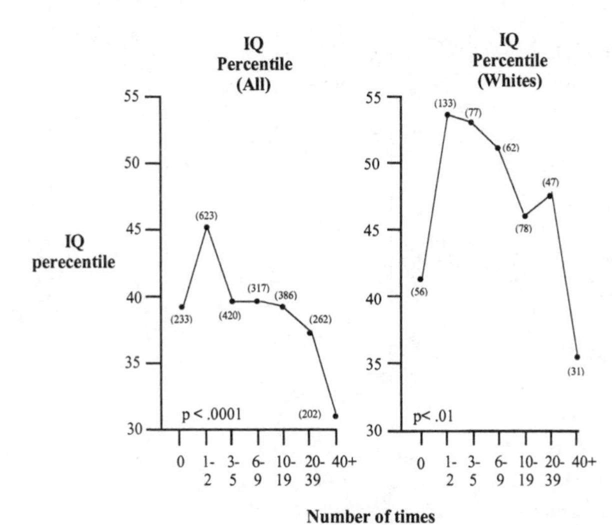

Figure 3. IQ percentile for subjects who smoked marijuana between 0 and 40+ times in the month prior to the interview. Figures in ( ) = number of subjects. P values are based on linear ANOVA. All = All races. Whites = non-Blacks and non-Hispanics. From the NLSY database.

Except for a blip for those smoking pot one to two times in the prior month, for all the subjects there was an accelerating trend toward lower IQ percentiles the more times that marijuana was smoked. The effect was stronger for the Whites. Here, except for a lower level for those who didn't smoke, there was an accelerating trend toward lower IQ percentiles as subjects smoked more and more marijuana. Thus, the mean IQ percentile was 52 to 54 for those smoking one to five times in the prior month, and dropped to the 36th percentile for those smoking forty or more times in the prior month.

The results for the use of barbiturates and other sedatives are shown in Figure 4. For the total group, the age at the birth of their first child was 22.5 years for those beginning barbiturate use between 8 to 14 years, and progressively increased to 25.8 years for those beginning later than 21 years of age.

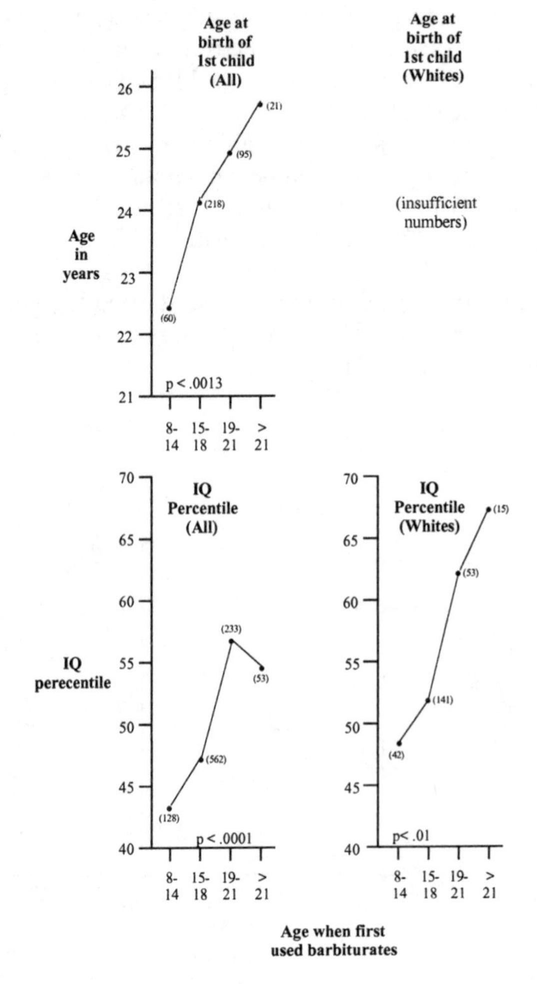

Figure 4. Age at birth of first child and IQ percentile for subjects first beginning to use barbiturates or other sedatives between ages 8 and > 21 years. Figures in ( ) = number of subjects. P values are based on linear ANOVA. All = All races. Whites = non-Blacks and non-Hispanics. From the NLSY database.

The numbers were insufficient for analysis for Whites only. For the total group, the average IQ percentile was 44 for those beginning to use barbiturates at ages 8 to 14, and increased to the 55.0-55.5th percentile for those beginning at 19 years of age or later. For the Whites only, again the trends were even more prominent, with an increase in IQ percentile from 48.8 for those beginning the use of barbiturates at ages 8 to 14 years, and progressively increasing to 67.3 for those starting after 21 years of age.

The results for the number of times cocaine was used in the month prior to the interview versus IQ percentile (Figure 5, next page) showed a pattern somewhat similar to that for the number of times marijuana was used (Figure 3).

Thus, as with marijuana, the IQ percentile was lower (47) for those with minimal use (1-9 times) than for those who had used cocaine 10-39 times in the prior month (54.5). However, from then on, as the number of times used increased, there was a progressive decrease in IQ percentile to 38 for those using 1000+ times in the prior month. The pattern for Whites was unique, with a modest progressive rise in IQ percentile (55.2 to 62) for those using cocaine 1 to 999 times, followed by a dramatic drop to 35.7 for those using cocaine more than 1,000 times.

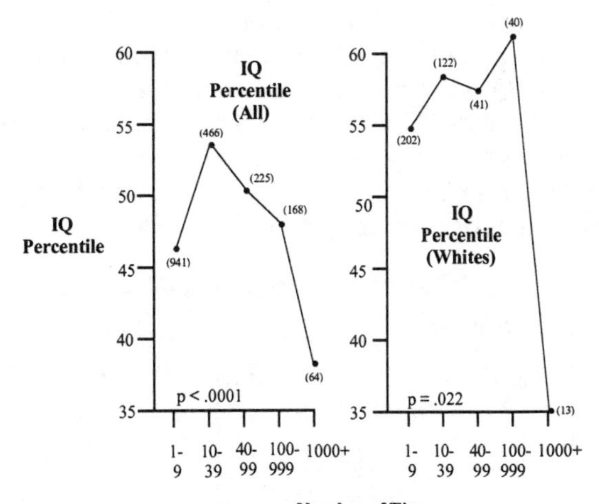

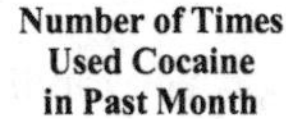

Figure 5. IQ percentile for subjects who used cocaine between 1 and 1000+ times in the month prior to the interview. Figures in ( ) = number of subjects. P values are based on linear ANOVA. All = All races. Whites = non-Blacks and non-Hispanics. From the NLSY database.

The results for age when subjects first used cocaine (Figure 6) were more uniform.

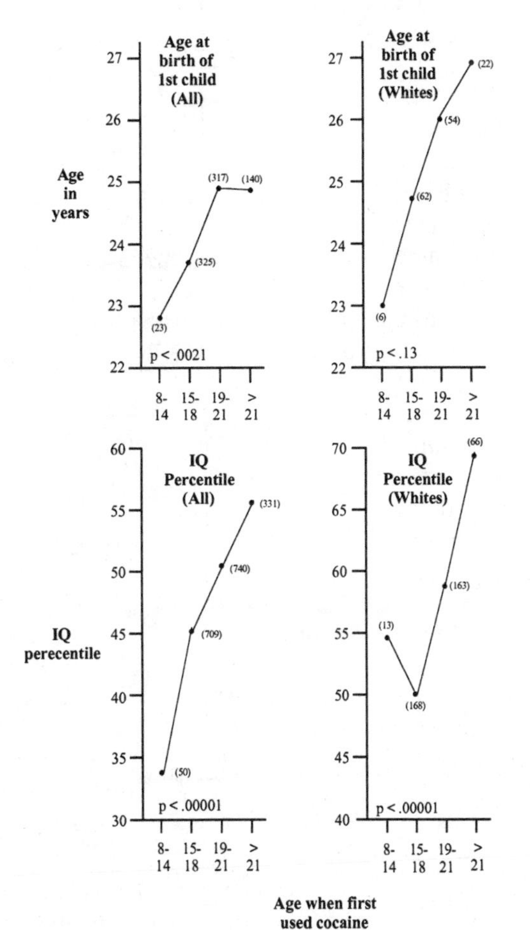

Figure 6. Age at birth of first child and IQ percentile for subjects first beginning to use cocaine between ages 8 and > 21 years. Figures in ( ) = number of subjects. P values are based on linear ANOVA. All = All races. Whites = non-Blacks and non-Hispanics. From the NLSY database.

For both the total group and Whites only, the age at the birth of their first child was lowest for those beginning to use cocaine very early (8-14 years) with a progressive increase as the age of first use increased. The ranges were greatest for Whites (23.1-27). For the total group, the IQ percentile was lowest for those beginning to use cocaine at 8 to 14 years (34.5) and progressively increased to the 56th percentile for those starting when they were older than 21 years of age. The Whites showed a similar pattern, except for a modestly higher IQ percentile for those starting at 8 to 14 years versus those starting at 15 to 18 years. The average IQ percentile was 70 for those Whites starting to use cocaine when they were older than age 21.

There were so few subjects in the NLSY survey that used heroin, that the figures were only valid for the IQ percentile for the total group (Figure 7).

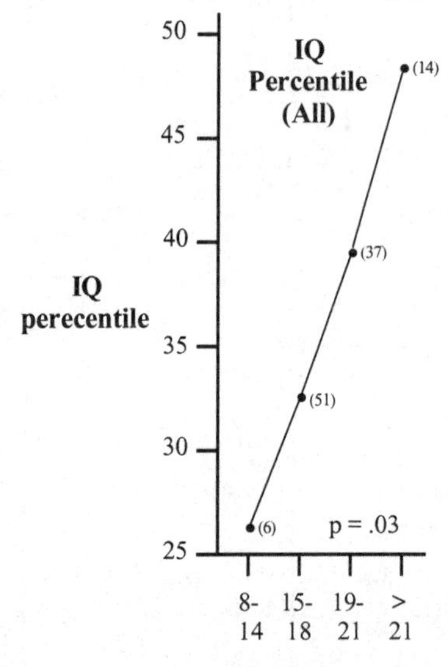

Figure 7. Age at birth of first child and IQ percentile for subjects first beginning to use heroin between ages 8 and > 21 years. Figures in ( ) = number of subjects. P values are based on linear ANOVA. All = total database. Whites = non-Blacks and non-Hispanics. From the NLSY database.

There was a progressive increase in the IQ percentile from the 26.5th percentile for those starting at age 8 to 14 years, to the 49th percentile for those starting after 21 years of age.

It is of interest that the IQ ranges were different for different drugs. Thus, for the results about the age when they first started using the drugs for the total group, the IQ range was highest (44th to 58th percentile) for those using barbiturates, next highest for those using cocaine (34th to 56th percentile), and lowest for those using heroin (26th to 48th percentile) and marijuana (26th to 41th percentile).

In addition to that fact that those with drug abuse problems began having children earlier, a second dysgenic effect would be present if they also had more children. The results were significant for the number of times drugs were used in the past month, for cocaine and tranquilizers, and for the age of initial use of drugs for barbiturates. The latter is illustrated in Figure 8.

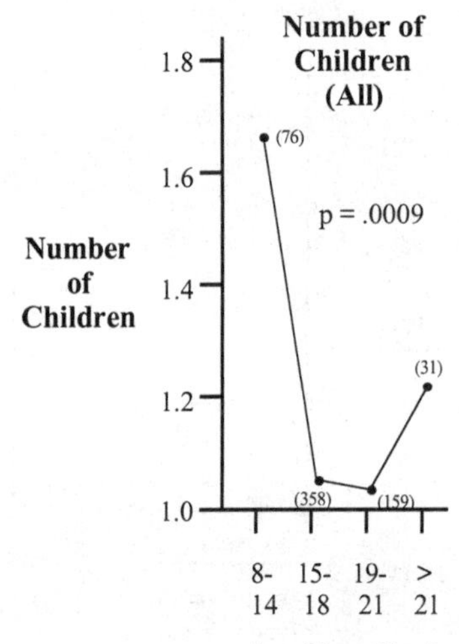

Figure 8. Number of children versus age first used barbiturates. Figures in ( ) = number of subjects. P values are based on linear ANOVA. All = all races. From the NLSY database.

The trends were in a dysgenic direction for most of the rest of the drugs, but were not statistically significant.

It should not come as a surprise that individuals who start using drugs at a very early age would also tend to end up being the heaviest users. To test this, the correlation between the age of onset of drug use and the number of times the drug was used in the month prior to the interview was determined for the different drugs. The results are given in Table 1.

Table 1. Correlation coefficients between age of beginning to use drugs and number of times the drugs were used in the prior month (all races).

Drug	Correlation Coefficient	p value
Amphetamines	-.254	<.01
Barbiturates	-.256	<.01
Cigarettes	-.160	<.01
Cocaine	-.121	<.01
Heroin	-.274	<.01
Marijuana	-.245	<.01
Tranquilizers	-.254	<.01

Each was significant at $p < .01$. Most clustered around a coefficient of -.25, except for cigarettes, which was lower, and cocaine, which had the lowest correlation coefficient of the drugs examined. Heroin had the highest correlation. The coefficients were negative because the lower the age of initiation of drug use, the greater the eventual use of the drug. While social and environmental factors clearly play a role in this correlation, it also reflects the fact that the greater the genetic loading for genes contributing to addictive behaviors, the younger the subjects will begin to use drugs and the more often they will use drugs.

Since the above graphs suggested greater correlations among the Whites only group for some of the drugs, the coefficients were also determined for this group. This showed similar results to those for the total group, except for a higher correlation coefficient for barbiturates (-.322) and a lower figure for cigarettes (-.063). It was also of interest to determine the degree to which IQ was correlated with the how often the drugs were used. Table 2 shows the results.

Table 2. Correlation coefficients between the IQ percentile and the number of times the drugs were used in the prior month (all races).

Drug	Correlation Coefficient	p value
Amphetamines	-.018	N.S.
Barbiturates	-.117	<.01
Cigarettes	.024	N.S.
Cocaine	-.007	N.S.
Heroin	-.053	N.S.
Marijuana	-.095	<.01
Tranquilizers	-.076	<.05

The negative correlations indicated the lower the IQ, the greater the number of times the different drugs were used. The correlations were significant for barbiturates, marijuana, and tranquilizers. The results were similar when the Whites only group was examined. The two major differences were higher and significant correlation coefficients for amphetamines (-.144, $p < .01$) and for cigarettes (-.132, $p < .01$).

The other important question to examine was if there was a correlation between IQ and the age of initiation of drug use. These results are shown in Table 3.

Table 3. Correlation coefficients between the IQ percentile and the age of initiation of drug use (all races).

Drug	Correlation Coefficient	p value
Amphetamines	.058	<.01
Barbiturates	.155	<.01
Cigarettes	.039	<.01
Cocaine	.163	<.01
Heroin	.156	<.01
Marijuana	.060	<.01
Tranquilizers	.153	<.01

Here all the correlations were significant and positive, meaning that the higher the IQ, the later in life the initiation of drug use, or conversely, the lower the IQ, the earlier the initiation of drug use. When the Whites only group was examined, the results were very similar, with the two major differences being a higher correlation coefficient for cocaine (.243) and marijuana (.199). In every case, the correlations between IQ and age of initiation of drug use were greater than the correlations between IQ and the number of times the drugs were used. Thus, individuals in the lower IQ percentiles were at significantly greater risk to begin early use of drugs than those in the higher IQ percentiles.

Summary – The NLSY data showed a significant dysgenic effect associated with drug abuse. Subjects with the most severe drug abuse problems, whether defined as a very early age of initiation of drug use, or high frequency of drug abuse, had a significantly lower IQ, had children at an earlier age, and, in several cases, had significantly more children than individuals with minor or no drug abuse problems. The earlier age of having children and, to a lesser extent, the greater number of children, selects for the genes associated with drug abuse and drug addiction. The correlation with IQ, a trait known to be at least 50% genetically determined, independently confirms the role of genes in drug addiction.

Chapter 23

NLSY – Alcoholism

Since some degree of alcohol use is very common in the general population, identification of any possible dysgenic effects would require the examination of those individuals with more severe alcohol abuse problems. For example, of 10,597 NLSY subjects asked "Have you ever had a drink of an alcoholic beverage?" 95% responded "yes." However, many other measures of the severity of alcohol abuse or dependence were also asked, allowing the identification of those with significant drinking problems. To accomplish this, an alcoholism score was calculated for each subject based on a number of questions relating to alcohol abuse or dependence.

Each positive response was counted as 1. The following are some examples: *Have you ever had a drink of an alcoholic beverage? Found it difficult to stop drinking once you had started? Driven a car after having too much to drink? Done things while drinking that could have caused you or someone else to be hurt? Found that the same amount of alcohol had less effect than before? Had a spouse or someone you lived with threaten to leave you or actually leave you because of your drinking? Wanted to cut down or stop drinking and found you couldn't? Found yourself sweating heavily or shaking after drinking or the morning after? Needed a drink so badly you couldn't think of anything else? Stayed away from work because of drinking? Lost ties to friends or family because of drinking? Continued to drink alcohol even though it was a threat to your health? Heard or seen things that weren't really there after drinking or the morning after? Taken a drink to keep yourself from shaking?* The maximum score was 28.

Figure 1 (next page) shows the correlation between the alcoholism score and IQ for the total group and for Whites only. For the whole group this showed an increase in IQ percentile from 39.8 for those with an alcoholism score of 0-3, to 42 to 46 for those with a score of 4-12. The IQ percentile then progressively decreased to the 20.7th percentile for those with a score of more than 24. For the Whites only group, the IQ percentiles were 51 to 55 for those with an alcoholism score of 0-12. They then progressively decreased to a mean of 16.5 for those with an alcoholism score of 22 or greater.

For the total group, there was an increase in the age at the birth of the first child from 23.1 years for those with an alcoholism score of 0-3, to 24 for those with a score of 4-12. It then dropped to 22.7 years for those with scores of 13-18, and to 22 years for those with scores of 22-28 ($p = .007$). There were minimal differences in the number of children born to subjects with the different alcoholism scores.

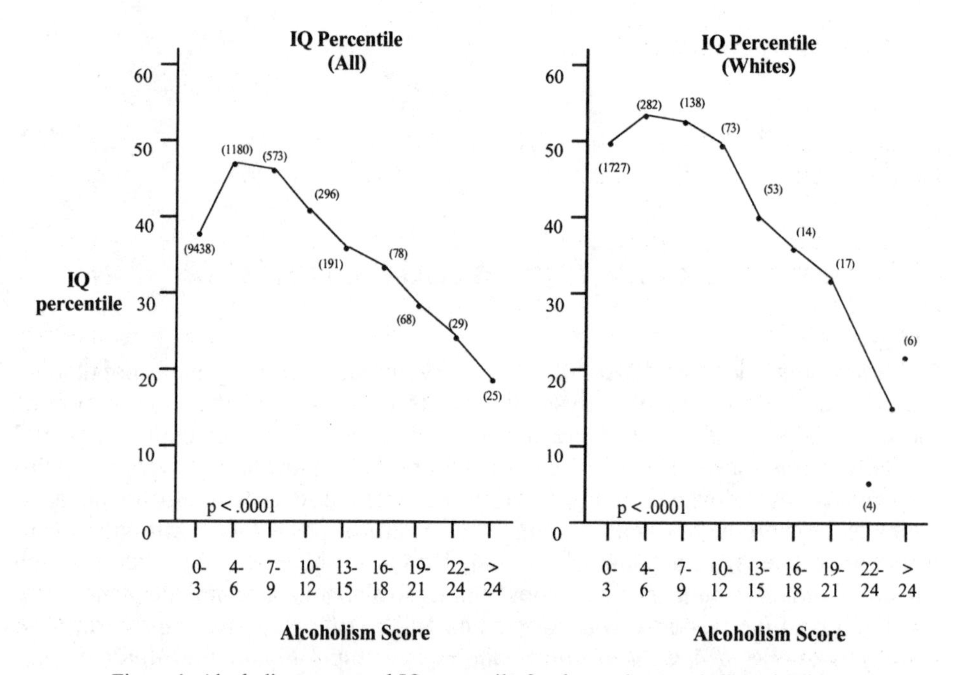

Figure 1. Alcoholism score and IQ percentile for the total group (All) and Whites only. Figures in ( ) = number of subjects. p values determined by linear ANOVA. Since the numbers for the Whites only in the 22-24 and > 24 alcoholism scores were small, the line was drawn to an average of the two. From the NLSY database.

Since the study included information on the presence of alcohol abuse or dependence in the relatives, it was possible to determine if a positive family history was associated with any dysgenic trends. Table 1 shows the results for those without and with a family history of alcoholism.

Table 1. Family History of Alcoholism.

	Neg Fam. Hx.		Pos. Fam. Hx		
	N	Mean	N	Mean	p
IQ percentile	9189	41.7	2689	38.4	<.000001
Age at birth of first child	4541	23.6	1590	22.6	<.000001
Number of children	6800	1.34	2211	1.55	<.000001

All three variables showed modest dysgenic trends with lower IQs, lower ages at the birth of the first child, and higher numbers of children for those with a first-degree relative (mother, father, brother, sister) with alcoholism.

It was informative to look at both the presence or absence of alcohol problems and the presence or absence of a family history of alcoholism. Thus, individuals with an alcoholism score of less than 10 and a negative family history of alcoholism were in group 1. Those with an alcoholism score of 10 or more and a negative family history of alcoholism were in group 2. Those with an alcoholism score of 10 or more and a positive family history of alcoholism were in group 3. The IQ percentile results are shown in Table 2.

Table 2. IQ percentile for Group 1 to 3.

	N	**IQ %tile**	**p**
Group 1 (Alc. score < 10, - Fam Hx.)	8778	41.8	
Group 2 (Alc. score ≥ 10, - Fam.Hx.)	411	40.0	
Group 3 (Alc. score ≥ 10, + Fam.Hx.)	276	34.1	.0001

Thus, the IQ percentile was significantly lower for those with both a history of alcohol problems and a family history of alcoholism, i.e. the genetic form of alcoholism. There was a modest difference in age at the birth of the first child of 23.6 years for those in group 1 versus 22.5 years for group 3, and of the number of children from 1.35 for those in group 1 to 1.44 in group 3.

Summary – While less striking than the results for drug abuse, some dysgenic trends were also present in subjects with problem drinking, especially if there was also a positive family history of alcoholism. IQ scores were significantly lower for subjects with the most severe drinking problems and there was a modest trend for those with a positive family history to have children earlier and have more children than whose without a positive family history. The greatest effect on lower IQ was seen in those with the genetic form of alcoholism.

Chapter 24

NLSY – Sexual Behavior

A major aspect of the hypothesis outlined in the *Introduction* is that one of the reasons there is selection for the genes for disruptive and addictive behaviors is because individuals carrying these genes tend have their first child earlier than those without these genes. This could be due, in part, because they become sexually active earlier and may show less responsibility in the use of contraception. The ultimate proof of this rests on the identification of the genes involved and the demonstration that they are more common in such individuals. While that is not yet possible, the NLSY subjects were asked a number of questions that allowed some of the predictions of this hypothesis to be tested.

The Use of Contraception

One of the relevant questions, asked only in 1982, was "Many couples in the United States do something to plan the number of children they have, or the time when they have them. Could you please tell me which of the items on this card best describes your current situation?"

There were 1,720 "yes" responses to the situation, "No birth control, not celibate," and 3,048 "yes" responses to the situation of, "Not using any method of birth control – not having sexual intercourse." The remaining situation was having sex and using birth control, and a listing of the possible types of birth control used was provided. Figure 1 shows the average IQ percentile for these three groups of individuals.

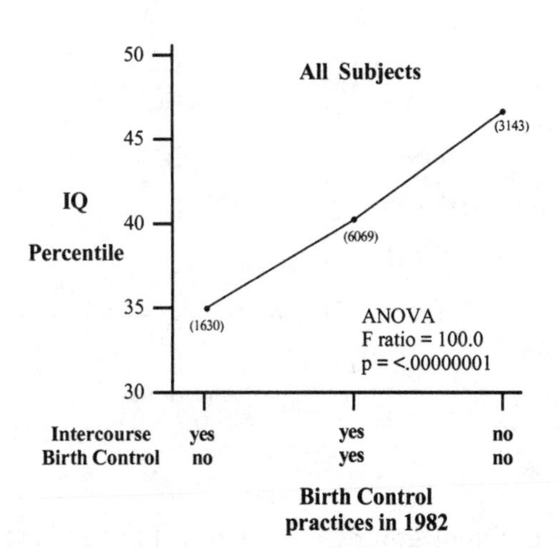

Figure 1. Average IQ percentile for individuals who have sex but don't use contraception, versus those who use contraception, versus those who were not having sex at the time of the 1982 interview. From the NLSY database.

This showed that those that were having sex without using contraception had a significantly lower average IQ than those who used contraception, and both were lower than those not having sex. The average age of the subjects at the time the question was asked was 21.0 years for those not using contraception, 21.2 years for those using contraception, and 20.0 for those not having sex – not remarkably different.

For the IQ difference to have a dysgenic effect, it would require a significant difference in either the age at the birth of the first child, or number of children. The age at the birth of the first child, based on information obtained ten years later, is shown in Figure 2.

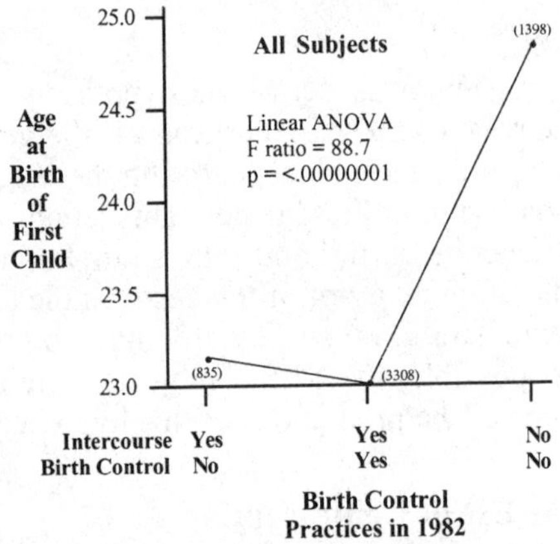

Figure 2. Age of the birth of the first child (1992 information) for subjects who in 1982 had intercourse without using contraception, intercourse with contraception, or no intercourse. From the NLSY database.

This showed that, at this age, those having intercourse, whether contraception was used or not, tended to have their first child at a significantly earlier age (23.0 to 23.2 years) than those who were not having sex in 1982. On the surface this many not seem surprising, since those who were not having sex at age 20, when the question was asked, could not have had a child yet, and this would tend to push upward the average age at the birth of their first child. However, as shown in Figure 1, these individuals also had a significantly higher IQ, and thus there would be selection against the genes for a higher IQ.

The total number of children born to subjects in these three groups, based on the 1993 interview, is shown in Figure 3.

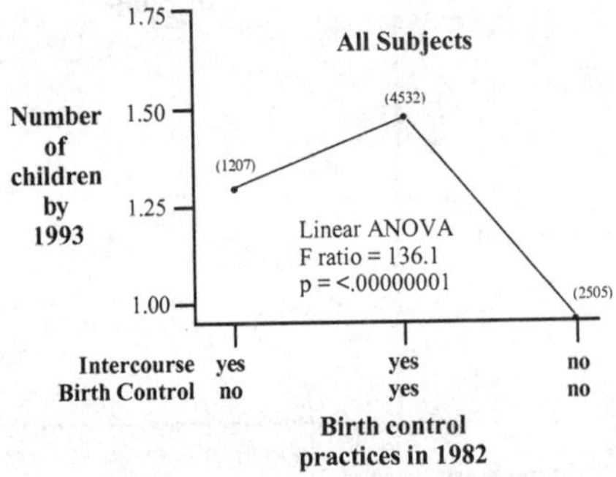

Figure 3. Total number of children the three groups had by 1993, based on whether they had sex without using contraception, sex with contraception, or no sex, in 1982. From the NLSY database.

Again, those having sex in 1982, whether they were using contraception or not, tended to have had significantly more children by 1993 than those not having sex in 1982.

The number of siblings for the different groups is shown in Figure 4 (next page).

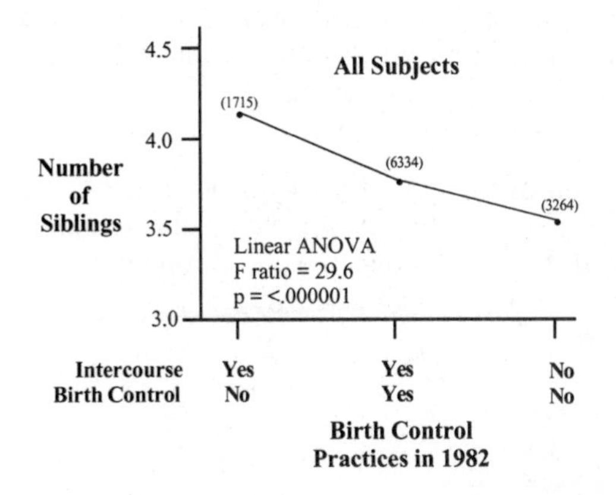

Figure 4. The number of siblings for the three groups based on whether they had sex without using contraception, sex with contraception, or no sex in 1982. From the NLSY database.

Those having sex and not using contraception in 1982 tended to come from larger families (4.2 siblings) than those not having sex (3.6 siblings).

When Whites only were examined, the results for all of these variables were similar to those for the whole group, indicating that race played little role in these trends.

To determine if there was any correlation between contraceptive practices and drug use, the average number of times marijuana was used in the prior month was determined for the three groups. The results are shown in Figure 5.

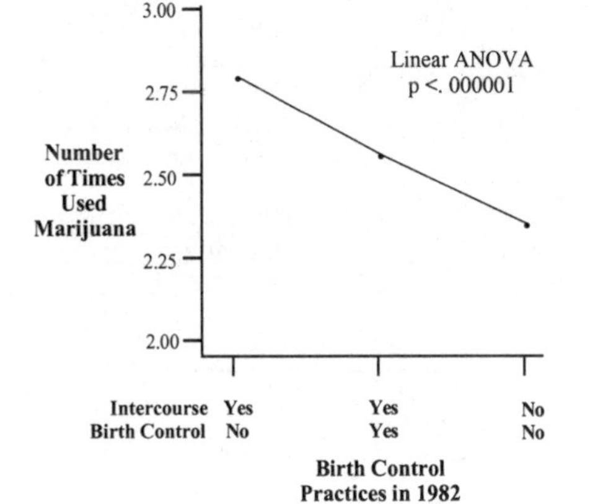

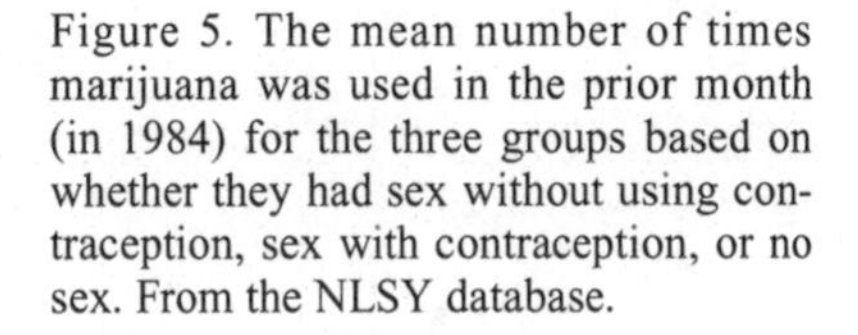
Figure 5. The mean number of times marijuana was used in the prior month (in 1984) for the three groups based on whether they had sex without using contraception, sex with contraception, or no sex. From the NLSY database.

To determine if alcohol use was associated with contraception practices, the mean alcoholism score (see Chapter 23) was determined for the three groups. The results are shown in Figure 6.

Figure 6. The mean alcoholism score for the three groups based on whether they had sex without using contraception, sex with contraception, or no sex. From the NLSY database.

There was a progressive decrease in the alcoholism score from those who had sex but did not use birth control, to those using birth control, to those not having sex.

Age at First Intercourse

The data also allowed an examination of the age of subjects when they first began having sex. Since the data were given separately for males and females, the results for each sex will be presented. Figure 7 shows the correlation between age of initiating sex and IQ.

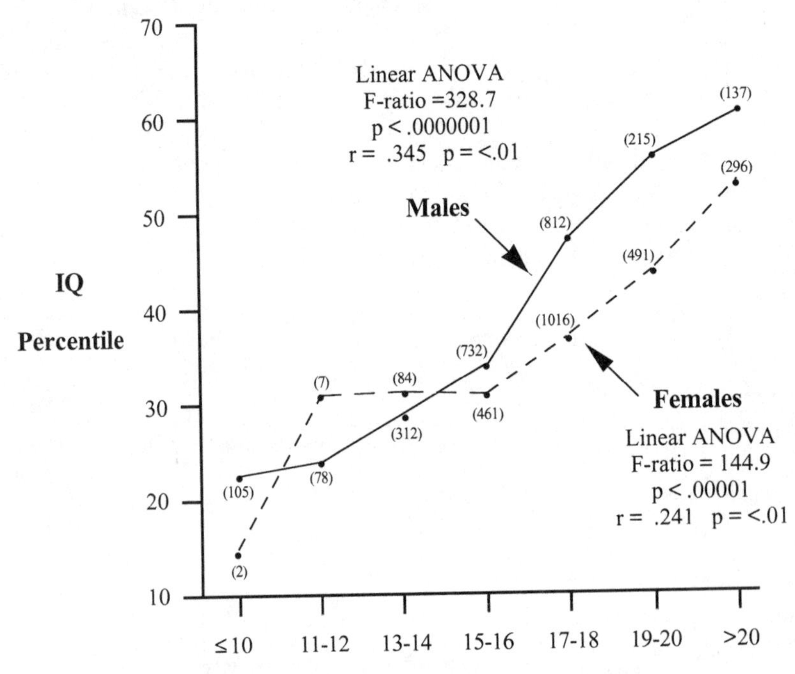

Figure 7. Relationship between age at first sexual intercourse and IQ percentile for males and females. From NLSY database.

For both sexes, the IQ percentile was significantly lower for those becoming sexually active at less than 16 years of age compared to those initiating sexual activity later than age 16. The correlation coefficient (r) between age at first intercourse and IQ percentile was positive and significant for both sexes. This has been widely reported by many other studies.[160,168,206,235]

The relationship between the age at first intercourse and age at the birth of the first child is shown in Figure 8 (next page).

This was significant for both sexes, but much more dramatic for females. Here, those becoming sexually active between 11 to 14 years of age had their first child between 18.4 and 19.4 years of age. There was then a progressive increase in age at the birth of the first child to 26.5 years for those becoming sexually active after 20 years of age. The correlation coefficient between age at first intercourse and age at the birth of the first child was .487, showing a high degree of correlation. While, to a certain extent, these correlations would be expected, the fact that average age at the birth of the first child was up to six years later than the average age of first intercourse suggests that factors relating to inherent behaviors and attitudes are involved, other than simply the issue of timing. This is supported by the correlation between age at first intercourse and IQ, illegal activities, drug and alcohol use (see below).

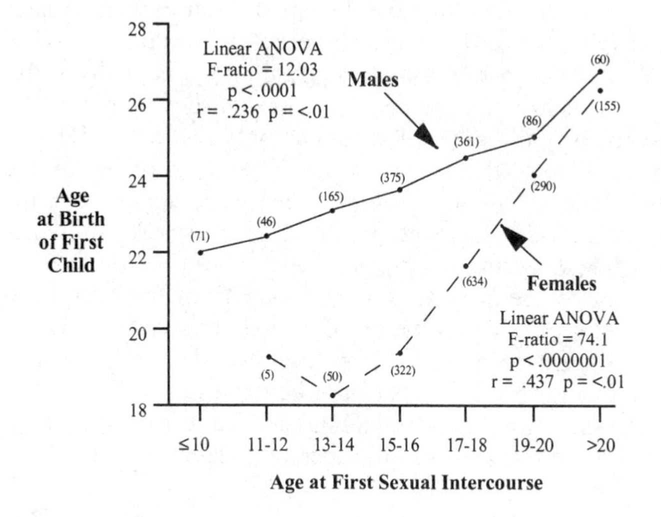

Figure 8. Relationship between age at first intercourse and age at the birth of the first child for males and females. From the NLSY database.

The relationship between age at first sexual intercourse and eventual number of children is shown in Figure 9. Those subjects, especially females, who became sexually active at an early age had more children over a lifetime than those beginning sexual activity at a later age. The male and female rate converged for those initiating sex after age 20. A figure of 2.1 children is considered to be the replacement rate. Only women initiating sex at 16 years of age or earlier had a replacement rate of fertility.

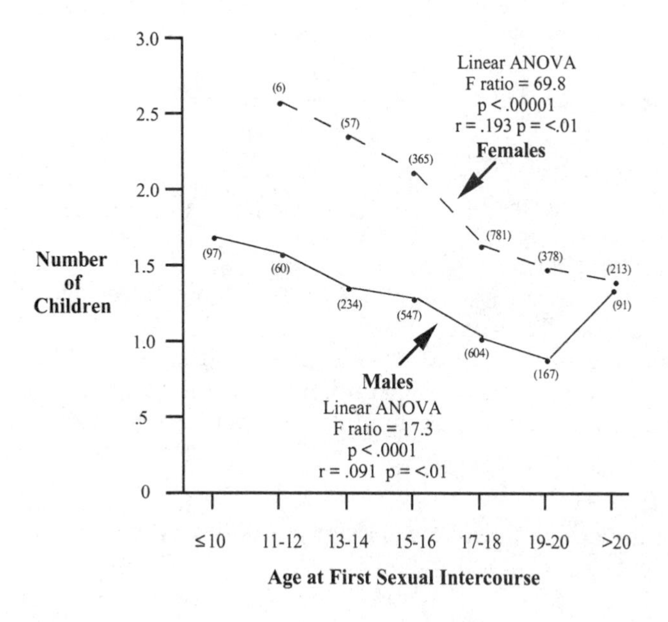

Figure 9. Relationship between age at first sexual intercourse and the total number of children by 1993 for males and females. From the NLSY database.

Those initiating sex early also came from significantly larger families. For example, for males, the number of siblings averaged 4.1

to 4.5 for those beginning to have sex at age 12 or less, versus 3.2 siblings for those beginning sex at 20 or later.

Disruptive and Addictive Behaviors

So far, all that has been demonstrated is that subjects who become sexually active early show a significant tendency to have their first child earlier, to have more children, and to come from larger families. These trends would have a dysgenic effect only if the subjects also had some less desirable genetically influenced behaviors. One of these is low IQ, and the above results show a highly significant correlation between early initiation of sexual activity and lower IQ. But what about other behaviors? One of the aspects of the hypothesis of Jessor and Jessor[140] was that there would be a connection between early sexual activity, teenage pregnancy, delinquent behavior, smoking, and drug use. As shown in Figure 8, the females who began having sex between ages 11 and 16 on average had their first child before they were 20. This supports the part of the hypothesis suggesting an association between early initiation of sexual activity and teenage pregnancy. But what about other aspects of the hypothesis? Is there any association between having sex without using birth control or being sexually active at an early age, and delinquency, smoking, drug or alcohol use? To test this, the relationship between contraceptive use and the number of illegal activities was examined. The results are shown in Figure 10.

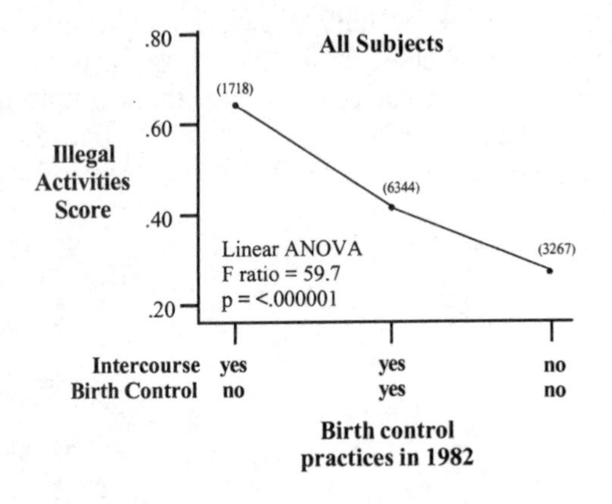

Figure 10. The illegal activities score the three groups had by 1993, based on whether they had sex without using contraception, sex with contraception, or no sex, in 1982.

This showed that those subjects having sex and not using contraception were more likely to engage in a number of illegal activities than those having sex and using contraception and those not having sex. Figure 11 (next page) shows the relationship between age of early initiation of sex and the illegal behavior score.

The illegal behavior score was significantly higher the earlier individuals began initiating sex, and the correlation coefficient between the illegal behavior score and the age of initiating intercourse was significant for both sexes.

Drug and Alcohol Use

The results for age of initiation of smoking and age at first sexual intercourse are shown in Figure 12 (next page).

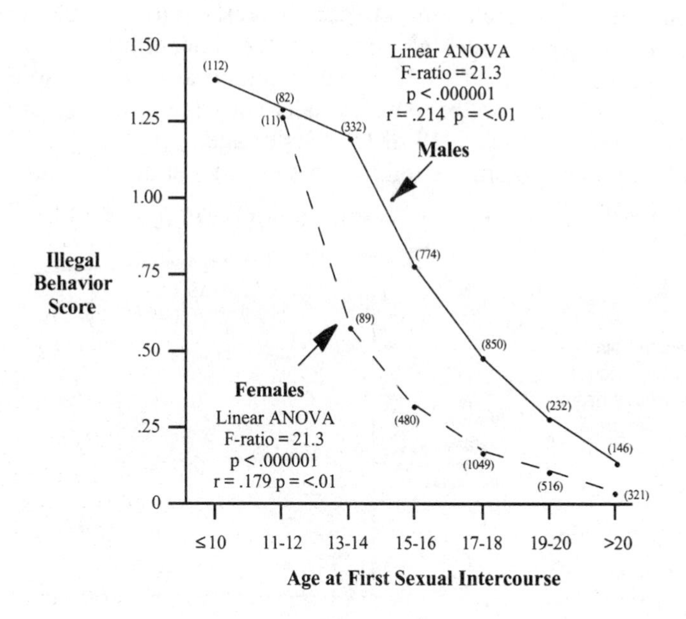

Figure 11. Relationship between age at first sexual intercourse and illegal behavior score. From the NLSY database.

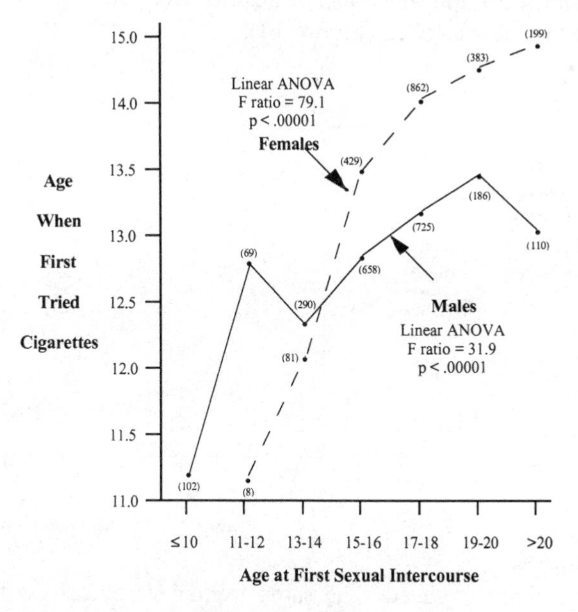

Figure 12. Relationship between age at first sexual intercourse and the age of beginning to smoke cigarettes. From the NLSY database.

These was a striking correlation between the early initiation of sex and the early initiation of smoking, especially for females. The small number of girls who began having sex at age 11 to 12 began smoking at a comparable age. By contrast, those girls who did not become sexually active until age 17 or older did not begin smoking until they were 14 to 15 years of age.

The age of beginning to smoke marijuana showed a similar association with the age of initiating sex (Figure 13).

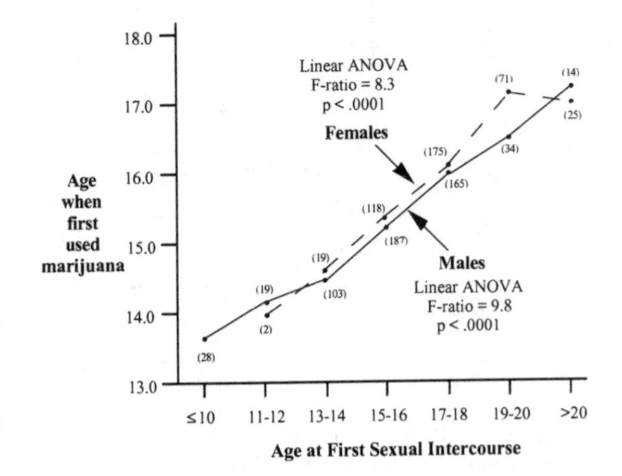

Figure 13. Relationship between age at first sexual intercourse and age when first used marijuana. From the NLSY database.

The results for males and females were virtually identical, with those who were sexually active by 12 years of age initiating pot smoking around age 14, while those who were not sexually active until age 19 or more did not begin smoking pot until around age 17. The results were similar for the average number of times marijuana was used (Figure 14).

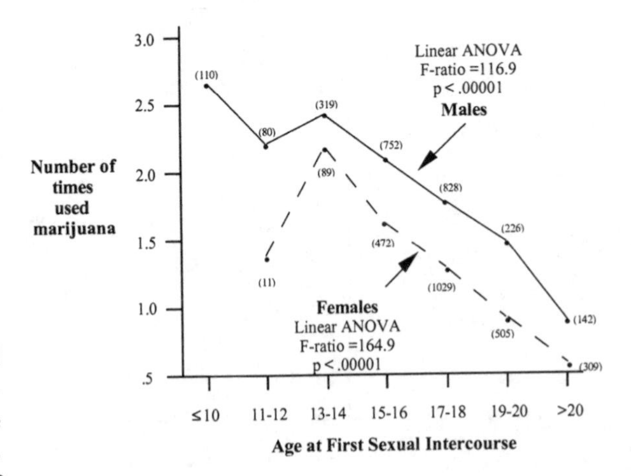

Figure 14. Relationship between age at first sexual intercourse and number of times marijuana was used in the prior month. All subjects. From the NLSY database.

For both males and females there was a progressive decrease in the number of times marijuana was used as the age of becoming sexually active increased. The same trends were seen with the use of other drugs, including amphetamines, barbiturates, cocaine, tranquilizers, and others. The results were essentially identical among the White only group, indicating race was not a significant factor. For example, the results for the number of times marijuana was used versus the age of initiation of sexual activity for Whites only is

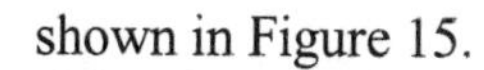
shown in Figure 15.

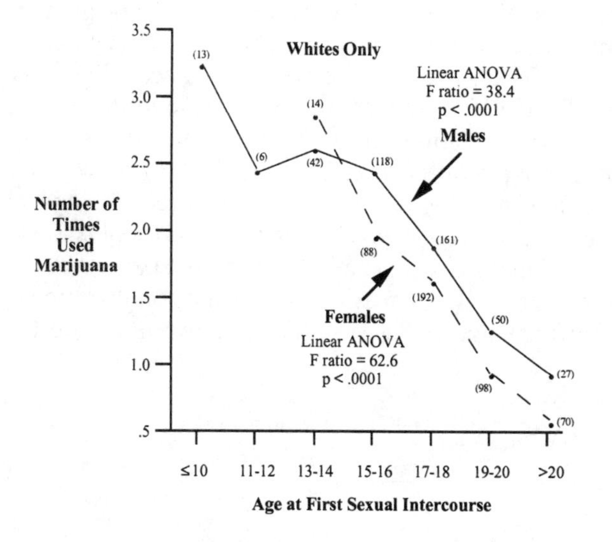

Figure 15. Relationship between age at first sexual intercourse and number of times marijuana was used in the prior month, for Whites only. From the NLSY database.

Comparing this to Figure 14 shows that, except for the smaller numbers and absence of females smoking pot and initiating sex at less than 13 years old, the results were very similar.

Alcohol use also showed a significant correlation with age at first intercourse. For example, in Chapter 23, subjects were divided into those with an alcoholism score of less than 10 and a negative family history of alcoholism, those with a score 10 or greater and no family history of alcoholism, and those with a score of 10 or more and a positive family history of alcoholism. The frequency of these three groups according to the age of first intercourse is shown in Figure 16.

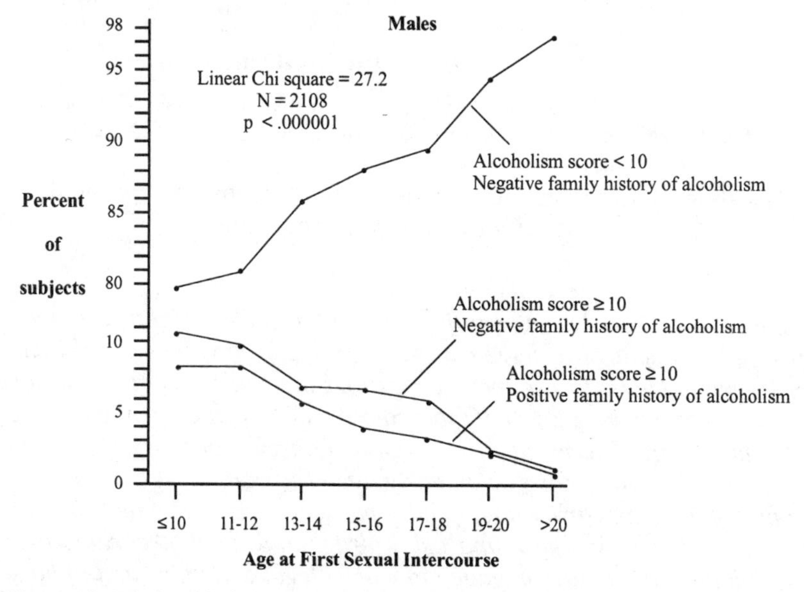

Figure 16. Relationship between age at first sexual intercourse and the presence or absence of alcoholism or a family history of alcoholism. All subjects. From the NLSY database.

Among those who initiated sexual activity at age 20 years of age or greater, 97.6% were non-alcoholics and had no family history of alcoholism. This percentage progressive decreased with earlier and earlier ages of initiating sexual activity to 80.2% for those initiating sex at age 10 or earlier. The percentage of those with alcohol problems showed a reciprocal relationship, with 19.7% of those starting sex at age 10 or less having alcohol problems, either with (8.6%) or without (11.1%) a family history of alcoholism. This progressively decreased as subjects began having sex later and later to 2.4% (0.8% positive family history plus 1.6% negative family history) for those initiating sexual activity after age 20.

This relationship was also examined by comparing the mean alcoholism score across groups initiating sex at different ages. These results are shown in Figure 17.

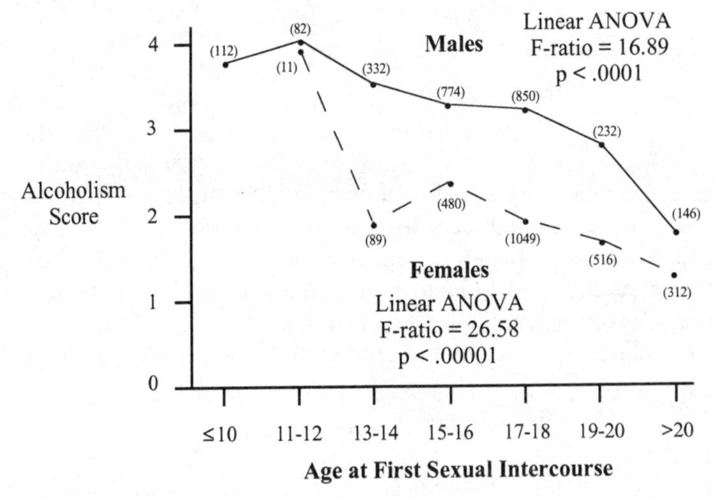

Figure 17. Relationship between age at first sexual intercourse, for males and females, and the mean alcoholism score. All subjects. From NLSY database.

This showed a progressive decrease in the alcoholism score with increasing age of initiating sexual activity for both males and females.

Summary – There was a significant trend for those subjects who had sex without using birth control to have children earlier, to have more children, to come from larger families, to use more drugs and alcohol, to engage in more illegal activities, and to have a lower IQ than those using birth control or not having sex at 21 years of age. There was also a significant trend for those males and females who became sexually active at an early age (14 years of age or earlier) to have their first child at an earlier age, to have more children, to come from larger families, to have begun smoking and using drugs and alcohol earlier, to use more drugs and alcohol, to have engaged in more illegal activities, and to have a lower IQ than those becoming sexually active later. To the extent that genetic factors are involved in IQ, smoking, drug and alcohol use or abuse, and illegal behaviors, these trends will result in the selection of those genes.

Chapter 25

NLSY – Marriage

Teenage pregnancies are dramatically more common now than they were in the 1950s.[181] Such pregnancies, especially those occurring in unwed mothers, produce a considerable cost to society in terms of Aid to Families of Dependent Children (AFDC) and other welfare costs. While these are of social and political concern, the issue of importance to this book is whether they also have a genetic cost to society. The NLSY data allowed an examination of several aspects of this potential problem.

For comparison purposes, four different classes of subjects were examined: 1.) unmarried pregnant teenagers, including fathers of pregnant girlfriends, where teenage is defined as individuals 18 years of age or younger, 2.) Married pregnant teenagers or husbands of pregnant wives, 3.) Married, non-pregnant teenagers, and 4.) Non-married, non-pregnant teenagers. Group 1 was defined as those subjects who had their first child prior to age 19 and were married after their first child was born. Group 2 was defined as those subjects who had their first child prior to age 19 and were married prior to the birth of that child. Group 3 was defined as those who were married prior to age 19 but had their first child after age 19. Finally, group 4 was defined as those who were married after age 18, or not married, and either had no children or had a child after age 18.

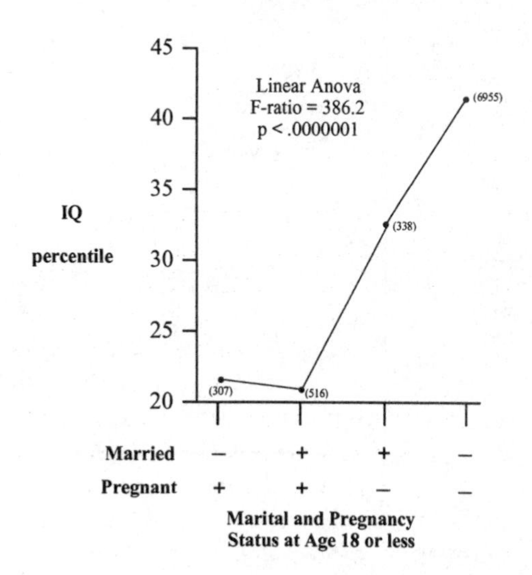

IQ

One of the first concerns relating to a possible dysgenic effect of teenage pregnancy was whether there was a difference in the average IQ percentile for those who got pregnant in teenage years compared to those who did not. The results are shown in Figure 1.

Figure 1. Correlation of pregnancy or marriage in teen years with IQ percentile. From the NLSY database.

These results showed a dramatically lower IQ for pregnant teenagers, whether they were married or not; thus, the mean IQ percentile was less than 22 for both the unmarried and married pregnant teenagers. It increased to 33 for those who were married at age 18 or younger but were not pregnant, and to 41.8 for those who were neither married nor pregnant as teenagers.

Schooling

One of the problems of teenage marriages is that they tend to interfere with schooling. To examine this, the highest grade completed was determined for each of the four groups. The results are shown in Figure 2.

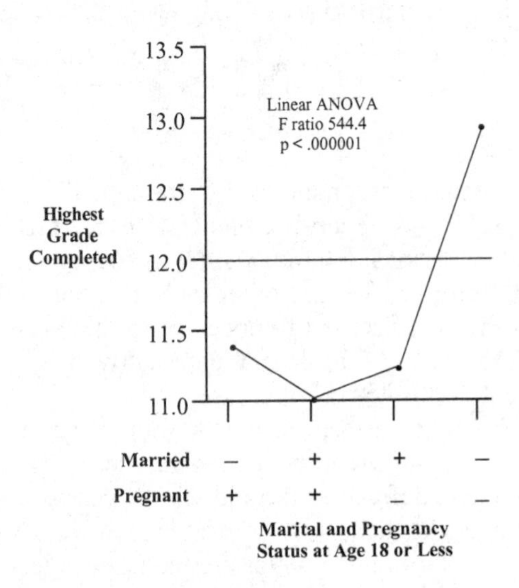

Figure 2. Correlation of pregnancy or marriage in teen years with highest grade completed. The horizontal line at 12 represents graduation from high school. From the NLSY database.

This showed, not unexpectedly, that both teenage pregnancy and teenage marriage, whether associated with pregnancy or not, were correlated with dropping out of school before graduation from high school; thus, the average highest grade completed was less than 11.5 for all three groups. By contrast, those who were neither married nor pregnant as teenagers, on average, completed high school and one year of college.

Age at Birth of the First Child

To determine the potential dysgenic effect of teenage pregnancy or marriage, it was necessary to examine the correlation with the age at the birth of the first child. This is shown in Figure 3.

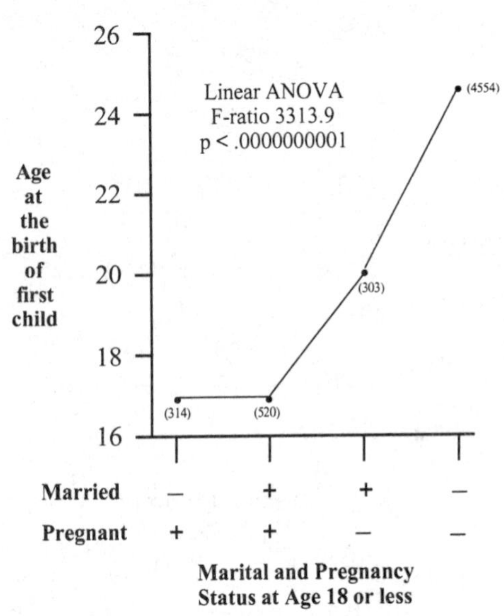

Figure 3. Correlation of pregnancy or marriage in teen years with age at the birth of the first child. From the NLSY database.

Clearly, a teenage pregnancy would be associated with a birth of the first child in teenage years, as shown. This figure shows the magnitude of the difference for those pregnant or married in their teens, versus those waiting until after age 18 to marry. The average age at the birth of the first child was 24.7 years for those marrying after age 18, versus 17 years for those getting pregnant as teenagers. Those who were married but not pregnant as teenagers, had their first child at an intermediate age of 20.5 years.

The correlations between age of marriage and the age at the birth of the first child, for those with and without an illegitimate first child, are shown in Figure 4.

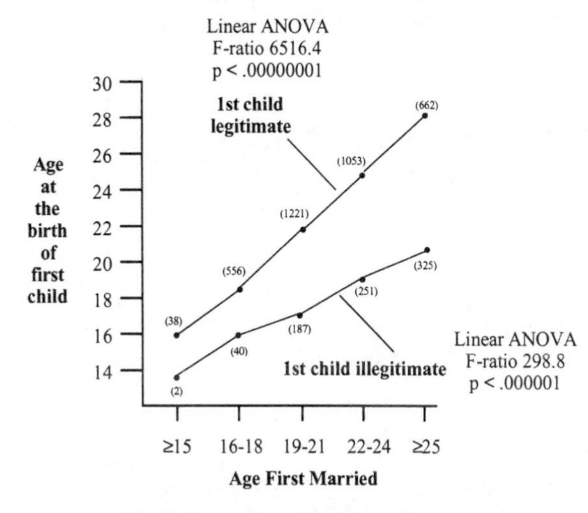

Figure 4. Correlation of the age when first married and age at the birth of the first child for those with an illegitimate or legitimate first child. From the NLSY database.

Not surprisingly, the earlier individuals got married, the earlier they had children. This correlation was much greater for those who had their first child after they were married than those whose first child was illegitimate. In addition, those individuals whose first child was illegitimate tended to have their first child much earlier, regardless of their age when they got married. This is likely to have a dysgenic effect if both having an illegitimate child and getting married early are associated with a behavior with a genetic component, such as IQ. The results for this analysis are shown in Figure 5.

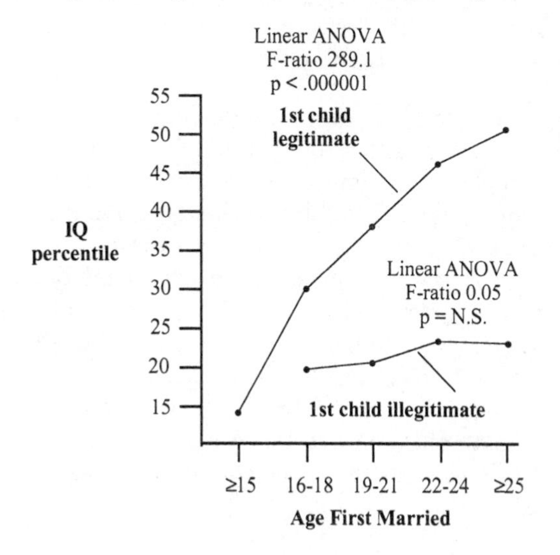

Figure 5. Correlation of age first married and IQ percentile for those with an illegitimate or legitimate first child. From the NLSY database.

For those whose first child was legitimate, there was a progressive increase in IQ percentile the older they were when they got married. This increased from the 15th percentile for those marrying at age 15 years or earlier to the 53rd percentile for those marrying at age 25 or older. By contrast, those whose first child was illegitimate tended to uniformly have a low IQ percentile, in the 20 to 23 range, regardless of the age at which they were eventually married.

To further explore the interrelationship between marriage, illegitimate pregnancy, and education, the correlation between the age at the birth of the first child and the highest grade completed was examined for those with and without illegitimate children. The results are shown in Figure 6.

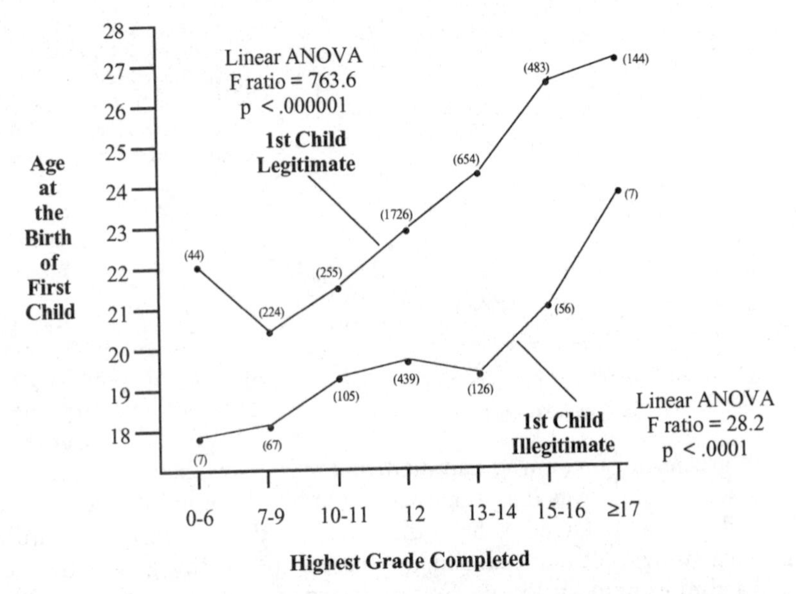

Figure 6. Correlation between highest grade completed in high school and the age at birth of the first child, for those with an illegitimate or legitimate first child. From the NLSY database.

For those whose first child was legitimate, there was a high degree of correlation between the highest grade completed and the age at the birth of their first child. This ranged from age 21 for those dropping out in junior high to age 27 or more for those graduating from college. For those whose first child was illegitimate, the age at the birth of the first child was much lower, ranging from 18 years for those dropping out in grade school to 19.9 years for those completing high school and one or two years of college. There were relatively few individuals whose first child was illegitimate and who also completed college or went to graduate school. For those who did, the birth of the first child ranged from 21 to 24 years.

Illegitimacy could also be associated with a dysgenic effect for the genes involved if individuals with illegitimate children had more children over their lifetime than those who did not have illegitimate children. The results of examining this are shown in Figure 7 (next page).

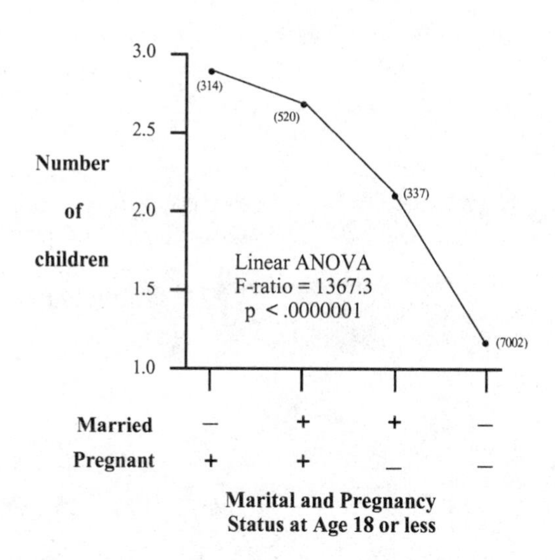

Figure 7. Correlation between teenage marriage and pregnancy and the average number of children. From the NLSY database.

Over their lifetime, unwed pregnant teenagers had more children than all other groups. Those who were neither pregnant nor married in their teenage years had the fewest children over their lifetime. The figures for the number of siblings for subjects in the four groups were 5.2, 4.3, 3.9, and 3.6 respectively ($p < .0001$).

The relative value of the above correlations were the same when restricted to Whites only.

Summary – Individuals who were pregnant as teenagers (girls), or caused a pregnancy as teenagers (boys), had a significantly lower average IQ percentile than those who were married as teenagers but not pregnant, or those who were neither married nor pregnant as teenagers. There was a strong association between teenage pregnancy and/or marriage and dropping out of school before completing high school. By contrast, those who were neither pregnant nor married as teenagers, on average, completed high school and one year of college. As expected, those who were pregnant as teenagers were significantly younger when they had their first child. In addition, they had more children than those who had no teenage pregnancy. Having an illegitimate child was associated with a lower IQ, a shorter time in school, having children earlier, having more children, and coming from larger families than those who had legitimate children, regardless of when they were married. These observations indicate that having illegitimate children and getting married in the teenage years will select for the genes associated with a lower IQ and other genetically influenced behaviors in this group of individuals.

Chapter 26

NLSY – Welfare

Many of the pregnant teenagers described in the previous chapter were supported by welfare. There were several variables in the NLSY that allowed an examination of the characteristics of the subjects who were on one or more welfare programs. The most relevant program was Aid to Families with Dependent Children, or AFDC. Here I simply examine the question of whether those who receive AFDC tend to have their first child earlier, or have more children, or have a lower IQ, than those not receiving AFDC. In its initial years the AFDC program was set up to provide financial support for widows with children; thus it was meant only for married women whose husbands had died. However, over time, as it began to be used instead for unwed mothers, the program began to rapidly expand. By its very nature it is obvious that mothers on AFDC would tend to have their first child earlier than those not on AFDC. The real question is, What is the degree to which this occurs? The results for these questions are shown in Table 1.

Table 1. Characteristics of mothers on AFDC or Food Stamps (F.S.) for the year 1984.

	On AFDC N=500 Mean	Not on AFDC N=2810 Mean	p
Age at birth of first child	18.61	23.20	<.000000001
Number of children	2.67	1.45	<.000000001
IQ percentile	21.62	42.66	<.000000001
Number of siblings	5.16	2.56	<.000000001
	On F.S. N=650 Mean	**Not on F.S. N=2670 Mean**	**p**
Age at birth of first child	19.22	23.33	<.000000001
Number of children	2.58	1.41	<.000000001
IQ percentile	23.24	43.42	<.000000001
Number of siblings	5.01	3.69	<.000000001

For uniformity the analysis was restricted only to the NLSY women who had children. It is not surprising that the average age of those on AFDC was significantly lower than for those women not on AFDC, since this is a program particularly designed to help unmarried teenage mothers. However, the same results were seen for mothers on food stamps, a welfare program less directed only to younger women. A second effect was the significant increase in number of children, or number of mother's siblings, for both those on AFDC and food stamps. Compounding the effect was the dramatic difference in IQ percentile for mothers on AFDC (21.6 percentile) versus those not on AFDC (42.7 percentile).

Summary – Significant effects were present for mothers utilizing two welfare programs – Aid to Families of Dependent Children (AFDC) and food stamps. In both cases, mothers in the program had their first child an average of four years earlier, had more children, more siblings, and a lower average IQ than those not in these programs.

Chapter 27

NLSY – Suspended or Expelled from School

In addition to sexual behavior, a second critical variable in determining the age at which individuals begin having children, and thus the degree to which the genes they carry are selected for, is school performance. The longer individuals stay in school, the later they begin having children. Staying in school depends on many factors, one being whether or not one has been expelled from school. This factor is related to delinquency, since one of the reasons for expulsion is delinquent behavior. It can also be related to academic abilities, since a second factor is school performance and grades. In this chapter I will examine various dysgenic factors related to being suspended or expelled from school.

A few points of note. First, a minority of NLSY subjects were suspended from school (25%) and an even smaller number were expelled (4.4%). In the figures, the variable being examined may show wide fluctuations as the number of subjects in a group decreases. However, it is the trends that are important. Second, despite the different numbers, the results for being suspended from school were essentially identical to being expelled from school. Thus, for each variable, the results for only one or the other will be shown.

For the sake of continuity with the previous chapters, I will first examine the relationship between suspension from school and sexual behavior (Figure 1).

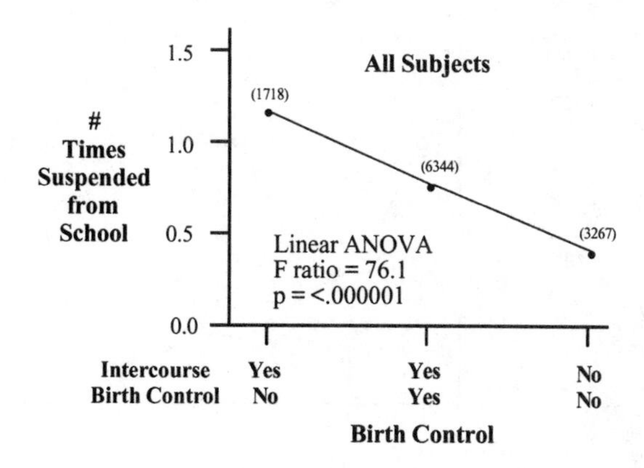

Figure 1. The average number of times suspended from school by birth control practices at age 21. From the NLSY database.

Individuals who were having sex but not using birth control were suspended from school more often than those using birth control. Both were suspended more often than those not having sex.

The relationship between suspension from school and age of initiation of sexual activity is shown in Figure 2.

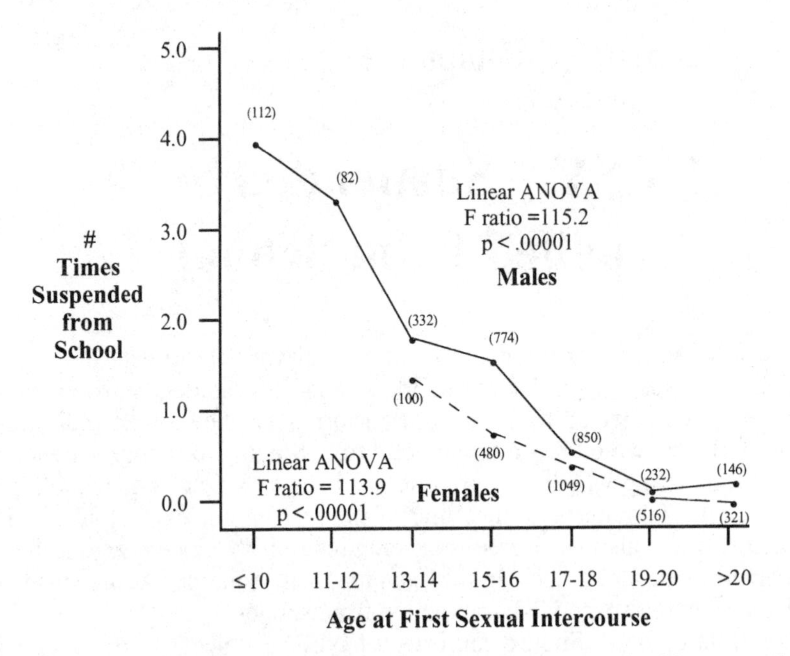

Figure 2. The average number of times suspended from school by age at first sexual intercourse. From the NLSY database.

The earlier males began having sex, the more often they were suspended from school. This was particularly dramatic for those who began having sexual intercourse before age 13. This supports a correlation between delinquency, school problems, and precocious sexual activity. For females the graph is truncated because of the small number of females initiating sex prior to age 13. Nonetheless, the correlations at later ages were similar to that for males.

The correlation between IQ and being expelled from school is shown in Figure 3.

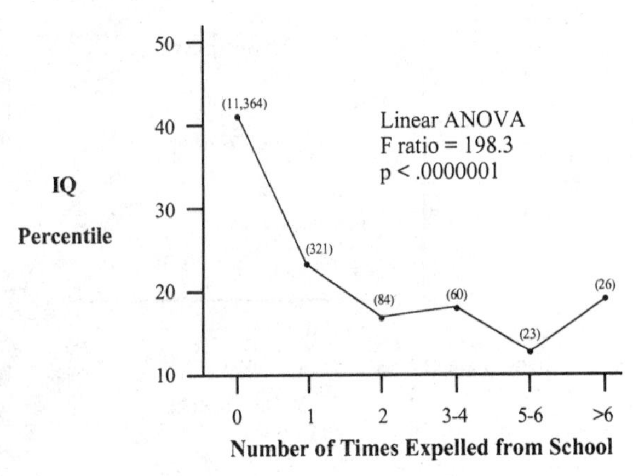

Figure 3. The average IQ percentile versus the number of times suspended from school. From the NLSY database.

There was a marked drop in average IQ percentile for those never expelled from school to those ever expelled, whether it was only once or more than 6 times. To the degree that poor academic performance contributes to expulsion from school, this is not unexpected. However, as shown previously (Chapter 21), delinquent behavior is also associated with a lower average IQ, and this also plays a role in school expulsion.

There was also a correlation between the use of marijuana and the number of times expelled from school (Figure 4).

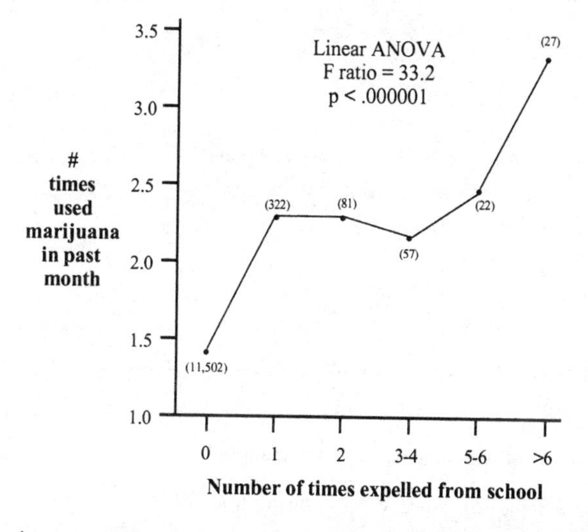

Figure 4. The average number of times used marijuana in the prior month versus the number of times expelled from school. From the NLSY database.

The average frequency of use of marijuana in the prior month increased from 1.45 times for those never expelled from school, to 2.3 to 2.5 times for those expelled 1 to 6 times, to 3.45 times for those expelled more than 6 times. There was also an association with alcohol use as measured by the alcoholism score (Figure 5).

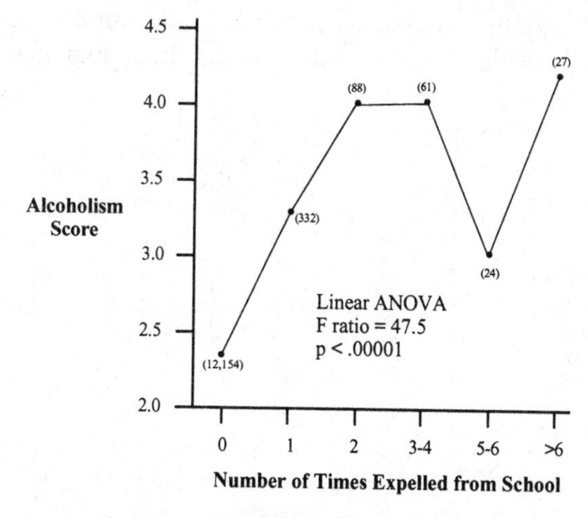

Figure 5. The average alcoholism score versus the number of times expelled from school. From the NLSY database.

Except for a dip for those expelled from school 5 to 6 times, there was a progressive increase in the mean alcoholism score from a low of 2.4 for those who were never expelled from school, to 4.3 for those expelled more than 6 times.

There was a remarkably progressive association between the number of times students were expelled from school the total number of illegal behaviors they had been involved in (Figure 6).

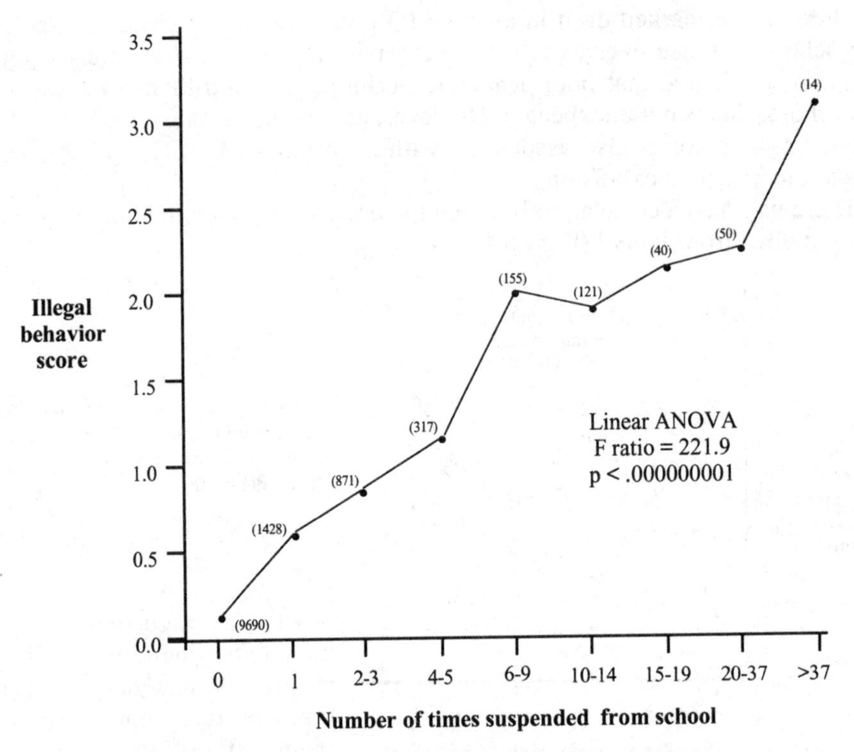

Figure 6. The average illegal behavior score versus the number of times expelled from school. From the NLSY database.

Thus, the majority of students were never suspended and this group averaged only 0.24 illegal behaviors. There was then a progressive increase in number of illegal activities with increasing number of times the students had been expelled from school, with an upper extreme of 3.3 for those suspended more than thirty-seven times.

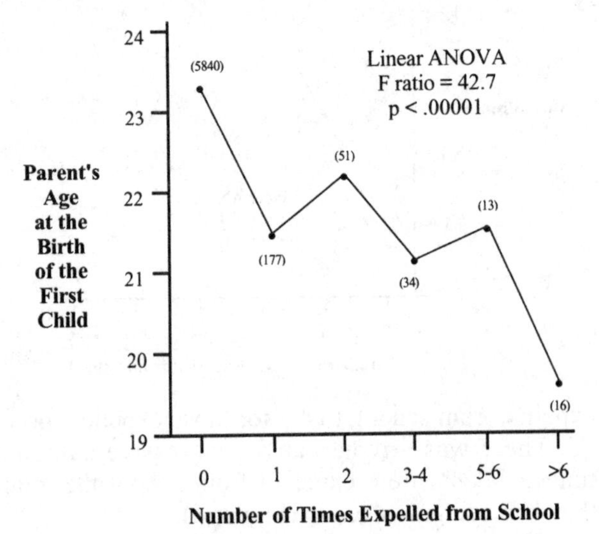

Figure 7. The parent's age at the birth of their first child versus the number of times expelled from school. From the NLSY database.

The degree to which suspension or expulsion is associated with a dysgenic effect depends upon whether it is also associated with having children earlier or having more children. The relationship between expulsion from school and the age at the birth of the first child is shown in Figure 7 (previous page).

The average age at the birth of the first child dropped from 23.4 years for those never expelled from school to 19.6 years for those expelled more than 6 times, with an irregular but progressive decrease for those expelled 1 to 6 times.

There was a modest and significant increase in the in the number of children from 1.4 for those who were never suspended to 2.15 for those suspended more than 37 times.

Summary – Students who were either suspended or expelled from school had a lower average IQ, were more likely to have sex without using birth control, started having sex earlier, were involved in more illegal behaviors, smoked more marijuana, had more problems with alcohol, had their first child earlier, and had more children than students who were not suspended or expelled. To the extent that being suspended or expelled from school, having a lower IQ, problems with drugs and alcohol, and delinquent behavior, are controlled by genes, the earlier age of the birth of the first child and increased number of children will result in a selection for those genes.

Chapter 28

NLSY – Education

A major theme of this book is that education, more than any other factor, is the primary culprit in generating the selective force resulting in the selection of genes for addictive and disruptive behaviors and learning disorders. As shown in Chapter 38, at the beginning of the twentieth century the educational playing field was level, with only 2% of the population attending college. By the latter half of the century this had ballooned to over 35% of the population attending college. The hypothesis assumes there is a direct correlation between the years of education, IQ, and older age of having the first child, and a negative correlation with the total number of children. The more years spent in school, the greater the delay in initiating childbearing and the fewer the total number of children. Since this delay in childbearing was drawn almost exclusively from individuals with a higher IQ, this results in a selection against the higher IQ genes. Finally, there was also an inverse correlation such that the higher the IQ, the lower the frequency of learning disorders and the number of disruptive and addictive behaviors (see previous chapters). This educationally-driven delay in childbearing provides the driving force behind the selection for genes for disruptive and addictive behaviors and learning disorders.

This chapter uses the NLSY to test whether the above thoughts are purely the fantasy of a demented geneticist, are true but to a trivial degree, or are true to an alarming degree.

One of the questions asked in the 1990 survey was, "What is the highest grade you have completed?" This was taken from the 1990 rather than from the 1992 or 1993 surveys simply because the largest number of subjects answered the question in this year. In the 1992 and 1993 surveys, many more subjects were skipped either because of loss to follow-up or because some segments of the original group were no longer being questioned. Since subjects averaged 31 years of age in 1990, this provided sufficient time for the vast majority to have completed whatever schooling they were going to complete.

The first question of importance was, "What is the correlation between number of years of school completed and IQ?" Logically, the correlation should be positive, but to what degree? Figure 1 (next page) shows the results.

This shows an almost perfect correlation between the IQ percentile and the number of years of school attended. The skeptic might think this was due to the effect of education on IQ testing since the environment does play some role in IQ. However, the IQ tends to be fixed by 4 to 7 years of age and tends to be relatively independent of education. Only the most intense interventions for children exposed to extreme environmental deprivations have resulted in significant increases in IQ.[246] The F ratio of 6,586.3, shown below, was one of the highest of all the variables examined in this book. The correlation coefficient between IQ percentile and the highest grade completed was very high at .63 (p < .001).

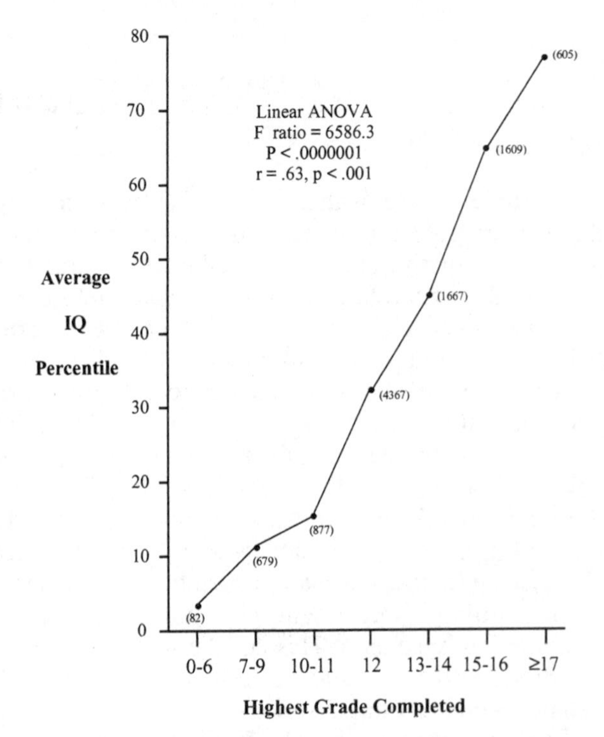

Figure 1. Highest grade completed by average IQ percentile. All subjects. From the NLSY database.

The correlation between the highest grade completed and age of first sexual intercourse is shown in Figure 2 (next page). Those males who initiated sexual activity at less than 10 to 13 years of age on average completed 11.7 years of schooling, i.e. did not complete high school. As males initiated sexual intercourse at progressively later years, they also completed progressively more years of schooling, to the point that those initiating sexual activity at 20 years of age or greater completed over two years of college. The correlation coefficient between the age of first intercourse and highest grade completed was .34 for females and .27 for males.

Presumably, if there was a positive correlation between the number of years of schooling completed and the age of initiation of sexual activity, there would also be a tendency for those who stayed in school longer to have their first child at a later age. Was this the case? Figure 3 (next page) shows the results.

Those subjects who completed only 7-9 grades of schooling had their first child, on average, at 20.6 years of age. Subsequent to this, the longer the subjects stayed in school, the older they were before they had their first child. This increased progressively to the point that those in graduate school (highest grade ≥ 17), on average, had their first child at 27.7 years of age. The correlation coefficient was .39. The trend slightly reversed itself (21.9 years) for the small number of subjects (62) who stopped their education in grade school. But, as shown below, this reversal did not hold for the number of children born to the different groups.

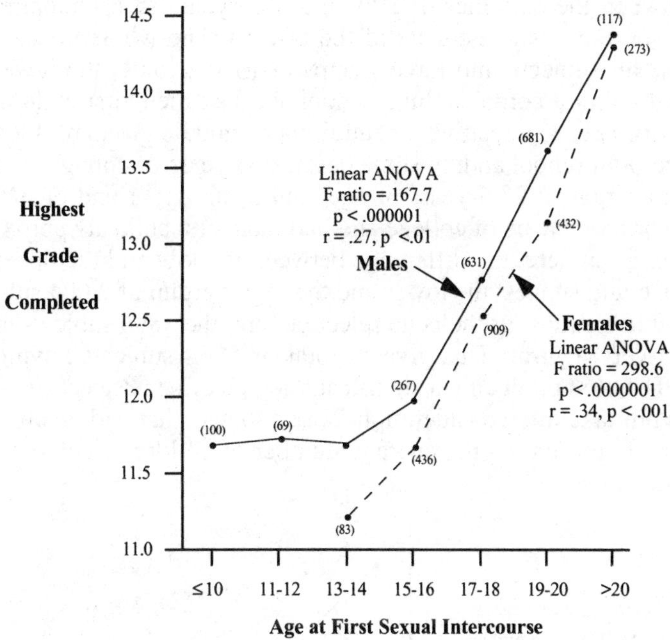

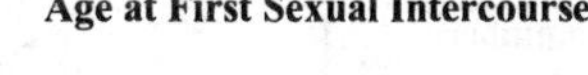

Figure 2. Age at first sexual intercourse versus highest grade completed. All subjects. From the NLSY database.

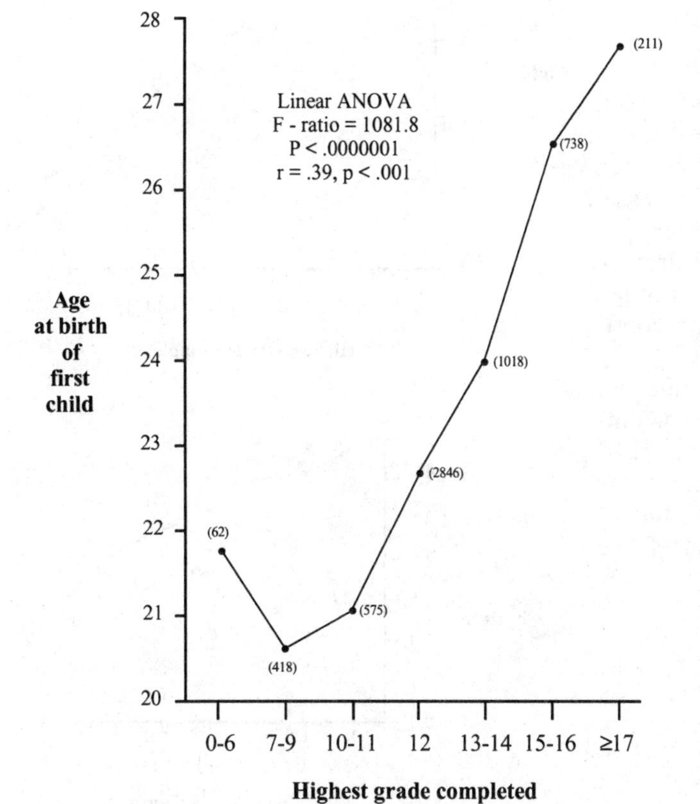

Figure 3. Age at the birth of the first child versus highest grade completed. All subjects. From the NLSY database.

The difference between the extremes of 20.6 and 27.7 years is a whopping 7.1 years. However, a more realistic estimate of the effect in the whole population would be to divide all subjects into three groups consisting of 1) the lower end of 1055 subjects who didn't complete high school and had their first child at approximately 21 years of age, 2) a relatively neutral major middle group of 3864 subjects who completed high school and, in some cases, two years of college, and had their first child on average at 23.5 years of age, and 3) an upper end of 949 subjects who had three or more years of college and had their first child at approximately 27 years of age. Even here, the difference between the low and the upper group was 6 years. One could suggest the lower and the upper group of 2,004 subjects would be responsible for most of the gene selection, and the 3864 subjects in the middle were selectively neutral. This gives a total of 5868 subjects having children with 34% (2004/5868) producing a significant degree of gene selection.

This analysis does not take into consideration those subjects that had no children. Figure 4 does this by presenting the average number of children (including none) by duration of education.

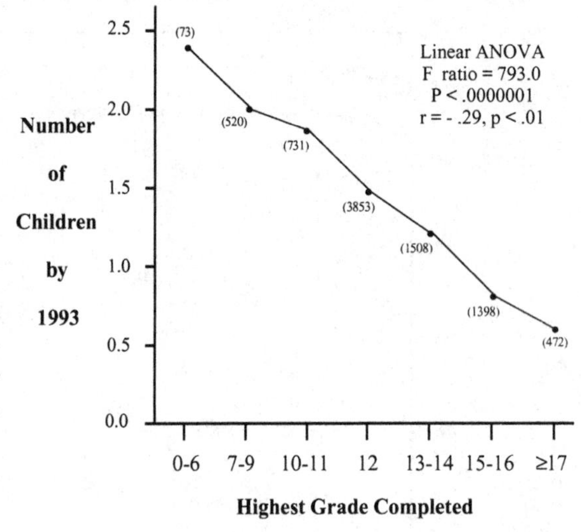

Figure 4. Number of children (by 1993) versus highest grade completed. All subjects. From the NLSY database.

Those subjects who completed 6 or less years of school had an average of 2.5 children. This progressively decreased with increasing years of schooling to .66 children by those who attended graduate school. The correlation coefficient was -.29.

These results are quite similar to those presented by Vining[256] based on the U.S. Bureau of the Census for women who were age 33 to 44 in 1980. These results are shown in Figure 5.

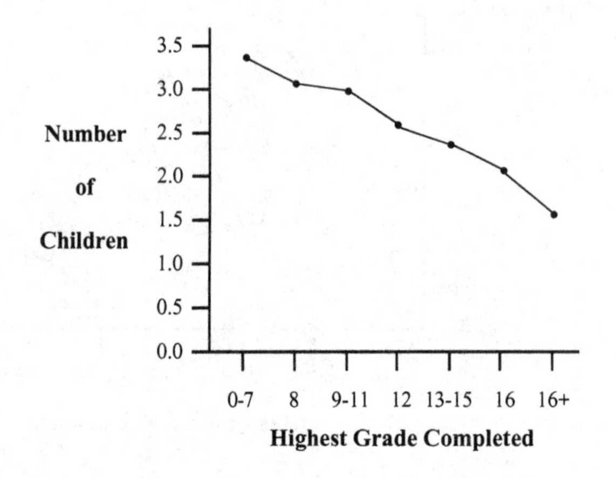

Figure 5. Number of children by highest grade completed based on U.S. Census reports for women aged 33-44 in 1980. From Vining, D. R. Jr. *The Behavioral and Brain Sciences* 9:167-216 256.

Comparison of Figure 5 with Figure 4, based on 1993 data, shows the trends remain the same, but the fertility rates are lower in 1993 than in 1980.

The relationship between education and age of childbearing shown by the NLSY data is supported by the U.S. census data as analyzed by Rindfluss and Sweet in their book, *Postwar Fertility Trends and Differentials in the United States*.[223] Two of the diagrams from that book, showing the timing of childbirth for women who dropped out of high school (education to grades 9-11) versus the timing for women with post-graduate schooling (education of 16+ years) for the year 1965, are redrawn in Figure 6.

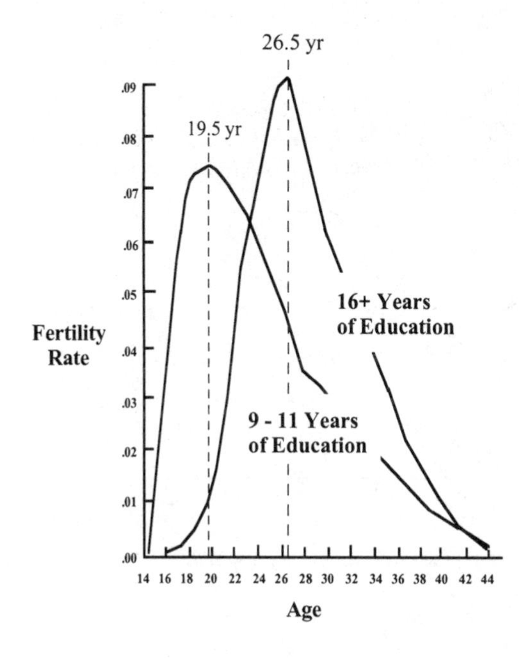

Figure 6. Fertility schedules for women with less than a high school degree (9-11 years of education) versus women with a post-graduate education (16+ years) for the year 1965. From Rindfluss and Sweet, *Post-war Fertility Trends and Differentials in the United States*, Academic Press, 1977.

This shows that for the women who completed 9 to 11 years of education, the peak of childbearing in 1965 was at age 19.5 years. By contrast, for the women with post-graduate schooling, in the peak of childbearing was at 26.5 years. The women with intermediate durations of education showed intermediate peak ages of fertility. The difference between the two groups shown was 7.0 years, virtually identical to the 7.1 years shown above for the NLSY data.

Back to the NLSY data. The correlation between the highest grade completed and the number of siblings is shown in Figure 7.

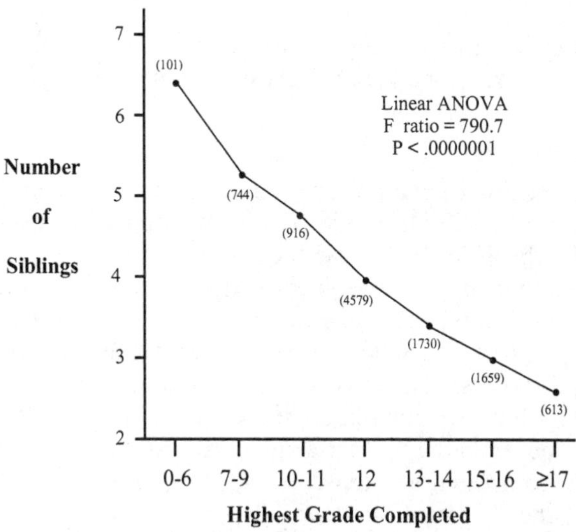

Figure 7. Number of siblings versus highest grade completed. All subjects. From the NLSY database.

The measure of the number of children in the previous generation is more significant than for the number of children of the NLSY subjects, and it decreased from 6.5 for those completing only 0 to 6 years of schooling to 2.6 for those completing 17 or more years of schooling.

For continuity with Chapter 27, the correlation between the number of times the subjects were suspended from school and the total number of years of education is shown in Figure 8. It should come as no surprise that those subjects who were most often suspended from school ended up leaving school earlier than those who were rarely suspended.

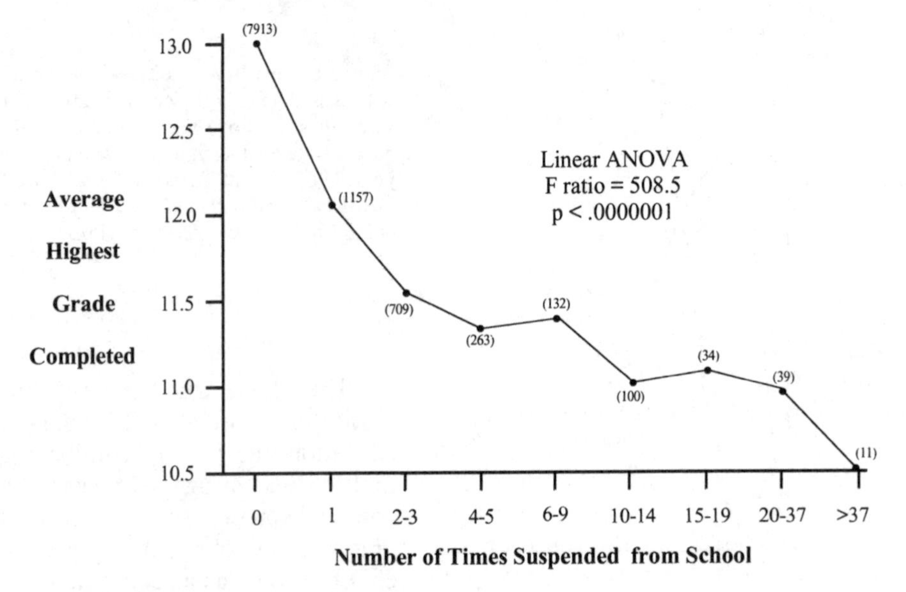

Figure 8. Highest grade completed versus number of times suspended from school. All subjects. From the NLSY database.

All of the above correlations were essentially identical when Whites alone were examined, indicating that race was not a factor in these results.

Multiple Regression Analysis

In many of the above relationships I have presented correlation coefficients, which are estimates of the degree to which two variables are related to each other. The highest and most significant of these was the correlation between IQ percentile and highest grade completed (r = .63). The next highest correlation was between IQ percentile and age at the birth of the first child (r = .38). However, for the purpose of trying to understand what forces are responsible for the selection of genes for learning disorders, disruptive and addictive behaviors, the correlations with the age at the birth of the first child are the most important. It is of note that here the correlation between the mother's age at the birth of her first child and the highest grade she completed is slightly higher (r = .392) than the correlation between her age at the first birth and her IQ percentile (r = .382). Multiple regression analysis is a statistical technique which weighs the relative degree of correlation of several variables. To determine the relative degree of correlation between the age at the birth of the first child and both the IQ percentile and the highest grade completed, both were simultaneously evaluated in a multiple regression analysis. The results showed that both were comparable, with *highest grade completed* taking the slight edge. In the following table the magnitude of the correlation is presented, and its significance (p).

Table 1. Multiple regression analysis between age at the birth of the first child and IQ percentile and highest grade completed.

Variable	Beta	T	p
Highest grade completed	.257	17.9	<.0001
IQ percentile	.233	16.2	<.0001

Thus, despite the high correlation between IQ and age at the birth of the first child, the correlation with the highest grade completed was even higher.

Summary – Not surprisingly, there was a high degree of positive correlation between the highest grade completed and IQ. Additional correlations showed that the fewer the number of grades completed, the earlier the age of first intercourse, the earlier the age at the birth of the first child, the greater the number of children, the greater the number of siblings, and the greater the number of times suspended or expelled from school. Thus, the longer subjects remained in school, the higher their IQ, the later they started having children, the fewer children they had, and the fewer siblings they had. To the extent that low IQ, learning disorders, and disruptive and addictive behaviors are genetic, these correlations indicate there will be a selection for these genes.

Chapter 29

NLSY– Grades

Learning disorders are a critical component in the spectrum of disorders that appear to be increasing in frequency in the latter half of the twentieth century. As described in previous chapters, the length of time that an individual remains in school is the chief determinant of how old they are when their first child is born. This, in turn, is a major factor in the selection for or against specific genes. Since the presence of learning disorders, or more generally, poor academic performance, is one of the single major determinants of low long children and young adults remain in school, it was important to examine the effect of this on the variables discussed in the previous chapters.

Since the NLSY did not specifically identify subjects with "learning disorders," I have instead examined academic performance, as measured by grades. This utilized questions about the grades obtained in five different classes in the ninth grade. The average was called the GPA, or "grade point average." For some comparisons it was necessary to place the GPA into categories. Thus, those with a GPA of 4, indicating an A in all five courses, were placed in the "A" group; those with a GPA of between 3.9 and 3.0 were in the "B" group, etc.

As with the variable *the highest grade completed,* it should not come as a surprise that there was also a high degree of correlation between IQ and the GPA. This is shown in Figure 1.

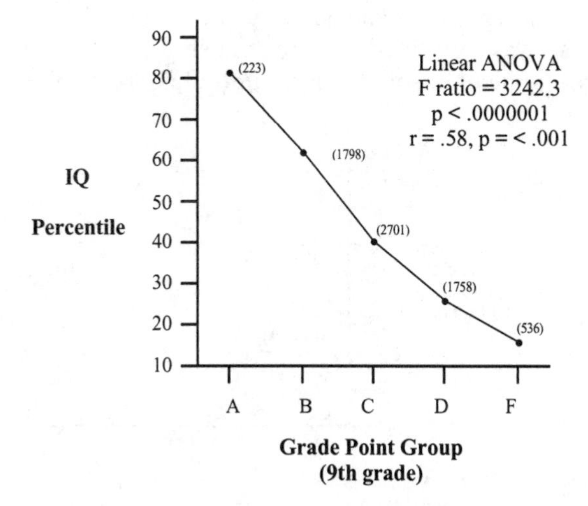

Figure 1. IQ percentile versus grade point average. All subjects. From the NLSY database.

This showed a highly significant correlation, with an average IQ percentile of 17.7 for those with an F average, to 82.9 for those with an A average. The correlation coefficient was 0.58, also second only to that between IQ and *highest grade completed.* It was not possible to obtain reasonable figures on the correlation between GPA and *highest average grade completed,* because none of the subjects who dropped out of school prior to ninth grade would be included in the averages. Another relevant school-related variable that could be examined was *number of times suspended from school* (Figure 2).

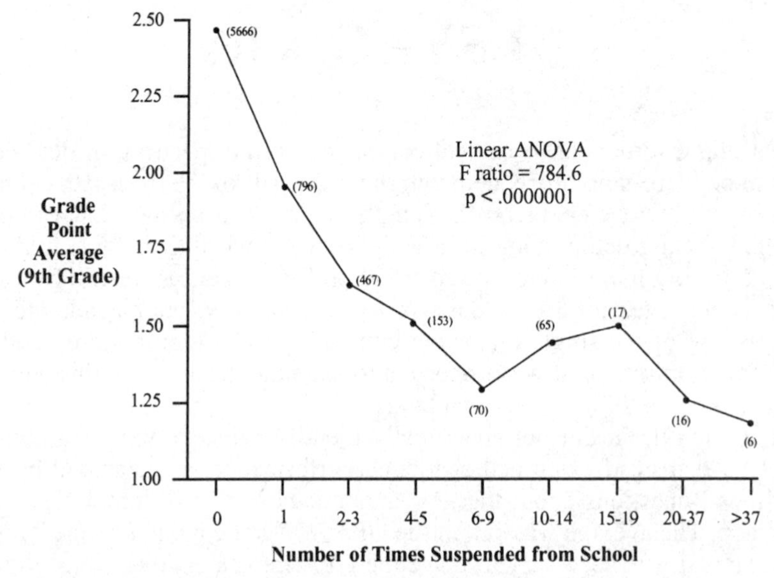

Figure 2. Grade point average for ninth grade versus number of times suspended from school. All subjects. From the NLSY database.

The GPA of those who had never been suspended from school was reasonably high, a C+. It dropped to a C average (2.0) for those suspended once, and even lower for those suspended more than once.

To examine the relationship between grades and addictive behaviors, the correlation with alcohol and drug use was examined. The relationship between the GPA and the presence or absence of alcoholism and/or a family history of alcoholism, is shown in Figure 3.

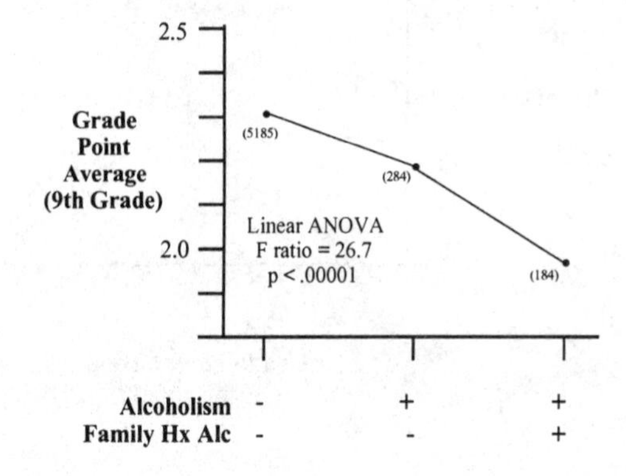

Figure 3. Grade point average for ninth grade versus presence or absence of alcoholism and/or a family history of alcoholism. All subjects. From the NLSY database.

The GPA was highest for those without alcoholism or a family history of alcoholism (2.32), next highest for those with alcoholism but a negative family history (2.20), and lowest for those with alcoholism and a family history of alcoholism (1.98). It should be noted that the GPA was based on ninth grade performance and many of the symptoms of alcoholism came on at a later age. This is consistent with the well-known clustering of learning disorders and ADHD as risk factors for the later development of alcoholism.

The correlation between the GPA and the number of times marijuana was used in the prior month, is shown in Figure 4.

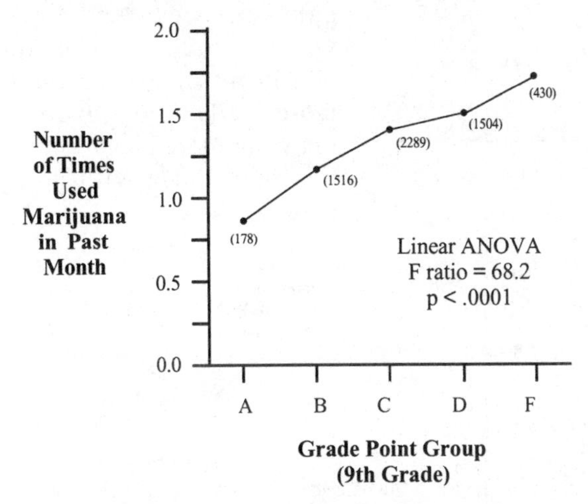

Figure 4. Number of times subjects used marijuana in the prior month versus GPA group. All subjects. From the NLSY database.

A decreasing GPA was associated with a progressive increase in the use of marijuana. This relationship was also apparent when the age of starting the use of marijuana was examined (Figure 5).

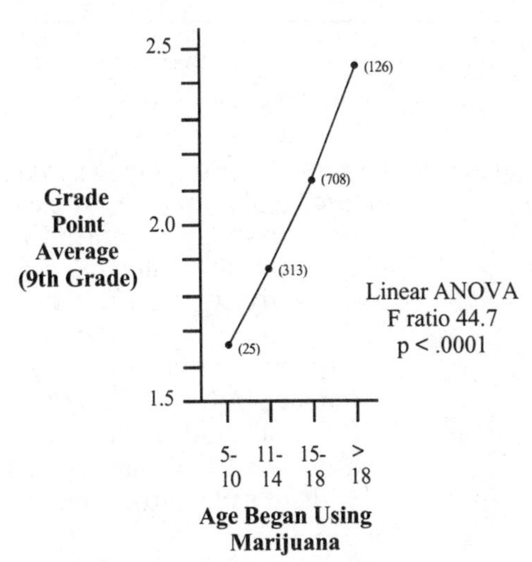

Figure 5. Grade point average in ninth grade versus age of beginning the use of marijuana. All subjects. From the NLSY database.

The GPA decreased progressively from 2.5 for those initiating the use of marijuana at 18 years of age or older, to 1.7 for those starting to use pot at age 11 or earlier. There was also a correlation between the GPA and the number of illegal acts the subjects had committed (Figure 6 - next page).

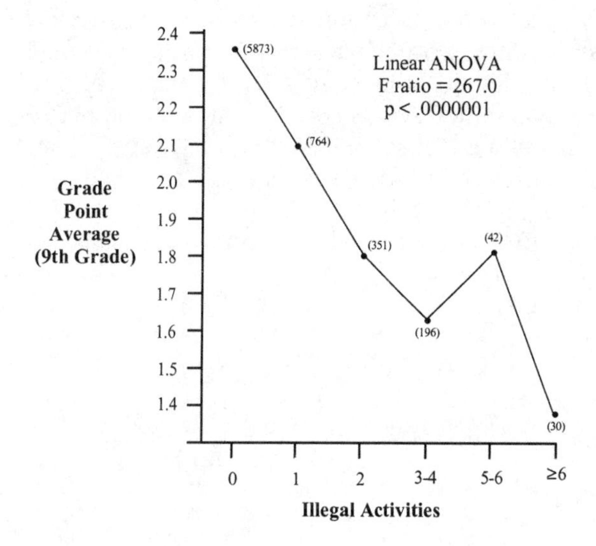

Figure 6. Grade point average in ninth grade versus number of illegal activities. All subjects. From the NLSY database.

The higher the GPA, the less the involvement in illegal activities, and vice versa. Students who do poorly in school are much more likely to be involved in two or more illegal activities than those with good grades.

The relationship between the GPA and age of initiation of sexual intercourse is shown in Figure 7.

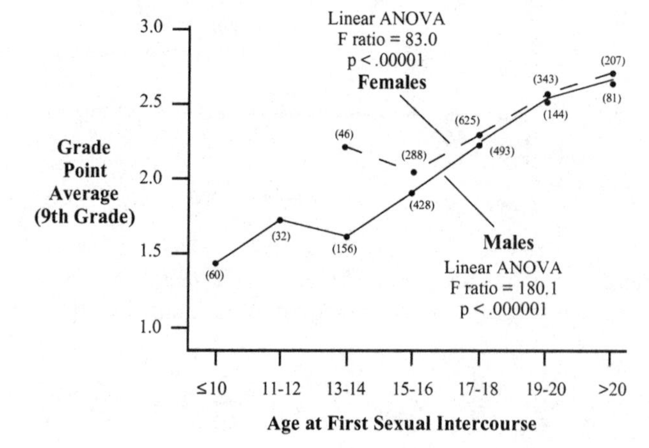

Figure 7. Grade point average in ninth grade versus age of initiation of sexual intercourse. All subjects. From the NLSY database.

The GPA for males who began having sex at 14 years of age or younger averaged 1.47 to 1.75. It then progressively increased to 2.67 for those initiating sexual activity at greater than 20 years of age. A similar pattern was seen for the females. As shown in Figure 8, poor grades were also associated with having sex without using birth control.

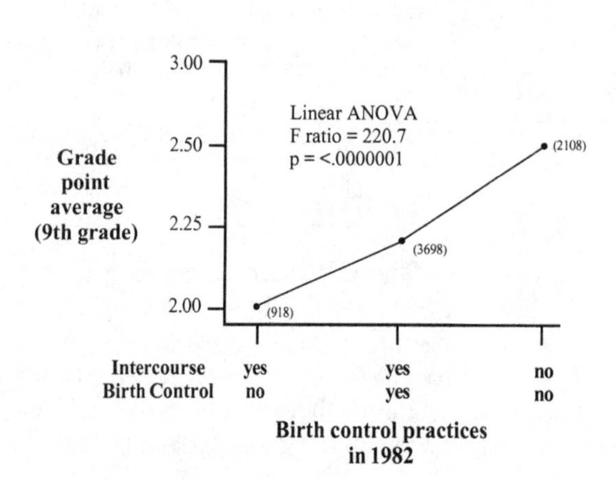

Figure 8. Grade point average in ninth grade versus birth control practices. All subjects. From the NLSY database.

Those who had sex without using contraception had the lowest GPA (2.09), those having sex and using contraception had a higher GPA (2.24), and those who were not having sex at age 21 had the highest GPA (2.56).

The above results indicate that a low GPA is associated with an increased risk of alcoholism, drug use, delinquent behavior, and an early initiation of sexual activity without the use of birth control. Determining if these factors had a dysgenic effect required examination of the correlation with age at the birth of the first child, number of children, and number of siblings. The results for the age at the birth of the first child are shown in Figure 9.

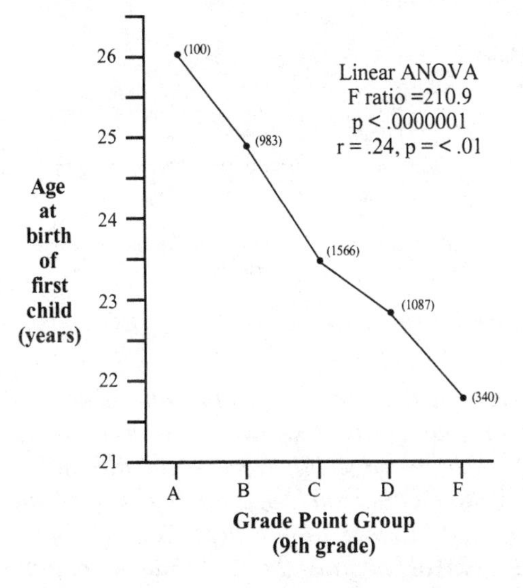

Figure 9. Age at the birth of the first child versus GPA group. All subjects. From the NLSY database.

As with the variable *highest number of grades completed*, there was also a highly significant association between the age at the birth of the first child and GPA group. Thus, those with a straight A average in ninth grade did not have their first child until an average of 26.3 years. This progressively decreased to 21.9 years for those with an F average. The difference between the two was 4.4 years. This was less than for the variable *highest number of grades completed*, suggesting that simply staying in school longer had a greater effect on when individuals began having children than the grades they received.

The relationship between GPA in ninth grade and the total number of children is shown in Figure 10.

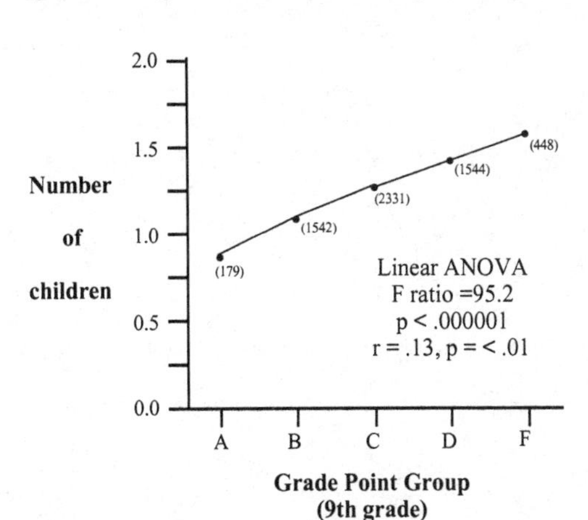

Figure 10. Number of children versus GPA group. All subjects. From the NLSY database.

The number of children (per individual) decreased from 1.62 for those with an F average to 0.94 for those with an A average. A more marked trend was present for the number of siblings (Figure 11).

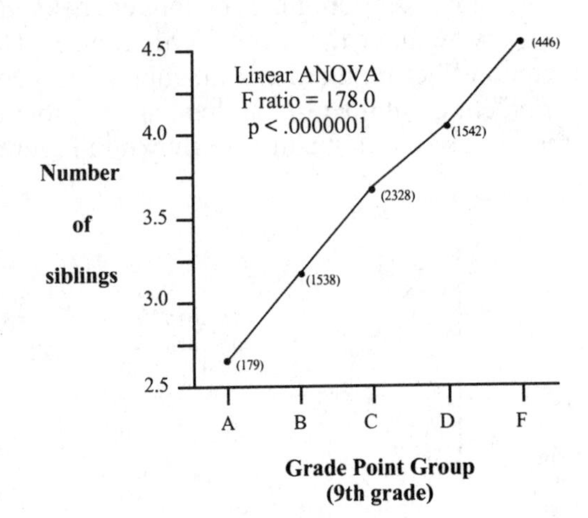

Figure 11. Number of siblings versus GPA group. All subjects. From the NLSY database.

The number of siblings progressively increased from 2.67 for those with an A average to 4.54 for those with an F average.

Summary – Using the grade point average for five courses in ninth grade as an indirect measure of cognitive or learning disabilities showed that individuals with poorer grades had a significantly lower average IQ, were suspended from school more often, were at increased risk for alcoholism, began using marijuana earlier, smoked more marijuana, participated in more illegal acts, began having sex earlier, were less likely to use birth control, began having children earlier, had more children, and came from larger families than those who got better grades. To the extent that these behaviors are controlled by genes, the latter three correlations would result in a significant selection for the genes involved.

Chapter 30

NLSY – School Dropouts

Closely related to the subject of education is the problem of school dropouts. Individuals who drop out before completing high school are placed at a distinct disadvantage in an increasingly complex society, where education is a critical requirement to gain entry to all but the most menial jobs. As shown in the previous chapter, there was a high correlation between the highest grade completed and IQ (r = .63), as well as with the age at the birth of the first child, number of children, and number of siblings. This combination would produce a strong selection for genes associated with learning disorders and lower, rather than higher, IQ. But would there also be a selection for any of the other behaviors of concern in this book? To investigate that, this chapter examines the association between dropping out of school early and various disruptive behaviors. Instead of using the entire seven-part range from the lowest to the highest grade completed, here they will be collapsed into four groups: 1.) dropouts, or those who failed to complete high school (n = 1,764); 2.) high school graduates (n = 4,588); 3. college students completing 1 to 4 years of college (n = 3,391); and 4.) post-graduates, or individuals completing any amount of post-graduate work (n = 615). While the comparison of the first two groups – dropouts versus high school graduates – is of primary interest, the addition of groups 3 and 4 allow additional comparisons. First, a look at expulsion from school.

Expulsion from School

It should come as no surprise that many individuals who were expelled from school would give up on education entirely and join the ranks of dropouts. Figure 1 shows the degree to which this was the case.

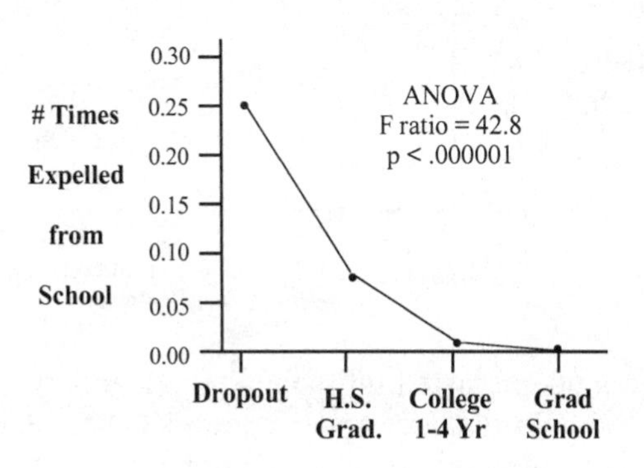

Figure 1. Correlation between the number of times expelled from school and becoming a school dropout. From the NLSY database.

In addition to the high correlation between being expelled from school and dropping out of school, this diagram also shows that expulsion from school does not inevitably lead to dropping out, since some of those expelled did eventually graduate from high school and some attended college. It is less inherently apparent whether there would be any correlation between dropping out of school and delinquent or illegal activities.

Illegal Activities

The association between the number of times individuals had stolen things worth more than $50 and being a dropout is shown in Figure 2.

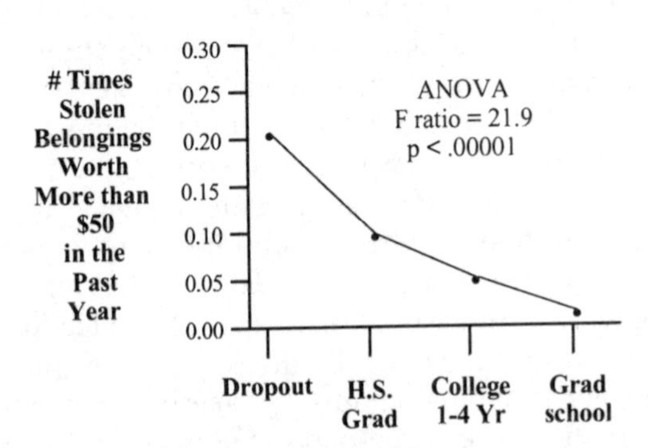

Figure 2. Correlation between the number of times individuals had stolen things and becoming a school dropout. From the NLSY database.

Those who dropped out before completing high school were twice as likely to have stolen something of value than high school graduates, and four times as likely as those attending college.

The correlation with getting into fights at school or work is shown in Figure 3.

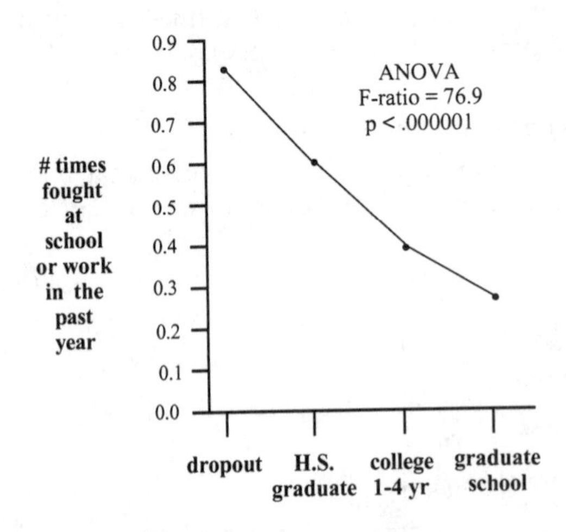

Figure 3. Correlation between the number of times individuals got into fights at school or work and dropping out of school. From the NLSY database.

This showed a highly significant and almost straight line negative correlation between the duration of education and the tendency to get into fights. Thus, the difference between college students and high school graduates was similar to the difference between high school graduates and dropouts. Since there was a high degree of correlation between the highest grade completed and the age at the birth of the first child, to the extent that resorting to fighting was, in part, controlled by an individual's genes, there would be selection for those genes.

Many additional behaviors, such as attacking with intent to kill, auto theft, and others, showed a similar trend. Rather than listing each, they can be summarized by showing the correlation with total number of illegal activities individuals were involved in (Figure 4).

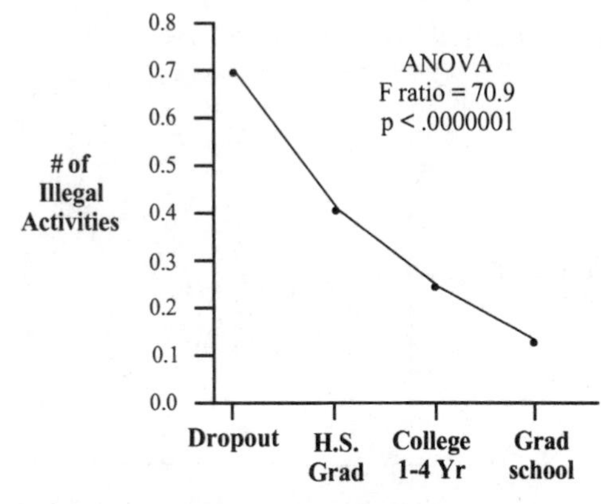

Figure 4. Correlation between the number of illegal activities subjects were involved in and dropping out of school. From the NLSY database.

The dropouts showed the highest involvement in illegal activities, followed by a progressive decrease with successively longer times spent in school. Another variable that tends to summarize such activities is the number of times an individual was convicted of an illegal activity (Figure 5).

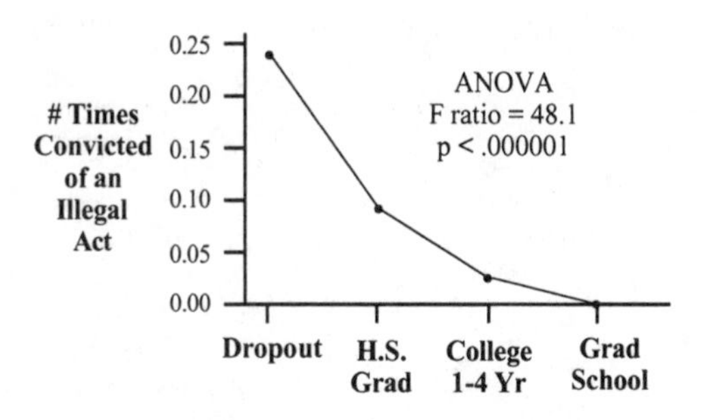

Figure 5. Correlation between the number of times subjects were convicted of an illegal act and dropping out of school. From the NLSY database.

The number of times individuals were convicted was 2.5 times higher for dropouts than for those who completed high school, and five time higher than for those attending college. A related variable is the number of times subjects were actually sent to an adult correctional institution (Figure 6).

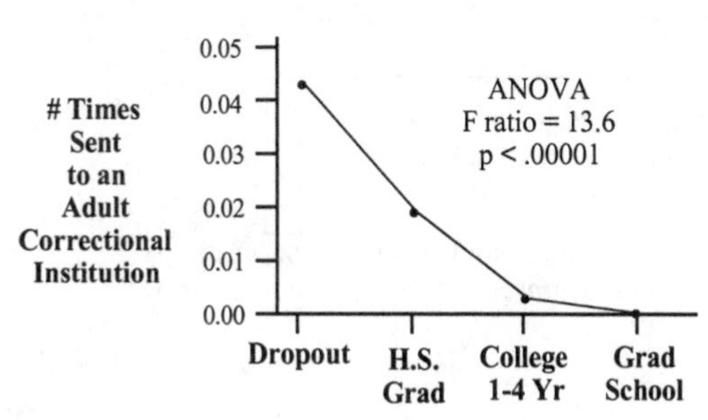

Figure 6. Correlation between the number of times subjects were incarcerated in an adult correctional institution and dropping out of school. From the NLSY database.

As with the number of convictions, the rate was more than twice as great for the dropouts compared to high school graduates, and nine times higher than for those attending college. The other issue of interest was the possible association between substance abuse and dropping out of school

Alcohol and Drug Abuse

The results for alcohol abuse, as indicated by the alcoholism score, are shown in Figure 7.

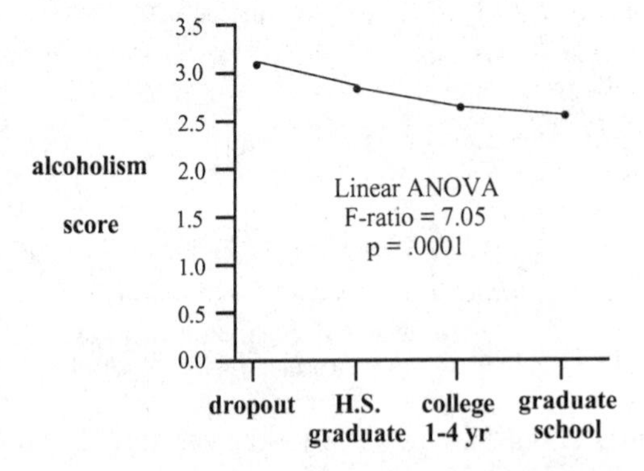

Figure 7. Correlation between the alcoholism score and dropping out of school. From the NLSY database.

While significant, the correlation between dropping out of school and the alcoholism score was modest compared to the results for aggressive and delinquent behaviors. To examine drug abuse, the number of times subjects used marijuana in the year prior to the interview was examined (Figure 8).

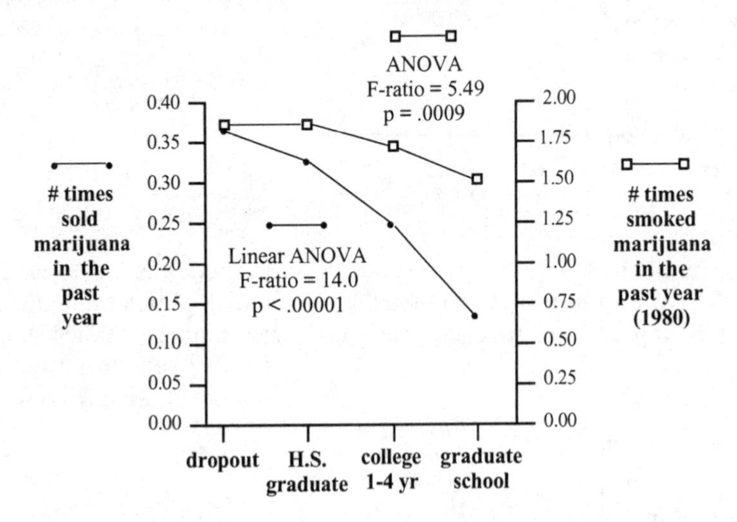

Figure 8. Correlation between number of times smoked marijuana in the past year, number of times sold marijuana, and dropping out of school. From the NLSY database.

Since, during this period, the use of marijuana in the young was rather ubiquitous, there was relatively little difference between those who were dropouts versus those who remained in school. However, when the variable of selling marijuana was examined, there was a much greater association with having dropped out of school.

Summary – Dropping out before completing high school was significantly associated with a wide range of delinquent and illegal behaviors, including stealing and fighting. This was reflected in a significant increase in the total number of illegal activities, number of arrests, convictions, and incarcerations in adult correctional institutions by dropouts. While there was also an association with alcohol and drug use, the association was greater with selling, rather than using, drugs. Previous chapters have shown a strong correlation between the highest grade completed and the age of the parents at the birth of their first child. Thus, to the degree that any of these behaviors are associated with genetic factors, there would be selection for the genes involved.

Chapter 31

NLSY Children – Disruptive Disorders

Between 1986 and 1992 the bi-yearly studies of the NLSY included the administration of an extensive set of assessments of the children of the NLSY women. A portion of this assessment included childhood behavioral problems. Scores for four of these – peer conflict, hyperactivity, antisocial behavior, and anxious/depressed – seemed to best represent the behaviors relevant to the hypothesis. The peer conflict score had a range from 0 (no peer conflict) to 3 (major peer conflict). The hyperactivity and anxious/depressed scores ranged from 0 to 5, and the antisocial behavior score ranged from 0 to 6.

These evaluations provided results analogous to the Berkeley data, where information was also available on both mothers and their children. There were several questions I wanted to explore in relation to the hypothesis. The first and most relevant was whether the mothers of the most severely affected children started having had their first child at an earlier age than mothers of the less affected children. The second, also relevant to gene selection, was whether the mothers of the more severely affected children had more children than mothers of the less affected children.

Because of the importance of IQ in relation to a wide range of variables important in gene selection, this was also charted. Finally, as an estimate of socioeconomic factors, the parent's income was examined. Since the parent's income during a single year might not be representative, it was based on the yearly net family income averaged over twelve different years from 1979 to 1993. The data will be presented as a diagram for each behavior score, showing the correlation with each of the above four variables.

Peer Conflict

The results for the child's peer conflict score are shown in Figure 1 (next page). The average of the mother's age at the birth of her first child progressively decreased from 21.6 years for children with a 0 peer conflict score to 20.0 years for those with a score of 3. The reverse trend was seen for the number of children ranging from 2.29 when the child's score was 0, to 2.61 for mothers of children with a score of 3. The mother's IQ percentile decreased from 35 when the child's peer conflict score was 0, to 24 when it was 3. The parent's income showed a comparable trend.

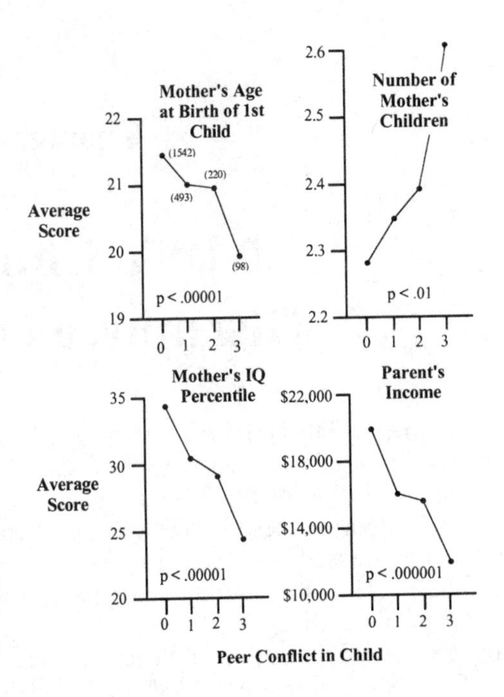

Figure 1. Peer conflict scores of the children of the NLSY mothers in relationship to the mother's age at the birth of her first child, the number of children the mothers had, the mother's IQ percentile, and the parent's income. The p values = the significance by linear ANOVA. From the NLSY database.

Antisocial Behavior

The antisocial behavior score most closely mimics conduct disorder, and here these results were more dramatic than for any of the four behaviors. These results are shown in Figure 2.

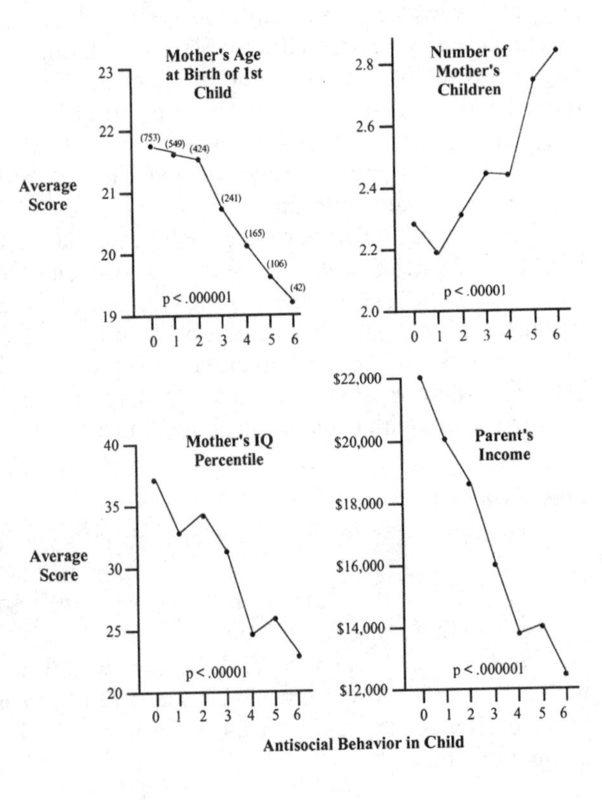

Figure 2. Antisocial behavior scores of the children of the NLSY mothers in relationship to the mother's age at the birth of her first child, the number of children the mothers had, the mother's IQ percentile, and the parent's income. The p values = the significance by linear ANOVA. From the NLSY database.

There was a highly significant decrease in the mother's age at the birth of her first child and an increase in the number of children the mother had, with a progressive increase in the antisocial behavior score. Again there was a parallel decrease in mother's IQ percentile and decrease in parent's income with an increase in the score.

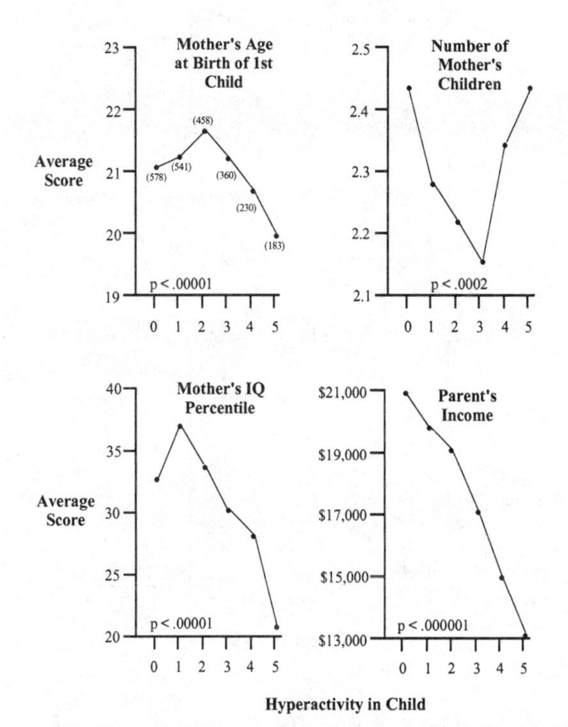

Hyperactivity

The hyperactivity score gave somewhat similar results, except for the number of the mother's children (Figure 3).

Figure 3. Hyperactivity scores of the children of the NLSY mothers in relationship to the mother's age at the birth of her first child, the number of children the mothers had, the mother's IQ percentile, and the parent's income. The p values = the significance by linear ANOVA. From the NLSY database.

The average mother's age at the birth of her first child increased modestly to 21.9 from a hyperactivity score of 0 to 2 and then dropped to 20.1 at a score of 5. The number of the mother's children showed a biphasic drop from 2.45 at a score of 0, to 2.15 at a score of 3, and back to 2.45 at a score of 5. The mother's IQ percentile and the parent's income again paralleled each other. The mother's IQ dropped from the 38th percentile when her children had a score of 1 to the 21st percentile for mothers of children with a hyperactivity score of 5.

Anxious/depressed

Since anxiety and depression tend to represent somewhat the antithesis of hyperactivity, peer conflict, and antisocial behavior, the anxiety/depression score could potentially produce much different results than those seen for the disruptive behavioral disorders. However, given the results presented in Part I showing an increase in the frequency of depression and anxiety in the past thirty years, it was not surprising that the results were more similar than different (Figure 4, next page).

There was a progressive decrease in the average mother's age at the birth of her first child, except at the score of 5, where it increased somewhat. The number of the mother's children progressively increased from 2.2 at a score of 0 to 2.53 at a score of 5. Again the mother's IQ and the parent's income paralleled each other, even to the point of both increasing at an anxious/depressed score of 5.

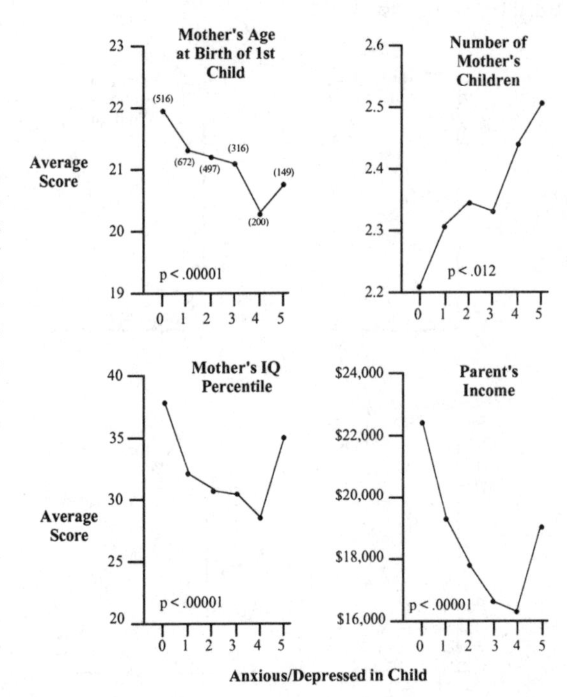

Figure 4. Anxious/depressed scores of the children of the NLSY mothers in relationship to the mother's age at the birth of her first child, the number of children the mothers had, the mother's IQ percentile, and the parent's income. The p values = the significance by linear ANOVA. From the NLSY database.

Summary – The assessment of the children of the NLSY mothers allowed an examination of whether the mothers of children with the most severe symptoms of antisocial behavior, hyperactivity, peer conflicts and anxious/depressed symptoms tended to have the their first children earlier or to have more children than mothers of children with minimal or no symptoms. This turned out to be both true and significant for all four of the behavioral scores examined. To the extent that these behaviors have a genetic component, there would be a selection for the genes involved. There was a parallel relationship between the generally progressive decrease in the mother's IQ percentile and the family income, with increases in the behavioral scores.

Chapter 32

NCDS – The National Child Development Study

The previous chapters showed that teenage mothers often had problems with conduct disorder. There are few studies that allow us to ask, do the children of these mothers also have conduct and behavioral problems, including teenage pregnancies? The most accurate way to examine this is to perform a longitudinal study, examining a large number of children, some illegitimate and born to younger mothers, and some legitimate and born to older mothers. The National Child Development Study allowed this to be done.

In 1958, all of the 17,000 babies born in England over a one week period were entered into one of the largest longitudinal studies ever undertaken. The first part was designed to study problems at birth and was called the Perinatal Mortality Survey. After that there were follow-up studies at 7, 11, 16, 23, and 33 years of age. Information was obtained from a number of sources, including parents, teachers, physicians, and the children themselves. Maughan and Pickles[180] used this data to examine the outcome of three groups of children – those born out of wedlock but not adopted and eventually taken care of by one or more parent (the *illegitimate* group, n = 363), those born out of wedlock and adopted (the *adopted* group, n = 180), and those born of married parents (the *legitimate* group). Since the latter was much larger, a representative sampling of 1,435 legitimate children was used in the study. The mothers of all the illegitimate children – that is, the mothers of both the illegitimate and adopted groups – were younger than the mothers of the legitimate children.

At ages 7, 11, and 16, the children were assessed with a behavioral rating scale. Figure 1 (next page) shows the results for the total score, normalized such that the scores for the legitimate children were 0.0.

At all three ages, the scores for the illegitimate children were higher than for the illegitimate but adopted children, and both were higher than for the legitimate children. This study uniquely allowed an examination of genetic and environmental factors in subsequent development. If the illegitimate children were more likely to inherit disruptive behavior genes from their parents than the legitimate children, it would be expected that, as they grew up, they would persistently display behavioral problems. This was the case. The total behavioral scores were significantly higher for the illegitimate children at all three ages. If these behaviors were due to disadvantaged socioeconomic conditions of mothers who had illegitimate children, these behaviors should not be present in those illegitimate children who were adopted into normal homes. The intermediate scores for these children indicated the behavior problems did not disappear in the adopted families. This persistence is consistent with a portion of the behavioral problems being genetically caused and resistant to elimination, despite growing up in more advantaged homes.

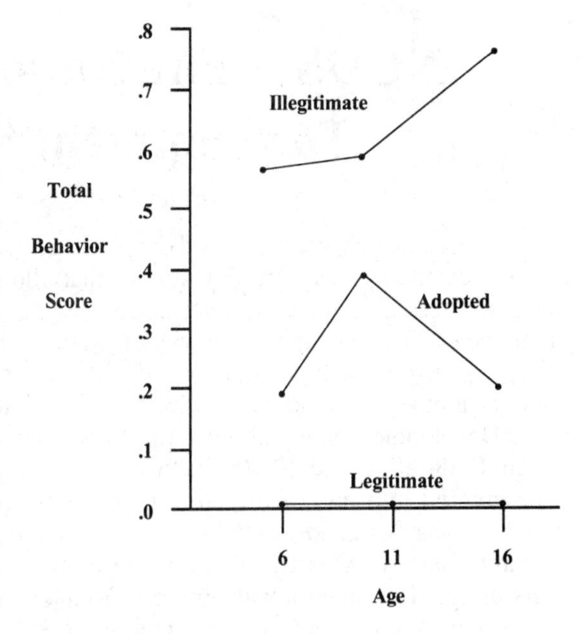

Figure 1. Total behavior score for adopted and illegitimate children normalized to legitimate children. From Maughan and Pickles in Robins, L.N. and Rutter, M. *Straight and devious pathways from childhood to adulthood.* Cambridge University Press, London, England, 1990.

The scores were also subdivided into individual behaviors, including ADHD (restless, hyperactive), antisocial and conduct problems, emotional difficulties, and problems with peer relationships. The subscores for ADHD, antisocial-conduct, and peer problems were all significantly higher in the illegitimate than legitimate children, and the adopted children were again intermediate.

Because of the length of the follow-up it was possible to also determine if the illegitimate children were also more likely to become teenage mothers themselves, i.e. have a teenage pregnancy. Figure 2 shows the results:

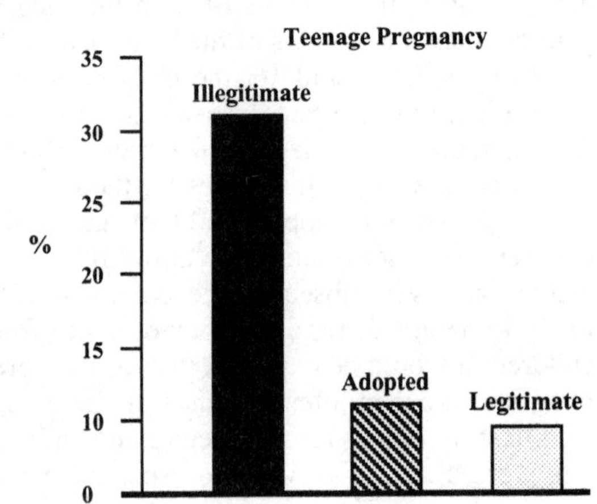

Figure 2. Percentage of teenage pregnancies for adopted, illegitimate, and legitimate children. From Maughan and Pickles in Robins, L.N. and Rutter, M. *Straight and devious pathways from childhood to adulthood.* Cambridge University Press, London, England, 1990.

There was a dramatically higher frequency of illegitimate children among the illegitimate children than among the illegitimate but adopted, or legitimate children. The fact that the percentage of adopted illegitimate children having a teenage pregnancy was only modestly higher than for the legitimate children, suggests that their improved environment played a role in decreasing the number of teenage pregnancies.

Summary – In a large longitudinal study in England, all children born in one week in 1958 were periodically examined as they grew up. Women who had illegitimate children were younger than those who had legitimate children. The frequency of behavioral problems, ADHD, conduct disorder, and peer problems were significantly higher in illegitimate children who were raised by one or both of the parents, and intermediate when the illegitimate children were adopted into other families. The illegitimate children were much more likely to have illegitimate children themselves. These results are consistent with a combined role of genetic and environmental factors in childhood behavioral disorders. In relation to the theme of this book, they are consistent with an increase in genetic behavioral problems in children born to younger out-of-wedlock teenage mothers.

Chapter 33

NCDS – IQ

As with the NLSY in the United States, I also obtained the database on the NCDS in England since it contained many of the variables required to examine the relationship between education, family size, substance abuse, delinquency, and other aspects in relation to the mother's age when she had her first child. This database was similar to the Berkeley study in two ways. First, since it focused on the child rather than the parent, the mother's age at the time of the first birth was related to behaviors in the child, not in the mother. Second, the specific variable relating to the age of the mother at the birth of her first child was not available. However, as with the Berkeley data, information relating to the year of birth of the child, mother's age, the total number of children she had, and the interval between her last child and the one born in 1958 was available. Thus, although the data was limited to those children born in 1958 who were only children, or had one prior sibling, there were enough subjects (over seven thousand) to provide accurate results. This made it possible to determine if the results obtained with the Berkeley and NLSY database were unique to the United States, or whether the same trends were present in other countries such as England.

This chapter was entitled NCDS – IQ to mirror the chapter on NLSY – IQ. Unlike the NLSY, no IQ tests *per se* were performed. However, there were tests of verbal and non-verbal general ability given to the children at 16 years of age that simulated IQ testing. For both tests, the subjects were divided into ten percentile groups based on whether they were in the lowest 10%, 11-20%, 21-30%, etc. These results in relation to the age of the mother at the time of her first birth, as shown in Figure 1 (next page).

There was a progressive and significant increase in age of first birth with increasing verbal and non-verbal performance of the child. Despite being a different study, in a different country, and involving tests of the child rather than the mother, the trends were similar to those of the NLSY (see Chapter 20).

The correlation between the age of the mother at the birth of her first child and the child's reading ability was also examined. Again the subjects were divided into percentile groups, except that the lowest and the highest groups were subdivided into two-five percentile subgroups to determine if the extremes showed an even greater spread (Figure 2, next page).

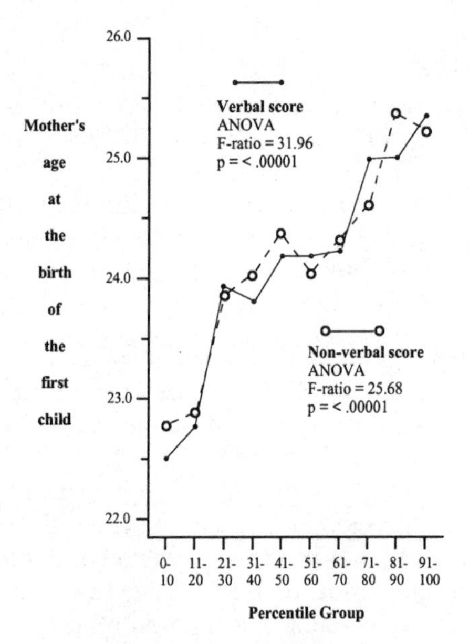

Figure 1. Age at the birth of their mother's first child versus verbal and non-verbal score in the child tested at age 16. (Variable for verbal score = N914 and for non-verbal score = N917). From NCDS, ESRC Data Archive, University of Essex, Colchester, Essex, England.

Reading comprehension is one of the most sensitive indicators of school academic performance and the presence or absence of learning disabilities. As shown in Figure 2, there was a remarkable correlation between the child's reading tests based on percentile group and the age at which the child's mother first began having children. The spread from the lowest 0-5th percentile (22.5 years) to the highest 96 to 100th percentile (26.8 years), was 4.3 years. The results with math ability were essentially identical.

While the spread in the NLSY data using IQ measurements was somewhat greater, 6.3 years, considering the greater ethnic homogeneity in England, and the differences in method of testing, and the fact that these are results on the children of the mothers rather than the mothers themselves, the trends

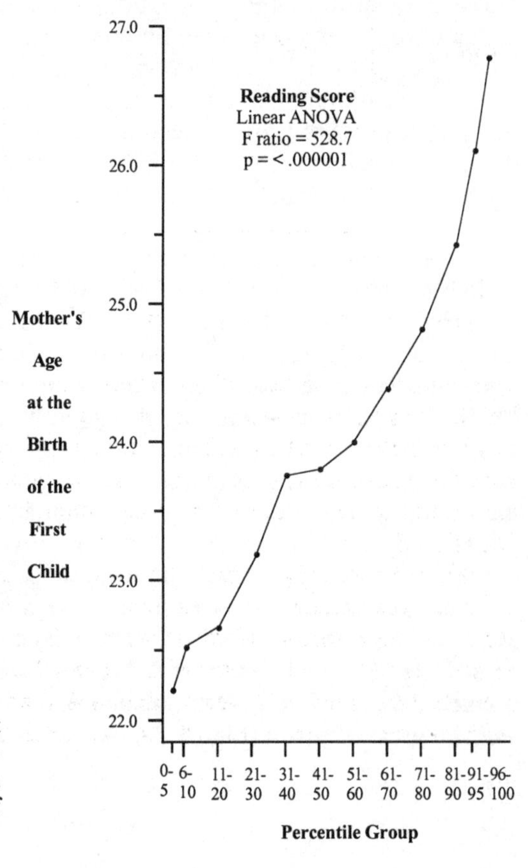

Figure 2. Age at the birth of the child's first child versus reading score in the child tested at age 16. (Variable = N923) From NCDS, ESRC Data Archive, University of Essex, Colchester, Essex, England.

were remarkably similar.

IQ and Second Generation Age at First Birth

By the time the subjects were 33 years of age, enough of them had children and the spread in ages of the births of their children was sufficiently broad to allow an examination of the relationship between the same reading score and the age at the birth of the child's first child. The results are shown in Figure 3.

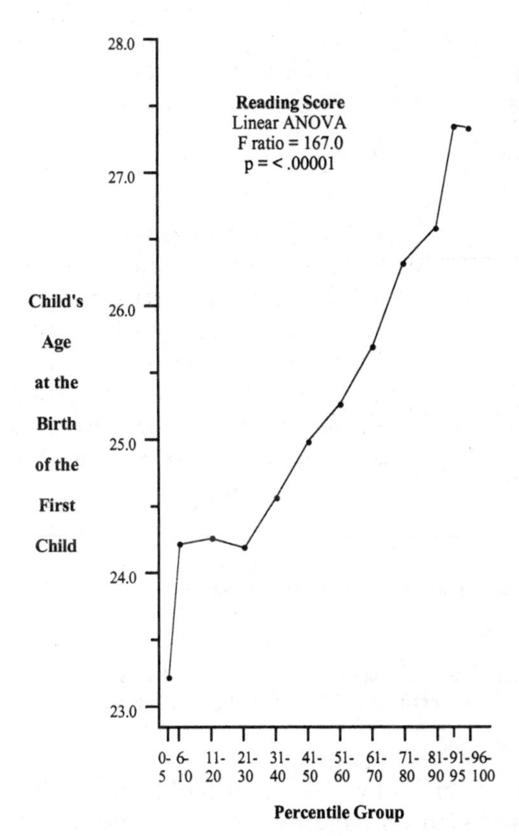

Figure 3. Age at the birth of the child's first child by reading score. From NCDS, ESRC Data Archive, University of Essex, Colchester, Essex, England.

The results were similar to those for the mother's age when the children were born. The spread was similar to that for their mothers.

To obtain an estimate of how significant a spread of four years was, it was necessary to obtain an accurate determination of the effective reproductive period of the women involved. This could be obtained with the NCDS data because women of all ages were randomly selected simply by virtue of having their first or second child in March of 1958. Since all such women were included, this provided an unbiased selection. The frequency distribution of the age at which these women had their first child is shown in Figure 4 (next page).

This showed that when the lower .8% and upper 1.6% were excluded, the effective reproductive period was 18 years, ranging from 18 to 35 years of age. Here, a 4-year spread represents 22% of the total reproductive span, and 2 years represents 11% of the span. If the lower 2.7% and upper 9.6% are excluded, the reproductive span is 11 years. Now a 4-year differential represents 36% and a 2-year differential represents 18% of the total reproductive span. Although these

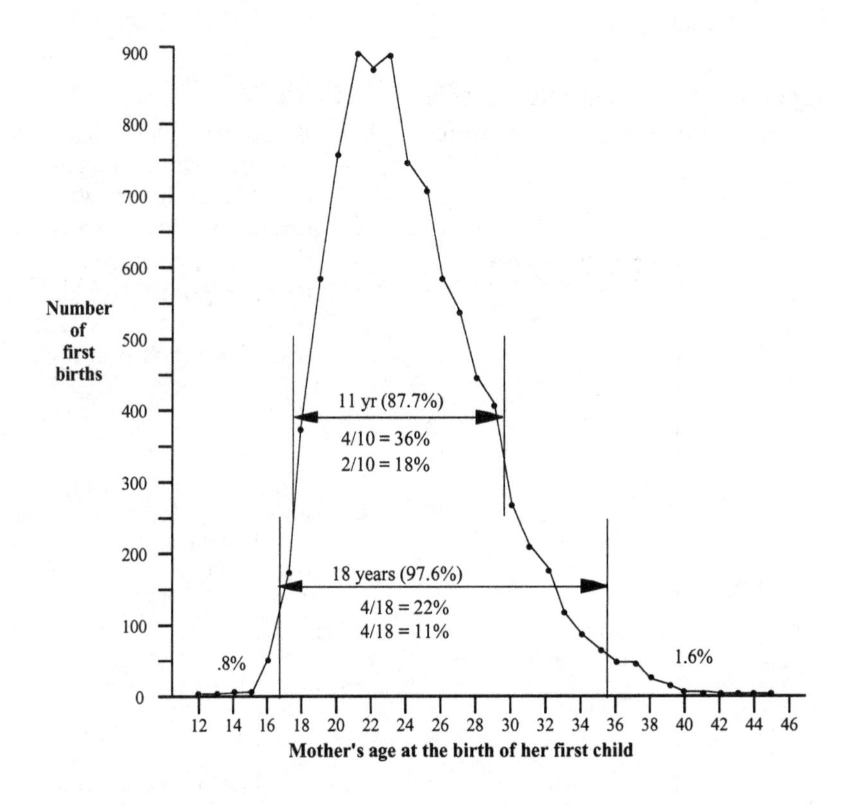

Figure 4. Distribution of the number of women having their first child by age. From NCDS, ESRC Data Archive, University of Essex, Colchester, Essex, England.

results are based on first births, this is the critical variable in regard to gene selection. These figures support the discussion on page 89 where a reproductive advantage of .25 per generation was used in the equation for estimating the rate of gene selection.

Summary – While specific IQ determination was not performed on the NCDS subjects, enough testing of academic performance was done to approximate IQ testing. As with the NLSY data, this showed that subjects with poorer cognitive skills had their first child at a significantly younger age than those scoring high on cognitive skills. The spread for both the mother and their children ranged from 3.0 to 4.3 years. Using the distribution of the number of first children born to the mothers of different ages indicated that the effective reproductive span for the majority of mothers ranged from 11 to 18 years. With such a span for first children, a difference in the age of first birth of 4 years for those with poor versus excellent cognitive skills represents 22 to 36% of this reproductive span – more than adequate to provide a strong selective drive for the genes involved.

Chapter 34

NCDS – Socioeconomic Status

The NCDS study included a variable that divided socioeconomic status (SES) into seven progressively decreasing classes. The results of the correlation between the mother's age at the birth of her first child and SES class are shown in Figure 1.

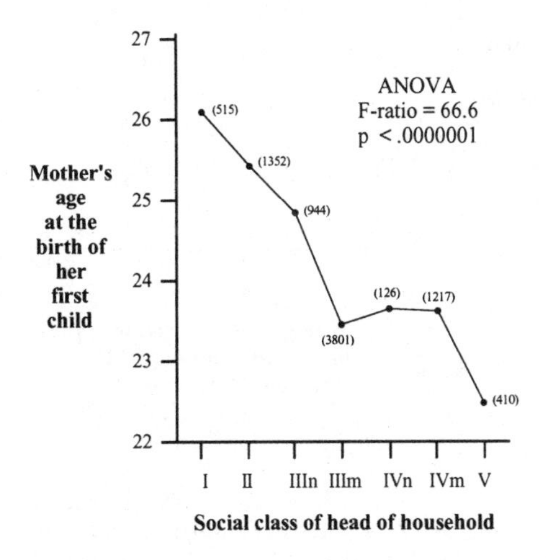

Figure 1. Mother's age at the birth of her first child versus the socioeconomic status of the male head of the household (variable 1P #N190). From NCDS, ESRC Data Archive, University of Essex, Colchester, Essex, England.

Those in class I were professionals, class II – semi-professional white collar workers, followed by progressively decreasing classes III through V, with III and IV divided into non-manual (n) and manual (m). For those in class I, the mean age of the mother at the birth of her first child was 26.1 years, with a significant and progressive decrease to 22.6 years for those in class V. Since individuals in class I and II had the most years of education, this was also an indirect measure of education.

Another variable that evaluated both socioeconomic status and family size was the number of children in the household under age 21, including those living away. While most of these were siblings, since the variable was simply total number of children, non-biological offspring were also included, thus serving as a measure of household population density and thus socioeconomic status. The results are shown in Figure 2 (next page).

This showed a highly significant correlation with the mother's age at the birth of her first child. For those households with only a single child this was 27.4 years. It dropped to 24.6 years for those with two children, and to 20.4 years for those with eight or more in the household.

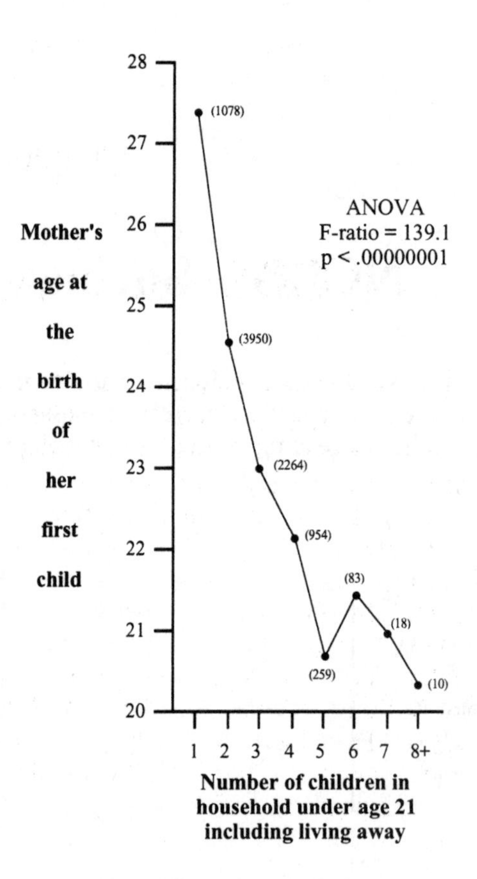

Figure 2. Mother's age at the birth of her first child versus number of children in the household under age 21, including those living away (variable 1P #N99). From NCDS, ESRC Data Archive, University of Essex, Colchester, Essex, England.

When the number of biological children alone was examined, i.e. the number of siblings, the mother's age at the birth of her first child progressively decreased from 24.9 years for those with no siblings to 19.5 years for those with eight or more siblings (F-ratio = 73.1, p = < .000001).

Summary – Socioeconomic class was measured by a direct measure and indirect measure based on the number of children living in the household. Both showed a marked correlation with the mother's age at the birth of her first child. This age was 25.5 to 27.4 years for those in the highest SES class and 20 to 22.5 years for those in the lowest classes, with intermediate values for intermediate classes.

Chapter 35

NCDS – Relevant Variables

The eighty-five variables in the NCDS study that seemed most relevant to the question of whether the mothers of children with potentially hereditary disorders related to impulsive, compulsive, addictive, and learning disorders have their first child earlier than average, are listed in Table 1 at the end of this chapter.

First, a few general comments. Some of the variables have a simple "yes" or "no" answer, while others have three potential answers "never," "sometimes," or "frequently." For each variable, the mean of the mother's age at the birth of her first child (MAFB) is given for both, or all three, of the answers. Variables with a P (1, 2 or 3) represent interviews of the parents, those with an S represent data from the teacher or school, those with an I represent interviews of the individual child, and those with an M (M-1 to M-24) are questions classified as "Malaise" questions. To evaluate the relative importance of these variables in relation to the mother's age at the birth of her first child, they were sorted in descending order of significance based on the F ratio. To evaluate the type of variable, each was assigned a letter which classified it according the general area that it related to. Thus A = ADHD, C = conduct and aggressive behavior, D = depression, F = family problems, X = anxiety, and O = other where it was difficult to clearly assign the variable to one of the previous categories.

For every variable except M-16 "Keyed up and jittery," the MAFB was lower for those who answered "yes" or "always" than for those answered "no." For the majority of the variables, this was highly significant with a p value of 0.0000, which means less than 0.0001. The variables that gave the highest F ratios were those relating to education, family problems, conduct or aggression, attention deficit disorder, and depression, while those relating primarily to anxiety and some in the "other" category were less significant. For example, of the fourteen variables relating to anxiety (type = X) only one was in the upper half, and five were among the ten variables that were not significant.

The top fourteen variables, in order, were *truant this year*; *school absences for trivial reasons*; *family difficulties, divorce, desertion*; *truant*; *fights with others*; *often tells lies*; *in trouble with the police*; *often disobedient*; *angry when corrected*; *cannot settle down*; *fights with children*; *family difficulties, domestic tension*; *restless, can't stay seated*; and *irritable, flies off the handle*. It is clear that these involve problems with school, aggression, conduct disorder, ADHD, and family discord. As an example, for those children who were never disobedient, the mean MAFB was 24.43 years. This decreased to 23.01 years for those who were sometimes disobedient, and 21.54 years for those who were always disobedient, a spread of 2.9 years. Those variables where the results were obtained by questioning the teachers or the school tended to be more significant than when the parent was interviewed. For example, of the twenty-five school questions, nineteen (76%) were in the upper half of significance, while of the twenty-seven parental questions, only eight (30%) were in the upper half.

While Table 1 provides insight into those variables that are the most significant, the presentation tends to be confusing. To make the relationship between the variables easier to see, in Table 1S they have been sorted by type. Several aspects of this presentation deserve comment.

Of the ADHD variables, the parents were asked about being squirmy and fidgety two different times, and the teachers were asked once. The results in all three cases showed a significant decrease in the mean MAFB for those identified as "always" being squirmy or fidgety.

Among the twenty-one conduct disorder and aggression variables, the mean MAFB was lower in all cases identified by a "yes" or "always' response, and this was significant for all but two. Thus, the mothers of children having behaviors such as fighting, lying, stealing, being disobedient, anger, being quarrelsome, rageful, and having temper tantrums tended to start having their children earlier than mothers of children who did not have these behaviors. The difference in MAFB ranged from 2 to 3 years.

Among the six variables identifying problems with depression, all showed a significantly lower mean MAFB for those identified by teachers, parents, or themselves as being miserable, tearful, and unhappy.

The highly significant decrease in mean MAFB for mothers answering "yes" to questions about problems with marital tension and divorce was of interest. This is consistent with our clinical experience showing high rates of discord and divorce in families in which one or more members have ADHD, TS, substance abuse, or other aggressive impulsive disorder.

The anxiety disorder variables tended to be less significant than those relating to aggressive, impulsive, hyperactive, and conduct disorder variables. Despite this, about half of them showed a significant decrease in the mean MAFB for those identified by a "yes" or "always."

All but three of the *other* variables were also significant, including three out of four relating to the presence of tics. These variables were included because they referred to features often observed in children with ADHD and TS. These included problems with sleep, night terrors, sleepwalking, bed-wetting, headaches, general aches and pains, biting nails, twitches, and mannerisms.

Education. As shown in the NLSY data, variables relating to education showed a significant correlation with the mother's age at the birth of her first child. Variables that related to truancy were especially significant. These variables cross over with conduct disorder, since one of the DSM-IV criteria for conduct disorder is "often truant from school, beginning before age 13 years."[91] To further examine educational problems, a series of questions not listed in the Table 1 and relating to the child's attitude toward school are shown in Figure 1 (next page).

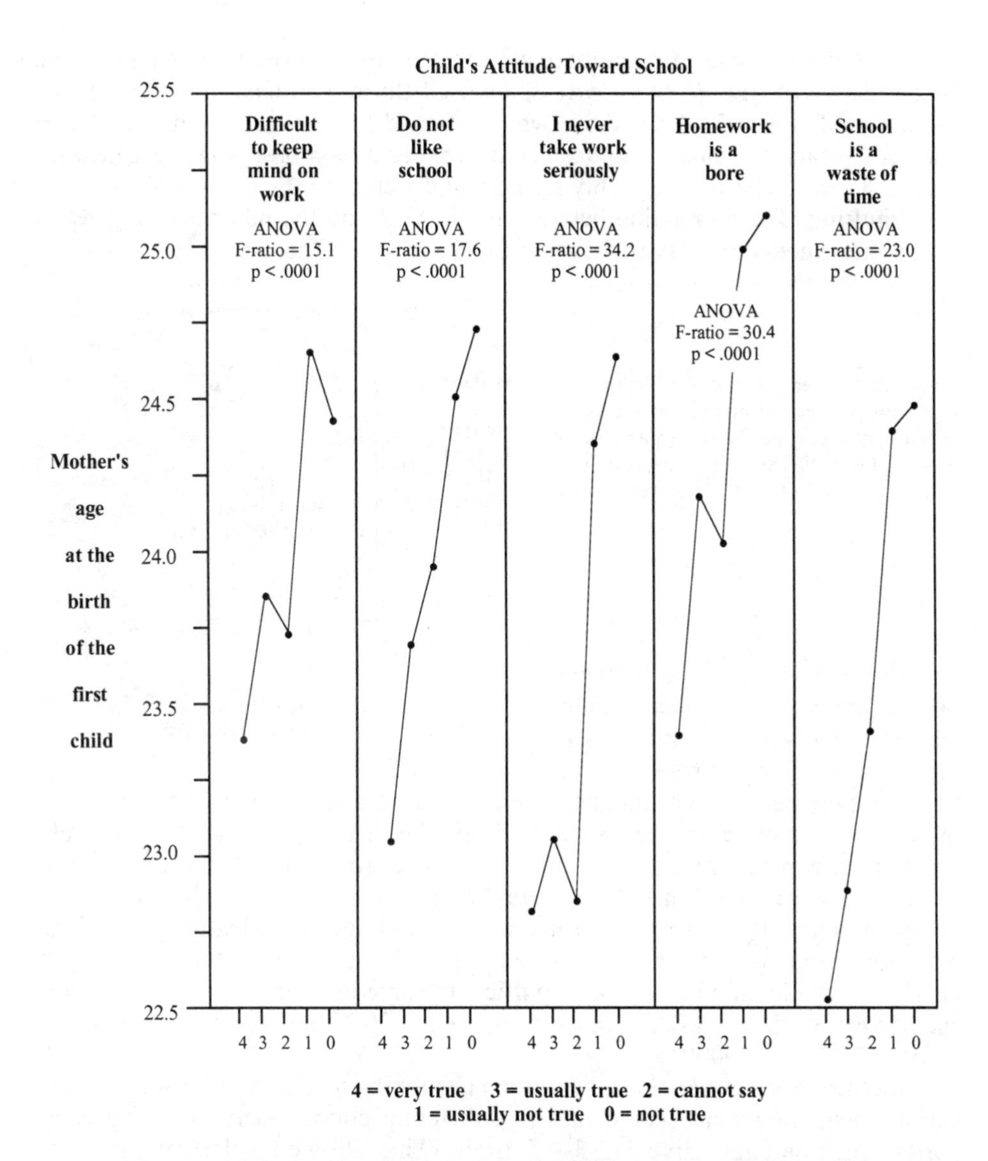

Figure 1. Mother's age at the birth of the first child versus the child's answers to a series of questions concerning his/her attitude toward school. (Variables = N2716, N2718-2721). From NCDS, ESRC Data Archive, University of Essex, Colchester, Essex, England.

While the first variable *Difficult to keep mind on work* also relates the ADHD symptoms, the other four are primarily related to a child's attitude toward school work. For the variable *Do not like school* there was a perfect linear increase in the mean MAFB from 23.1 years for those marked "very true," to 24.8 years for "never true." The two variables giving the highest F ratio were *I never take [school] work seriously* and *homework is a bore*. For the four variables relating to school attitude, the extremes in MAFB differed by 1.7 to 2.0 years.

A NCDS variable that approximated the number of years in school of the NLSY data was *Age likely to leave school.* Of those who answered "16 years," the mean MAFB = 23.3. Of those who answered "17 years," the mean MAFB was 24.6 years. Of those who answered "18 years or more," the mean MAFB was 25.8 years. These were highly significant differences (F-ratio = 176.8).

Smoking. The correlation between the MAFB and the number of cigarettes the mother smoked per day is shown in Figure 2.

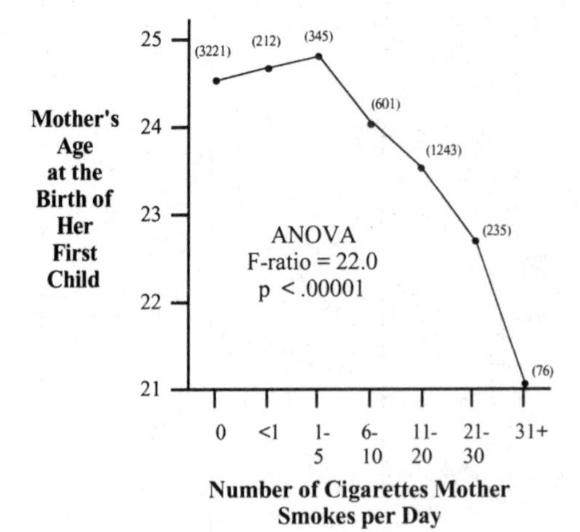

Figure 2. Mother's age at the birth of her first child versus the number of cigarettes mother smokes per day. (Variable = N2400). From NCDS, ESRC Data Archive, University of Essex, Colchester, Essex, England.

While the difference between the extremes was 3.4 years, there were relatively few individuals in the group smoking thirty-one cigarettes or more per day. The numbers were much more impressive for those smoking eleven or more cigarettes per day. While the difference in MAFB was only 1.6 years, a significant number of mothers, 26%, were in this category.

Behavioral Problems. One series of questions (N16648-N16652) was designed to identify a variety of problems in school, including learning disorders, behavior problems, temper tantrums, and lying. A total of 171 children of 4,142 queried were identified as having significant problems in these areas. The average MAFB for these was 22.8 years, compared to 26.3 years for the remainder (F ratio = 126.7, $p < .000001$).

Another series of variables, based on questioning teachers, was useful in that various behaviors were scored into five different continuously opposing categories, such as aggressive 1-2-3-4-5 timid. This allowed a determination of whether the intermediate scores gave intermediate MAFB values. These results are shown in Figure 3 (next page). As to the school's attitude toward the student, they represent the inverse of Figure 1, the student's attitude toward the school.

The results for the aggressive-timid and the impulsive-cautious variables suggested that the mean MAFB became progressively lower with the increasing magnitude of aggressiveness or impulsiveness. The most significant association was with lazy-hardworking (F-ratio = 47.9). By contrast, there was little difference in mean MAFB for withdrawn versus sociable.

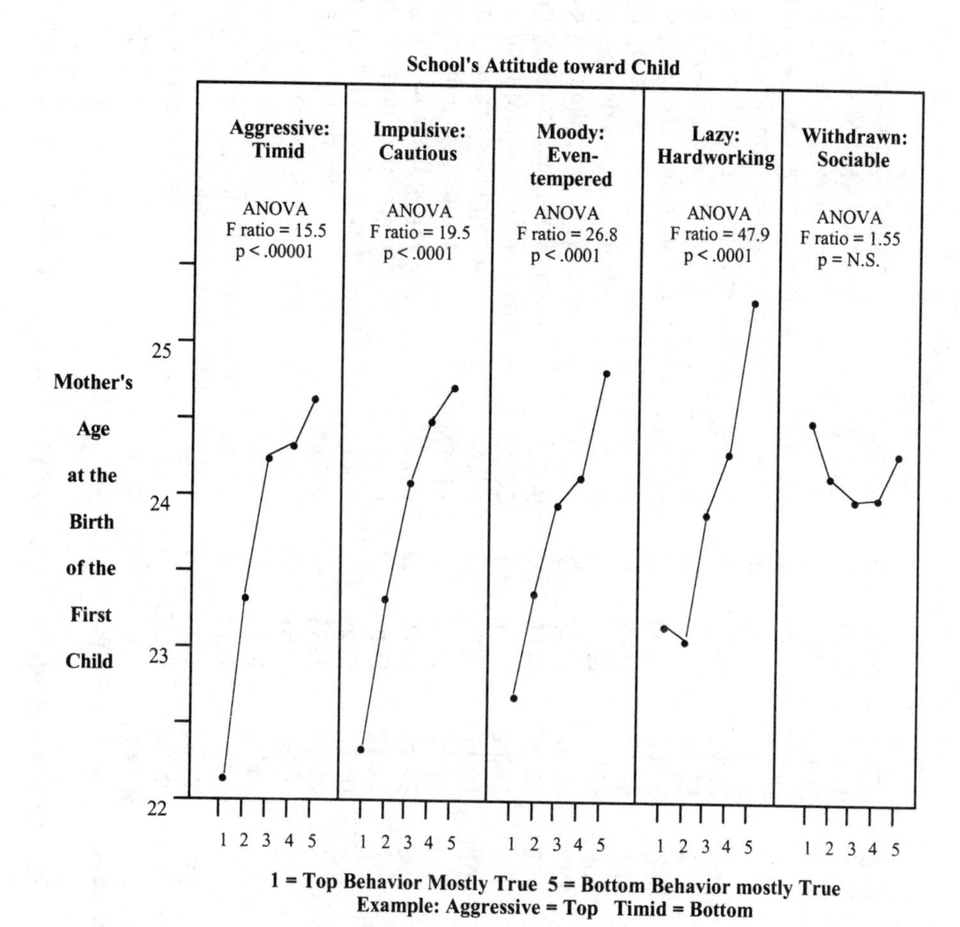

Figure 3. Mother's age at the birth of her first child versus teacher's evaluation of the child as a student. (Variables = N2326-N2331). From NCDS, ESRC Data Archive, University of Essex, Colchester, Essex, England.

Summary – An examination of eighty-five different variables from the NCDS database relating to a wide range of impulsive, aggressive, addictive, depressive, learning, school, and family problems indicated that the majority showed a significant decrease in the mean age at the birth of the first child for mothers of children identified as having these disorders. These observations indicate that the previous results based on the Berkeley and the NLSY data are not unique to the United States.

Table 1. NCDS Variables and Mean Mother's Age at the Birth of the First Child (sorted by F ratio)

			No or Never		Sometimes		Yes or Always			
Variable	**Var #**	**Type**	**N**	**MAFB**	**N**	**MAFB**	**N**	**MAFB**	**F ratio**	**p**
3I Truant this year	N2743	E	3284	24.78			2971	23.54	119.6	0.0000
3S School absences for trivial reas	N2309	E	5111	24.52	785	23.02	457	22.23	84.25	0.0000
1P Family diff., divorce, desertion	N321	F	7961	24.22			296	21.77	84.09	0.0000
3S Truant	N2297	E	5349	24.49	672	22.79	324	22.03	82.17	0.0000
2P Does child fight with others?	N1458	C	3687	24.74	3282	23.70	272	22.25	72.34	0.0000
3S Often tells lies	N2314	C	5558	24.42	628	22.69	178	21.62	71.86	0.0000
3S Child in trouble with police	N2293	C	5418	24.40			452	22.56	68.11	0.0000
3S Is often disobedient	N2310	C	5433	24.43	747	23.01	197	21.54	67.72	0.0000
3S Angry when corrected	N2320	C	5168	24.45	961	23.25	241	21.83	62.90	0.0000
3S Cannot settle down	N2311	A	5282	24.45	910	22.94	190	22.20	62.34	0.0000
1P Fights other children	N144	C	3623	24.64	4496	23.87	437	22.64	54.75	0.0000
1P Family diff, domestic tension	N322	F	7134	24.27			380	22.52	54.01	0.0000
3S Restless, can't stay seated	N2296	A	5329	24.42	836	23.09	207	22.11	53.96	0.0000
3S Irritable, flies off handle	N2304	C	5253	24.43	868	23.12	255	22.39	52.33	0.0000
S3 Unresponsive, apathetic	N2316	D	4721	24.48	1306	23.32	346	23.01	46.05	0.0000
3S Squirmy, fidgety	N2298	A	5451	24.39	777	23.08	138	22.23	41.60	0.0000
3S Destroys, damages property	N2299	C	6046	24.30	269	22.19	64	21.31	40.90	0.0000
1P Miserable or tearful	N137	D	4689	24.52	3501	23.78	401	22.96	40.74	0.0000
3S Frequently fights, quarrelsome	N2300	C	5899	24.32	394	22.73	87	21.47	38.50	0.0000
1P Family diff, in-law conflicts	N323	F	7241	24.26			151	21.98	36.88	0.0000
Currently smoking cigarettes	N5931	S	1937	24.53			2474	23.70	34.89	0.0000
M-7 Usually awake early	N6022	O	6022	24.43			908	23.40	34.85	0.0000
3S Has stolen 1x past year	N2315	C	6055	24.30	138	21.86	108	21.91	33.66	0.0000
3S Bullies other children	N2321	C	6005	24.29	285	22.55	89	21.76	32.83	0.0000

Variable	Var #	Type	No or Never N	No or Never MAFB	Sometimes N	Sometimes MAFB	Yes or Always N	Yes or Always MAFB	F ratio	p
M-14 Easily upset, irritated	N6029	C	5102	24.68			1440	23.71	31.66	0.0000
2P Child miserable or tearful	N1451	D	4128	24.48	2890	23.86	296	22.75	30.11	0.0000
3P Wet bed since age 5 yr	N2625	O	5638	24.24			436	23.06	28.03	0.0000
M-4 Have bad headaches	N6019	O	5776	24.41			755	23.49	27.46	0.0000
3S Often miserable, unhappy	N2305	D	5583	24.33	695	23.14	100	22.57	27.25	0.0000
M-2 Feel tired most of time	N6017	D	5507	24.42			1027	23.64	26.36	0.0000
1P Family contact probation off	N307	F	7733	24.25			71	21.49	25.98	0.0000
1P Generally destructive	N136	C	7319	24.28	1038	23.48	237	22.82	24.69	0.0000
M-6 Usually great diff sleeping	N6021	O	5925	24.39			614	23.45	23.97	0.0000
S3 Complains of aches & pains	N2317	O	5903	24.26	371	23.26	91	21.76	21.69	0.0000
1P Temper tantrums	N124	C	6180	24.29			2386	23.78	21.50	0.0000
2P Child irritable, quick-tempered	N1454	C	3314	24.48	2981	24.08	1025	23.51	19.58	0.0000
M-21 Heart often races	N6036	X	6151	24.37			383	23.31	19.46	0.0000
3S Bites nails	N2308	O	5635	24.30	476	23.10	181	23.23	19.37	0.0000
M-9 Often gets in violent rage	N6024	C	6191	24.35			347	23.27	19.31	0.0000
1P Bullied by other kids	N135	O	5397	24.33	2579	23.94	509	23.16	18.97	0.0000
3S Not much liked by other children	N2301	O	5475	24.30	788	23.51	108	22.48	18.36	0.0000
M-10 People annoy & irritate	N6025	C	4852	24.44			1687	23.90	17.28	0.0000
M-5 Often worried about things	N6020	X	3920	24.48			2618	24.03	15.11	0.0001
1P Irritable	N140	C	4339	24.38	3262	23.99	986	23.59	14.76	0.0000
Problems with writing, spelling	N4660	E	6049	24.37			490	23.55	14.64	0.0001
Problems with reading	N4659	E	6372	24.34			168	23.01	14.13	0.0002
1P Squirmy, fidgety	N138	A	4762	24.27	2810	24.17	1017	23.51	11.71	0.0000
1P Sleepwalking	N128	O	8257	24.18			313	23.30	11.32	0.0008
M-3 Often miserable, depressed	N6018	D	5724	24.37			798	23.80	11.05	0.0009
M-15 Scared to go out alone	N6030	X	6076	24.35			461	23.63	10.87	0.0010
1P Difficulty concentrating	N133	A	5799	24.28	2110	23.96	675	23.50	10.28	0.0000

Variable	Var #	Type	No or Never		Sometimes		Yes or Always		F ratio	p
			N	MAFB	N	MAFB	N	MAFB		
M-11 Twitching face, head, should	N6026	O	6076	24.30			464	23.68	9.22	0.0024
2P Difficulty settling to anything	N1447	A	4716	24.31	1867	24.09	705	23.56	8.92	0.0001
3S Twitches, mannerisms, tics	N2306	O	6207	24.21	144	22.93	27	22.11	8.44	0.0002
1P Family contact psychiat S.W.	N302	F	7745	24.21			89	22.89	8.39	0.0038
2P Bullied by other children	N1449	O	5394	24.24	1506	24.12	262	23.11	7.98	0.0004
1P Family diff, other	N326	F	7411	24.21			179	23.27	7.60	0.0058
1P Reluctance to go to school	N125	E	7708	24.18			859	23.75	7.09	0.0078
2P Child destroys thing	N1450	C	6829	24.24	420	23.53	79	23.20	6.83	0.0011
2P Does child bite nails	N1459	O	4891	24.29	1124	24.22	1299	23.79	6.08	0.0023
3S School phobia	N2318	X	6272	24.19	60	22.56	48	22.91	5.70	0.0034
3S Fearful of new situations	N2312	X	2312	24.27	1214	23.85	117	23.43	5.68	0.0035
1P Family diff, alcoholism	N325	F	7482	24.21			53	22.75	5.38	0.0204
M-12 Often suddenly scared	N6027	X	6047	24.34			493	23.86	5.10	0.0239
1P Difficulty getting to sleep	N127	O	6790	24.09			1776	24.36	5.00	0.0254
M-20 Things get on your nerves	N6035	X	6418	24.32			117	23.39	4.85	0.0277
M-13 Scared to be alone	N6028	X	5962	24.34			578	23.92	4.41	0.0356
2P Child has twitches face, eye	N1457	O	6719	24.22	432	24.02	156	23.21	4.08	0.0171
1P Bad dreams, night terrors	N126	O	7095	24.19			1420	23.93	4.04	0.0444
2P Child squirmy or fidgety	N1452	A	4384	24.24	2095	24.23	830	23.77	3.98	0.0189
2P Child worries about things	N1453	X	3078	24.20	3057	24.29	1092	23.87	3.68	0.0255
3S Stutter or stammer	N2319	O	6257	24.20	98	23.21	23	22.73	3.42	0.0327
1P Family diff, MR	N318	F	8030	24.18			81	23.27	3.16	0.0751
2P Child disobedient at home	N1460	C	3492	24.24	3634	24.17	201	23.48	2.67	0.0693
1P Family diff, mental	N317	F	7767	24.16			199	23.65	2.49	0.1142
M-8 Worry about health	N6023	X	6433	24.31			108	23.66	2.16	0.1415
3S Often worries about things	N2302	X	4039	24.23	2037	24.13	263	23.65	2.13	0.1188

Variable	Var #	Type	No or Never		Sometimes		Yes or Always		F ratio	p
			N	MAFB	N	MAFB	N	MAFB		
3S Child obese	N2338	O	5826	24.15	408	24.56	82	23.82	1.90	0.1520
1P Disobedient	N146	C	3379	24.21	4864	24.14	353	23.71	1.89	0.1511
1P Continually worried	N139	X	4101	24.09	3268	24.36	1144	24.03	1.64	0.1930
1P Mannerisms, twitches	N143	O	7844	24.17	542	23.93	195	23.85	1.15	0.3164
M-24 Had a nervous breakdown	N6039	X	6452	24.31			87	23.79	1.11	0.2915
2P Child sucks thumb	N1455	O	6739	24.17	302	24.46	280	24.24	0.59	0.5528
M-16 Keyed up and jittery	N6031	X	6313	24.29			225	24.46	0.29	0.5921
1P Family contact mental officer	N310	F	7733	24.20			72	23.90	0.25	0.6170

Table 1S. NCDS Variables and Mean Mother's Age at the Birth of the First Child (sorted by type)

			No or Never		Sometimes		Yes or Always			
Variable	Var #	Type	N	MAFB	N	MAFB	N	MAFB	F ratio	p
ADHD variables										
3S Cannot settle down	N2311	A	5282	24.45	910	22.94	190	22.20	62.34	0.0000
3S Restless, can't stay seated	N2296	A	5329	24.42	836	23.09	207	22.11	53.96	0.0000
3S Squirmy, fidgety	N2298	A	5451	24.39	777	23.08	138	22.23	41.60	0.0000
1P Squirmy, fidgety	N138	A	4762	24.27	2810	24.17	1017	23.51	11.71	0.0000
1P Difficulty concentrating	N133	A	5799	24.28	2110	23.96	675	23.50	10.28	0.0000
2P Difficulty settling to anything	N1447	A	4716	24.31	1867	24.09	705	23.56	8.92	0.0001
2P Child squirmy or fidgety	N1452	A	4384	24.24	2095	24.23	830	23.77	3.98	0.0189
Conduct disorder aggression variables										
2P Does child fight with others?	N1458	C	3687	24.74	3282	23.70	272	22.25	72.34	0.0000
3S Often tells lies	N2314	C	5558	24.42	628	22.69	178	21.62	71.86	0.0000
3S Child in trouble with police	N2293	C	5418	24.40			452	22.56	68.11	0.0000
3S Is often disobedient	N2310	C	5433	24.43	747	23.01	197	21.54	67.72	0.0000
3S Angry when corrected	N2320	C	5168	24.45	961	23.25	241	21.83	62.90	0.0000
1P Fights other children	N144	C	3623	24.64	4496	23.87	437	22.64	54.75	0.0000
3S Irritable, flies off handle	N2304	C	5253	24.43	868	23.12	255	22.39	52.33	0.0000
3S Destroys, damages property	N2299	C	6046	24.30	269	22.19	64	21.31	40.90	0.0000
3S Frequently fights, quarrelsome	N2300	C	5899	24.32	394	22.73	87	21.47	38.50	0.0000
3S Has stolen 1x past year	N2315	C	6055	24.30	138	21.86	108	21.91	33.66	0.0000
3S Bullies other children	N2321	C	6005	24.29	285	22.55	89	21.76	32.83	0.0000
M-14 Easily upset, irritated	N6029	C	5102	24.68			1440	23.71	31.66	0.0000
1P Generally destructive	N136	C	7319	24.28	1038	23.48	237	22.82	24.69	0.0000
1P Temper tantrums	N124	C	6180	24.29			2386	23.78	21.50	0.0000

Variable	Var #	Type	No or never N	MAFB	Sometimes N	MAFB	Yes or always N	MAFB	F-ratio	p
2P Child irritable, quick tempered	N1454	C	3314	24.48	2981	24.08	1025	23.51	19.58	0.0000
M-9 Often get in violent rage	N6024	C	6191	24.35			347	23.27	19.31	0.0000
M-10 People annoy & irritate	N6025	C	4852	24.44			1687	23.90	17.28	0.0000
1P Irritable	N140	C	4339	24.38	3262	23.99	986	23.59	14.76	0.0000
2P Child destroys thing	N1450	C	6829	24.24	420	23.53	79	23.20	6.83	0.0011
2P Child disobedient at home	N1460	C	3492	24.24	3634	24.17	201	23.48	2.67	0.0693
1P Disobedient	N146	C	3379	24.21	4864	24.14	353	23.71	1.89	0.1511
Depression variables										
S3 Unresponsive, apathetic	N2316	D	4721	24.48	1306	23.32	346	23.01	46.05	0.0000
1P Miserable or tearful	N137	D	4689	24.52	3501	23.78	401	22.96	40.74	0.0000
2P Child miserable or tearful	N1451	D	4128	24.48	2890	23.86	296	22.75	30.11	0.0000
3S Often miserable, unhappy	N2305	D	5583	24.33	695	23.14	100	22.57	27.25	0.0000
M-2 Feel tired most of time	N6017	D	5507	24.42			1027	23.64	26.36	0.0000
M-3 Often miserable, depressed	N6018	D	5724	24.37			798	23.80	11.05	0.0009
Educational variables										
3I Truant this year	N2743	E	3284	24.78			2971	23.54	119.6	0.0000
3S School absences for trivial reas	N2309	E	5111	24.52	785	23.02	457	22.23	84.25	0.0000
3S Truant	N2297	E	5349	24.49	672	22.79	324	22.03	82.17	0.0000
Problems with writing, spelling	N4660	E	6049	24.37			490	23.55	14.64	0.0001
Problems with reading	N4659	E	6372	24.34			168	23.01	14.13	0.0002
1P Reluctance to go to school	N125	E	7708	24.18						
Family difficulty variable										
1P Family diff, divorce, desertion	N321	F	7961	24.22			296	21.77	84.09	0.0000
1P Family diff, domestic tension	N322	F	7134	24.27			380	22.52	54.01	0.0000
1P Family diff, in-law conflicts	N323	F	7241	24.26			151	21.98	36.88	0.0000
1P Family contact probation off	N307	F	7733	24.25			71	21.49	25.98	0.0000

Variable	Var #	Type	No or Never		Sometimes		Yes or Always		F ratio	p
			N	MAFB	N	MAFB	N	MAFB		
1P Family contact psychiat S.W.	N302	F	7745	24.21			89	22.89	8.39	0.0038
1P Family diff, other	N326	F	7411	24.21			179	23.27	7.60	0.0058
1P Family diff, alcoholism	N325	F	7482	24.21			53	22.75	5.38	0.0204
1P Family diff, MR	N318	F	8030	24.18			81	23.27	3.16	0.0751
1P Family diff, mental	N317	F	7767	24.16			199	23.65	2.49	0.1142
1P Family contact mental officer	N310	F	7733	24.20			72	23.90	0.25	0.6170
Other variables										
M-7 Usually awake early	N6022	O	6022	24.43			908	23.40	34.85	0.0000
3P Wet bed since age 5 yr	N2625	O	5638	24.24			436	23.06	28.03	0.0000
M-4 Have bad headaches	N6019	O	5776	24.41			755	23.49	27.46	0.0000
M-6 Usually great diff sleeping	N6021	O	5925	24.39			614	23.45	23.97	0.0000
S3 Complains of aches & pains	N2317	O	5903	24.26	371	23.26	91	21.76	21.69	0.0000
3S Bites nails	N2308	O	5635	24.30	476	23.10	181	23.23	19.37	0.0000
1P Bullied by other kids	N135	O	5397	24.33	2579	23.94	509	23.16	18.97	0.0000
3S Not much liked by other children	N2301	O	5475	24.30	788	23.51	108	22.48	18.36	0.0000
1P Sleepwalking	N128	O	8257	24.18			313	23.30	11.32	0.0008
M-11 Twitching face, head, should	N6026	O	6076	24.30			464	23.68	9.22	0.0024
3S Twitches, mannerisms, tics	N2306	O	6207	24.21	144	22.93	27	22.11	8.44	0.0002
2P Bullied by other children	N1449	O	5394	24.24	1506	24.12	262	23.11	7.98	0.0004
2P Does child bite nails	N1459	O	4891	24.29	1124	24.22	1299	23.79	6.08	0.0023
1P Difficulty getting to sleep	N127	O	6790	24.09			1776	24.36	5.00	0.0254
2P Child has twitches face, eye	N1457	O	6719	24.22	432	24.02	156	23.21	4.08	0.0171
1P Bad dreams, night terrors	N126	O	7095	24.19			1420	23.93	4.04	0.0444
3S Stutter or stammer	N2319	O	6257	24.20	98	23.21	23	22.73	3.42	0.0327
3S Child obese	N2338	O	5826	24.15	408	24.56	82	23.82	1.90	0.0000
1P Mannerisms, twitches	N143	O	7844	24.17	542	23.93	195	23.85	1.15	0.3164

			No or Never		Sometimes		Yes or Always			
Variable	**Var #**	**Type**	**N**	**MAFB**	**N**	**MAFB**	**N**	**MAFB**	**F ratio**	**p**
2P Child sucks thumb	N1455	O	6739	24.17	302	24.46	280	24.24	0.59	0.5528
Substance abuse variable										
Currently smoking cigarettes	N5931	S	1937	24.53			2474	23.70	34.89	0.0000
Anxiety variables										
M-21 Heart often races	N6036	X	6151	24.37			383	23.31	19.46	0.0000
M-5 Often worried about things	N6020	X	3920	24.48			2618	24.03	15.11	0.0001
M-15 Scared to go out alone	N6030	X	6076	24.35			461	23.63	10.87	0.0010
3S School phobia	N2318	X	6272	24.19	60	22.56	48	22.91	5.70	0.0034
3S Fearful of new situations	N2312	X	2312	24.27	1214	23.85	117	23.43	5.68	0.0035
M-12 Often suddenly scared	N6027	X	6047	24.34			493	23.86	5.10	0.0239
M-20 Things get on your nerves	N6035	X	6418	24.32			117	23.39	4.85	0.0277
M-13 Scared to be alone	N6028	X	5962	24.34			578	23.92	4.41	0.0356
2P Child worries about things	N1453	X	3078	24.20	3057	24.29	1092	23.87	3.68	0.0255
M-8 Worry about health	N6023	X	6433	24.31			108	23.66	2.16	0.1415
3S Often worries about things	N2302	X	4039	24.23	2037	24.13	263	23.65	2.13	0.1188
1P Continually worried	N139	X	4101	24.09	3268	24.36	1144	24.03	1.64	0.1930
M-24 Had a nervous breakdown	N6039	X	6452	24.31			87	23.79	1.11	0.2915
M-16 Keyed up and jittery	N6031	X	6313	24.29			225	24.46	0.29	0.5921

Chapter 36

Data from New Zealand

In the preceding chapters databases covering individuals from California, the United States as a whole, and England have all shown a trend in which individuals with impulsive, disruptive, addictive, and learning disorders tend to have children earlier, have more children, and come from larger families than individuals without these disorders. As described in Chapter 12 on the genetics of conduct disorder, I reviewed the studies of Terri Moffitt and colleagues based on the Dunedin Multidisciplinary Health and Development Study.[191,194,195] This potentially allowed an examination of yet another country, New Zealand, to determine if these trends were indeed worldwide. In a response to questions about this data, Moffitt stated the problem well. "Those with ADHD + conduct disorder, i.e. severe impulse control problems, will be found in families that intergenerationally breed at a disproportionately high rate but invest disproportionately less in caring for the offspring." While there were some technical difficulties in examining the age at first birth of the mothers, she stated, "We may be able to do better than studying the boys' mothers' reproductive behavior, as we have collected extensive data on the boys' own reproductive behavior at the recent age 21 follow-up. The ADHD + conduct disorder boys have more often cohabited with a woman, fathered more babies, are living with fewer of those babies, are contributing financial support to fewer babies, are having more sex partners, and are more likely to refuse to use a condom, relative to the other males in the sample [unaffected, delinquent only, and ADHD only]. All of these findings are significant by ANOVA analysis."

Moffitt's studies and those of others[112,171] indicate that one of the best predictors of later criminal activity was the presence in childhood of a combination of two or more of the following: ADHD, conduct disorder, aggression, learning disorders, and adverse family circumstances.

Summary – Young adults in New Zealand with ADHD and conduct disorder tend to be irresponsible about the use of contraception, father more children, and show less responsibility for their support than other males.

Chapter 37

National Center for Health Statistics

Before progressing on to Part IV, there was one additional database that I wanted to examine. That was the information obtained by the National Center for Health Statistics (NCHS) in their Birth Cohort Linked Birth/Infant Death Data set. Since 1985 the NCHS and the Center for Disease Control and Prevention have made the above data sets easily available on CD discs. While the primary purpose of this data set was to monitor the frequency of various causes of infant mortality, enough variables were provided to examine the relationship between the mother's age when she had her first child and the number of years of education she had completed. As with several other data sets, the NCHS variables did not include the mother's age at the birth of her first child. However, since the interval to the previous liveborn child was given, restricting the results to mothers with only one or two children still provided a large number of cases to analyze. The results for the mothers are shown in Figure 1.

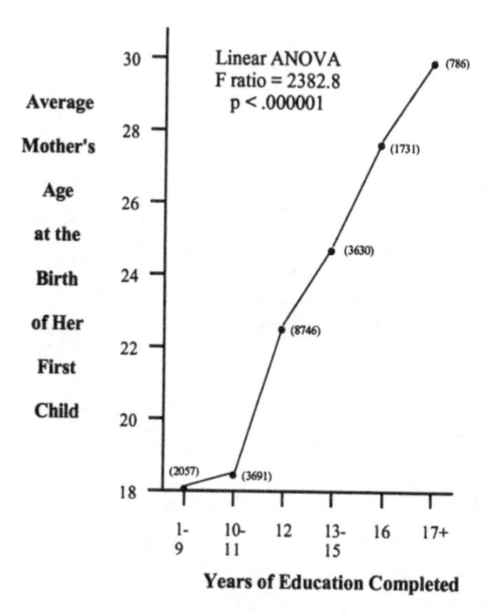

Figure 1. Age of the mother at the birth of her first child versus the number of years of education she completed. From the 1985 Birth Cohort Linked Birth/Infant Death Data Set. National Center for Health Statistics and Center for Disease Control and Prevention.

There were a total of 20,641 women in the survey. The average age at the birth of the first child for those who did not graduate from high school ranged from 18.1 to 18.8 years. This jumped to 22.7 years for those who finished high school, 27.8 years for those who finished college, and 30.2 years for those attending graduate school. Of the total, 27.8% did not graduate from high school, and 29.8% completed some college. The difference in average age at the birth of the first child for these two groups was approximately seven years. Since these two groups constituted 58% of the total group, together they would provide a significant force for gene selection.

The average number of years of education was correlated with the total number of children. Those with six or more children averaged 10.5 years of education, while those with one or two children averaged 12.2 years of education (F ratio = 79, $p < .000001$).

Summary – Data in the U.S. from the National Center for Health Statistics 1985 Birth Cohort Linked Birth/Infant Death Data set reverified the highly significant association between age of the mother at the birth of her first child and number of years of education she completed. These data were more striking than for any of the other data sets examined. For the women, 58% of the total group either failed to finish high school or completed at least one year of college. The difference in mean age at the birth of the first child for these two groups was 7 years. The combination of high percentage at the extremes and wide span in years for the age at the birth of the first child testify once again to the magnitude of the force for gene selection.

Part IV

Causes

Chapter 38

Education and Gene Selection

In previous chapters I have suggested that one of the major forces for the selection of genes associated with the disruptive, learning, and addictive behavioral disorders is a tendency for individuals carrying these genes to start having children earlier than those without these behaviors. In the earlier chapters it was noted that the average age that mothers with conduct disorder, drug abuse, or low cognitive skills had their first children ranged from 17 to 22 years of age, while the average age that mothers without these problems had their first children ranged from 23 to 30 years of age. This suggests that the separation between these two ranges is due as much to a tendency for individuals without disruptive behavioral disorders to have children later, as it is to a tendency for those with these behaviors to have children earlier. What is responsible for this?

College Education

One of the most potent life experiences that contributes to putting off childbearing is education. This could be a factor for gene selection in the twentieth century if there was a significant increase during this century in the percentage of the population that attended college. The following figure shows this was true.

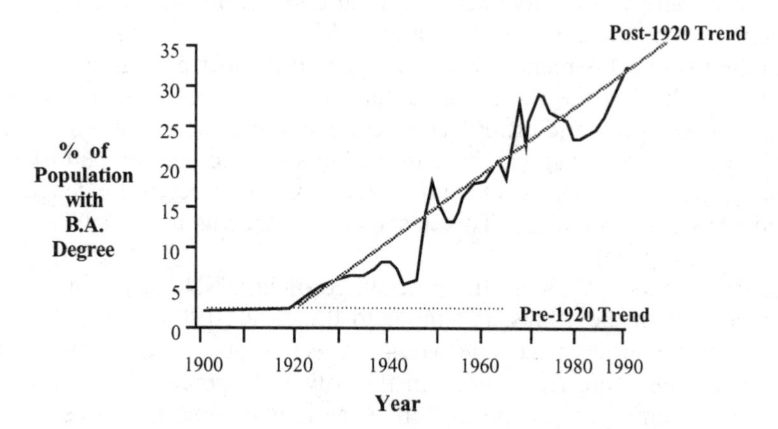

Figure 1. Percentage of the population obtaining a Bachelor's degree from 1900 to 1990. Redrawn from R. J. Herrnstein and C. Murray, *The Bell Curve*, The Free Press, 1994, p 32.

A static 2% of the population was obtaining a bachelor's degree from 1900 to 1920. This was followed by a dramatic progressive increase to over 35% in 1990. This, in of itself, would have no selective drive unless there was a genetic difference between those obtaining a B.A. degree versus those who did not. The most obvious difference would be in cognitive skills or IQ. The following diagram shows the relationship between the percentage of high school graduates going directly to college and IQ percentile.

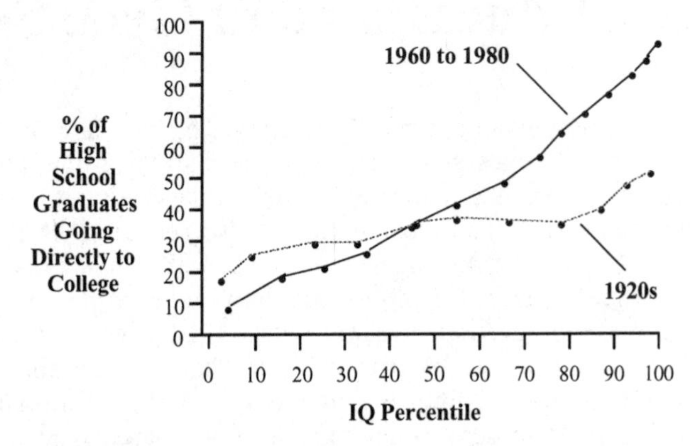

Figure 2. Percent of high school graduates going directly to college by IQ percentile for two time periods – 1920s and 1960s - 1980s. Redrawn from R. J. Herrnstein and C. Murray, *The Bell Curve*, The Free Press, 1994, p 35.

This shows that in the mid-1920s a stable 30% of high school graduates went directly to college, with little difference according to IQ. Thus, 20% of those in the lowest percentile went on to college, and there was no significant increase in the mean until reaching those in the 92nd to 98th percentile, when there was an increase to 50%. By contrast, in the period from 1960 to 1980, 10% of those high school graduates in the lowest IQ percentile went directly to college. There was then a progressive increase to almost 100% of those in the highest IQ percentiles. Thus, to the extent that going to college tends to delay childbearing (Chapter 28), in the 1920s there would have been very little selection for the genes for lower intelligence, since there was only a minor difference in college attendance for those in the lower IQ percentile versus those in the upper percentile. By contrast, by the second half of the twentieth century, the selective force was significant, since there was a dramatic difference in college attendance between those in the lowest IQ percentile (10%) and those in the higher IQ percentile (almost 100%).

During the college years, IQ continues to have an effect on the length of time that students stayed in college. This is shown for students in the 1980s, as shown in Figure 3 (next page).

The upper curve, showing the percentage of high school graduates directly entering college in the 1980s, is similar to the 1960s-1980s curve in Figure 2. The lower curve shows what one would expect – most of those who failed to complete college came from those in the lower IQ percentiles. Again, to the extent that remaining in college for longer periods of time increases a woman's age when she has her first child, there would be progressively greater selection for genes contributing to a lower IQ, because these genes will have a faster generational cycle.

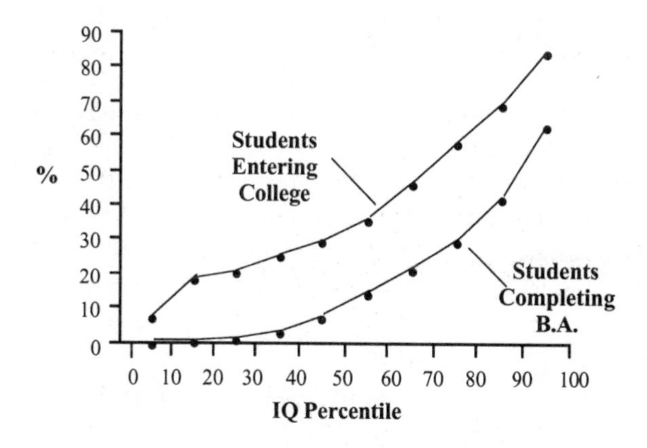

Figure 3. Comparison of the percent of students that entered college and the percent that completed college by IQ percentile. Redrawn from R. J. Herrnstein and C. Murray, *The Bell Curve*, The Free Press, 1994, p 37.

Heath et al. published an interesting article in *Nature*[130] entitled "Education policy and the heritability of educational attainment." They determined the education level of identical and fraternal twins from Norway, born before 1940, between 1940 and 1949, and after 1950. They found that at least in males, the genetic contribution to educational attainment increased from 45% for those born before 1940 to 67% -75% for those born after 1940. This is consistent with Figure 2 showing that IQs were much higher in subjects attending college after 1960 than in the 1920s. This, in turn, is related to the data in Figure 1. *As college became available to a much higher percentage of the population, admission requirements became more dependent on academic performance and IQ than on family wealth or socioeconomic status. As a result, the role of education in postponing childbirth became a stronger and stronger force in the selection against genes associated with a high IQ and an absence of learning disorders and disruptive behaviors.*

High School Education

These trends for a college education raise the question of whether the effect also extends a high school education. The following data indicates it does. The percentage of 17-year-olds who obtained a high school diploma rose dramatically over the duration of the twentieth century. This is shown in Figure 4.

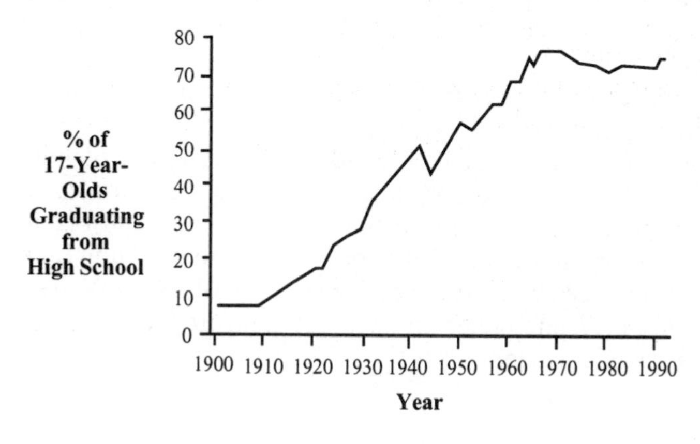

Figure 4. Percentage of the population of 17 year olds obtaining a high school diploma degree from 1900 to 1990. Redrawn from R. J. Herrnstein and C. Murray, *The Bell Curve*, The Free Press, 1994, p 144.

In the earliest part of the twentieth century less than 8% of the eligible 17 year olds were obtaining a high school diploma. Starting in 1910 this began a progressive and dramatic rise to over 70% by 1970, after which it showed a modest decline. As with the above data on a college education, this would have no selective drive unless there was a genetic difference between those obtaining a high school degree versus those who did not. Again the most obvious difference would be in cognitive skills or IQ. The following diagram (Figure 5) derived from the NLSY study, shows the relationship between the percentage of White youths who permanently dropped out of high school by IQ.

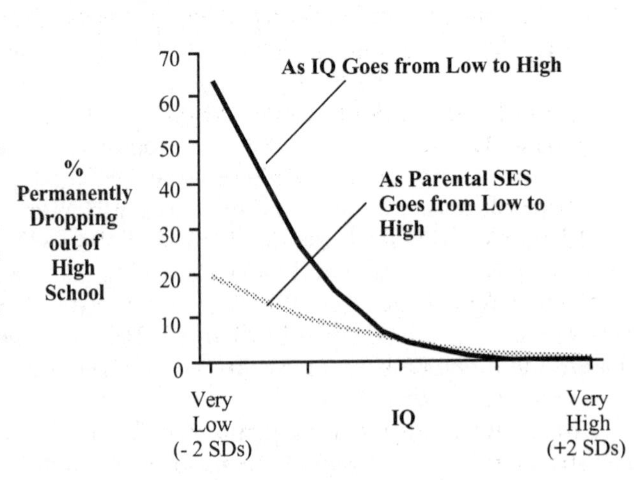

Figure 5. Percent of White students who permanently dropped out of high school by IQ measured in standard deviations. The gray line shows that socioeconomic status (SES) was much less of a factor than IQ. Redrawn from R. J. Herrnstein and C. Murray, *The Bell Curve*, The Free Press, 1994, p 149.

This shows the expected trend for individuals with lower cognitive skills to be the ones who dropped out of high school, and the converse that virtually none of those in the high IQ range dropped out of high school. Socioeconomic status was a relatively minor factor. As pointed out by Herrnstein and Murray,[131] "these numbers dispel the stereotype of the high school dropout as the bright but unlucky youngster whose talents are wasted because of economic disadvantage or a school system that cannot hold onto [them] – the stereotype that people have in mind when they lament the American dropout rate because it is frittering away the nation's human capital."

Older ages at first birth lead to longer interbirth intervals,[27] more effective contraceptive use,[265] and a preference for fewer children.[222]

Demographic studies of the percent of U.S. children who drop out of high school indicate that from 1968 to 1985 the figures were 1.7 to 2.3% for up to ninth grade, 3.9 to 5.5% for tenth grade, 4.3 to 6.5% for eleventh, and 6.2 to 8.9% for twelfth, for a total of approximately 20%, or 1 in 5 students, being high school dropouts.[160] The differences by race were Hispanics >> Blacks > Whites. For example, in 1985 17.6% of White males dropped out compared to 23.3% of Black males, and 38.3% Hispanic males.[160]

The National Longitudinal Studies, Education and Childbearing

The above and earlier results clearly show that obtaining or not obtaining a high school or college diploma can contribute to a dysgenic effect on IQ because those with lower IQ disproportionately drop out of school and begin having children earlier, and have more children, than those with higher IQ who remain in school.

A major theme of this book is that a significant degree of selection for genes for disruptive and addictive behaviors, learning disorders, and lower IQ has occurred primarily in the mid- and latter part of the twentieth century because the percentage of individuals attaining a higher education did not attain significant levels until this time. This is well-shown in Figure 1. In addition to the NLSY, the National Longitudinal Studies (NLS) also performed four surveys of older subjects. These provide further information on this point and are summarized in Table 1.

Table 1. The National Longitudinal Studies

Survey Group	Age Cohort	Sample Size	Initial/Latest Survey Year
Older Men	45-59	5020	1966/1990
Mature Women	30-44	5083	1967/1992
Young Men	14-24	5225	1966/1981
Young Women	14-24	5159	1968/1993
NLSY	14-22	12686	1679/1993
Children of NLSY	Birth-20	6509	1986/1993

Since the first four studies were initiated 11 to 13 years earlier than the NLSY, and since the first two involved older men and women, it was possible to determine if there was a change over time in the percentage of subjects obtaining a higher education (college or graduate school) and whether this resulted in a change over time in the age at which they first began to have children. Since the database only had information about age at the birth of first child in the Mature Women and Young Woman studies, it was possible to study this change only in women.

Mature Women Beginning College Between 1940 and 1955

Figure 6 (next page) shows the correlation between the age at the birth of the first child and the highest grade completed for the Mature Women who were born between 1922 and 1937. The average year of birth was 1929, so those who would be starting college at age 18 would be doing so between 1940 and 1955, with the average being in the year 1947.

This shows that, on average, in the late 1940s, women who dropped out before completing high school had their first child at an average age of 18.9 years. This rose to 20.8 years for those who completed high school, and 22.6 years for those who completed college. On average, the entire group had their first child at 19.8 years of age. The difference between the lowest and the highest was 4.0 years. While the age at the time of birth of the first child was greatest for those attending college, only 12.1% were in this category.

These results were for the entire study group consisting of 3,606 Caucasians, 1390 Blacks, and 87 other. When only the Caucasians were examined, the results were similar, with the age at the birth of the first child averaging 19.5 years for those who did not complete high school, 21.3 years for those who graduated from high school, and 22.5 years for those who graduated from college. The difference between the lowest and the highest was 3.0 years. The percentage of the total who attended college was 15.5.

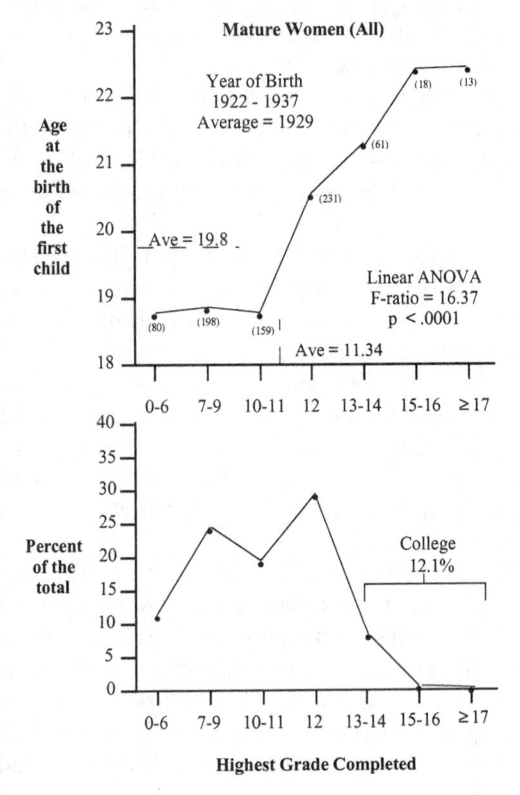

Figure 6. Top: Correlation between age at the birth of the first child and highest grade completed for the cohort of Mature Women. Bottom: percentage of the total for the highest grade completed. All = all races included. From the NLS Mature Women database.

Young Women Beginning College Between 1960 and 1971

The correlation between the age at the birth of the first child and the highest grade completed for the Young Women study is shown in Figure 7 (next page). These women were born between 1942 and 1953, with the average year of birth being 1947. Thus, those who would be starting college at age 18 would be doing so between 1960 and 1971, or in 1965, on average.

This shows that on average, in the mid-1960s, women who dropped out before completing high school had their first child at an average age of between 21.5 and 22.1 years. This rose to 23.5 years for those who completed high school and 25.3 to 26.1 years for those who completed college. On average, the entire group had their first child at 23.8 years of age. The difference between the lowest and the highest was 4.6 years. However, by this time almost three times as many women, 34.5%, were attending college.

Again, these results were for the entire study group, consisting of 3,638 Caucasians, 1,459 Blacks, and 62 other. When restricted to Caucasians, the results were similar. Here the average age at the birth of the first child for those who did not complete high school was 21.9 years. This increased to 23.7 years for those who completed high school, and 26.1 to 26.8 years for those completing college. The difference between the lowest and the highest was 4.9 years. The percentage of the total who attended college was 37.4.

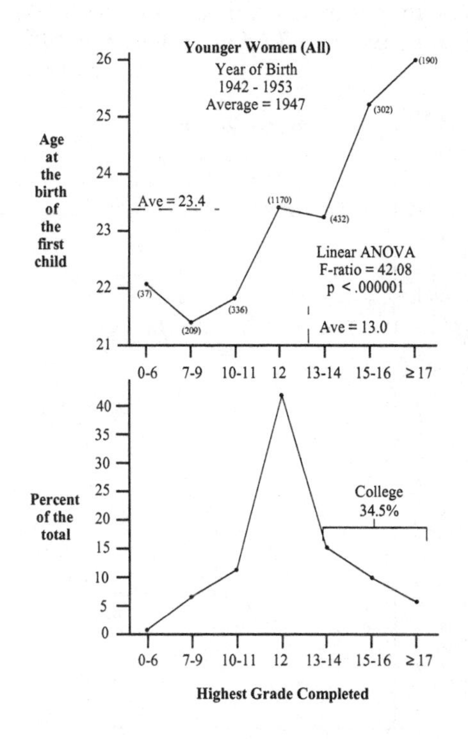

Figure 7. Top: Correlation between age at the birth of the first child and highest grade completed for the cohort of Young Women. Bottom: percentage of the total for the highest grade completed. All = all races included. From the NLS Young Women database.

A new aspect of the study of Young Women was that IQ testing was performed. This allowed an examination of the correlation between IQ score (where the average is 100) and the highest grade completed. This is shown in Figure 8.

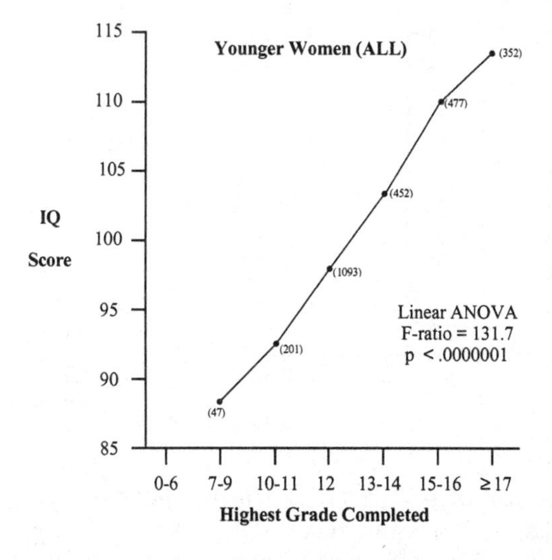

Figure 8. Correlation between the IQ score and the highest grade completed. All races. NLS Young Women database.

This showed a high correlation between IQ and highest grade completed.

NLSY Women Beginning College Between 1974 and 1982

The third cohort of women that the NLS provided was the NLSY. The results for the total set of males and females was shown earlier (Chapter 28). The correlation between the age at the birth of the first child and the highest grade completed for just the NLSY women is shown in Figure 9. These women were born between 1956 and 1964, with the average being 1960. Thus, those who would be starting college at age 18 would be doing so between 1974 and 1982, or in 1978 on average.

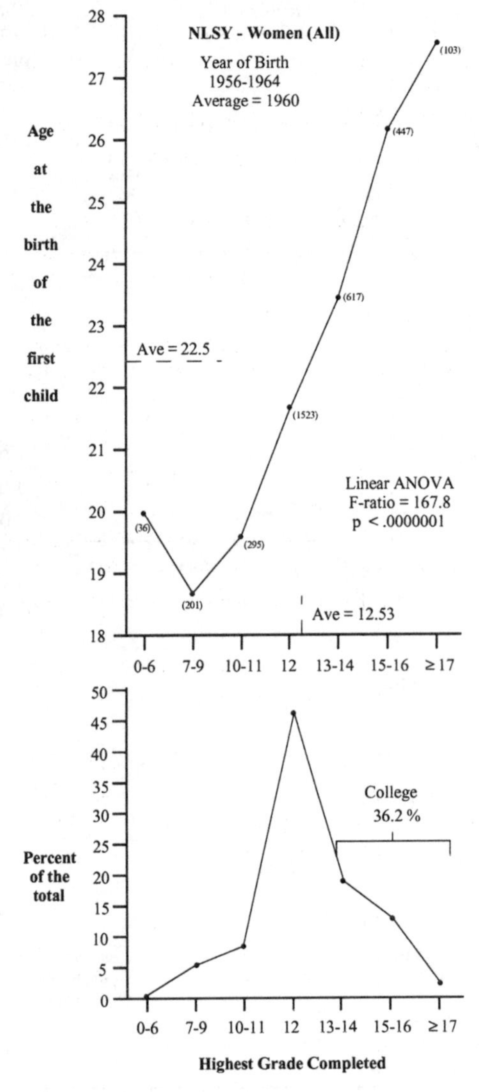

Figure 9. Top: Correlation between age at the birth of the first child and highest grade completed for the cohort of NLSY women. Bottom: percentage of the total for the highest grade completed. All races included. NLSY database.

This shows that, on average, in the late 1970s, women who dropped out before completing high school had their first child at an average age of between 18.7 and 20.0 years. This rose to 21.7 years for those who completed high school and 26.2 to 27.6 years for those who completed college. On average, the entire group had their first child at 22.5 years of age. The difference between the lowest and the highest age at the birth of the first child was 8.9 years, and, by this time, 36.2% were attending college. It takes only the briefest comparison of Figure 6, for the women in the late 1940s, to Figure 9, for the women in the late 1970s, to see a dramatic increase in both the spread between the youngest and oldest age at the birth of the first child, 4.0 to 8.9, an increase of 4.9 years, and in the percentage of women attending college, from 12.1 to 36.2%, an increase of 24.1%.

The results in Figure 9 were for females in the entire study group, consisting of 3,720 Caucasians, 1,561 Blacks, and 1,561 Hispanics. When restricted to Caucasians, the results were again very similar. Here the average age at the birth of the first child for those who did not complete high school was 19.7 years. This increased to 22.8 years for those who completed high school, and 27.2 to 28.0

years for those completing college. The difference between the lowest and the highest was 8.3 years and was 39.7% of the total who attended college. Again, comparison with the results for Caucasian women in the late 1940s shows a dramatic increase in both the spread between the youngest and oldest age at the birth of the first child – 3.0 years to 8.3 years, or an increase by 5.3 years – and in the percentage of women attending college of 15.5% to 39.7%, or an increase of 27.6%. These figures are summarized in Table 2.

Table 2. Childbirth Statistics for The National Longitudinal Studies (MAFB = mother's age at first birth)

Survey Group	Dates at Age 18	Ave. at Age 18	MAFB Youngest/ Oldest	A Spread in Years	B % in College	A x B
Mature Women	1940-1955	1947	18.9/22.6	4.0	12.1	48
Young Women	1960-1971	1965	21.5/26.1	4.6	34.5	159
NLSY women	1974-1982	1978	18.7/27.6	8.9	36.2	322

This shows that the spread in years between the age at the birth of the first child for those dropping out of school early, versus those remaining to go to college or beyond, increased from 4.0 years in 1947 to 8.9 years in 1978, and the percentage of women going to college or higher increased from 12.1% to 36.2%. Since, from a gene selection point of view, both of these are important, a crude estimate of the relative magnitude for both can be obtained as the product of the two (A x B). This increased from 48 in 1947 to 322 in 1978, a 6.7-fold difference.

While the NLS data did not allow an examination of earlier cohorts, it is easy to make an estimate. Thus, if the data from Figure 1 are used, in the period prior to the 1920s, only 2% of the population attended college. Leaving the spread in years for the age at first birth of those who did not complete high school versus those who attended college at 4.0 years gives a product of 8. This suggests that, in the twentieth century, the product of A and B in Table 2 increased from 8 to 322, a fortyfold increase, more than sufficient to produce and then accelerate the selection for genes that are more common in those who failed to finish high school. These relationships are diagrammed in the following figures. The first (Figure 10) shows the dramatic increase in the percentage of women attending college in from 1920 to 1978.

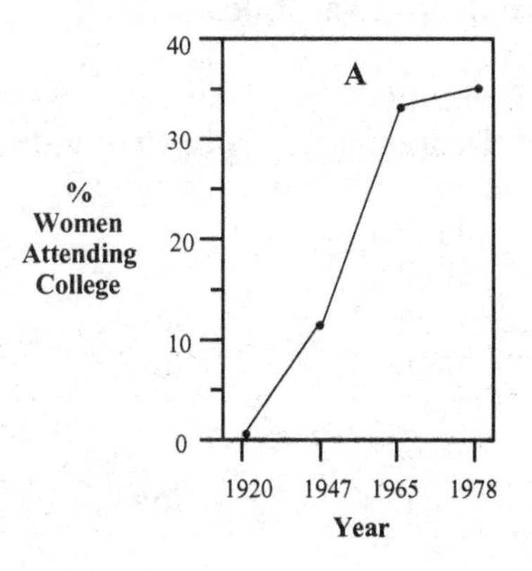

Figure 10. A. Percent of women attending college 1920 to 1978. From the NLS databases.

In 1920 less than 2% of women attended college. As documented in the National Longitudinal Studies, this progressively increased to 36.2% for women attending college in 1978. The increase in the span in years between the age of mother at the birth of her first child for women dropping out of high school versus those attending college is shown in Figure 11.

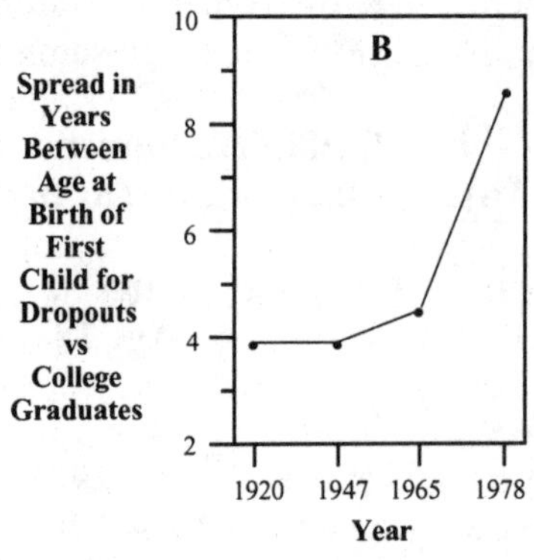

Figure 11. B. Span in years between the age at the birth of the first child for women dropping out of high school versus those attending college from 1920 to 1978. From the NLS databases.

This interval increased from approximately 4 years in 1920 and 1947 to 8.9 years in 1978. Since both of these variables are important from a selection point of view, their combined effect was estimated by multiplying them together (A x B) (Figure 12).

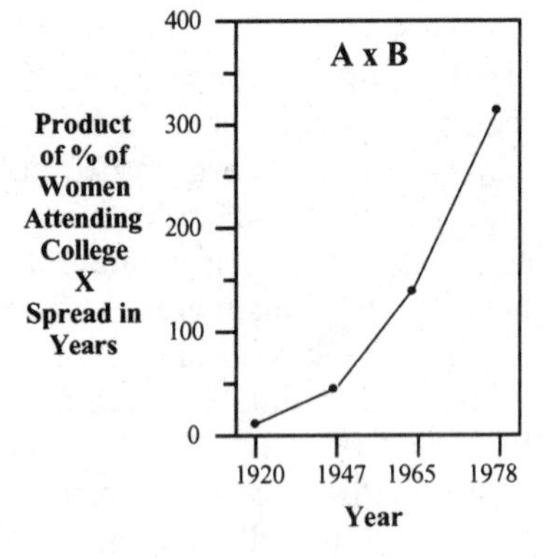

Figure 12. A x B. Product of the percent of women attending college times the spread in years between age at the birth of the first child for women dropouts versus those attending college, from 1920 to 1978. From the NLS databases.

This product increased from 8 in 1920 to 322 in 1978, a fortyfold increase. This illustrates the marked increase in selective effect that education has provided over the span of the twentieth century.

The extent to which education also plays a dysgenic role in addictive and disruptive behaviors depends upon the degree to which those who drop out of school show a higher frequency of substance abuse, conduct disorder, and other disruptive behaviors. The prior chapters can be summarized as follows:

Summary of School Dropouts

Behaviors	**Dropouts**	**Non-dropouts**
IQ	lower	higher
Incidence of alcohol abuse	higher	lower
Incidence of drug abuse and selling	higher	lower
Incidence of conduct disorder	higher	lower
Incidence of criminal record	higher	lower
Selective Effects		
Age of First Birth	lower	higher
Number of children	higher	lower

Despite these associations, it is clear that the act of dropping out of school *per se* is not responsible for these problems, but rather that individuals with learning, disruptive, and addictive behaviors tend to drop out of school earlier than those without these problems. While environmental factors clearly play a role, as reviewed in Part II, genetic factors are also very important. In addition, dropping out of school *per se* does not always lead to having children earlier. But again, as reviewed in previous chapters, there is a strong association between all of the factors leading to dropping out of school early, or not going to college, and early initiation of childbearing.

Independent studies by Bloom and Trussell[19] using data from the National Longitudinal Survey of Young Women, the National Survey of Family Growth, and the Current Population Survey of the Bureau of Census, also showed that the number of years spent in school was a major factor in both the delaying the age of birth of the first child and in not having children at all. They also reported an increase in the number of years between the age at first birth for those dropping out of school early versus those remaining in school, in the latter part of the twentieth century.

Intelligence and Family Size

In 1956 Anne Anastasi[9] reviewed the literature on intelligence and family size. Study after study showed a negative correlation coefficient, averaging about -.3, between intelligence and family size. In short, the lower the IQ, the larger the family. Many of these reports concluded that the average IQ of the country should be falling by 2 to 4 IQ points per generation. However, despite these many studies, there was no clear evidence that the IQ of the nation was decreasing, and several studies suggested it was increasing. Anastasi listed a number of possible explanations of this paradox. While she also pointed out that there was a tendency for individuals with a lower IQ to have children earlier, it was suggested that this was a relatively minor factor in gene selection. However, in most of the early studies the eugenic or dysgenic effect of the age at the birth of the first child was generally ignored, with family size being the focus of attention. While only two of the National Longitudinal Studies, the Young Women and the NLSY, contained sufficient information for comparison, there were intriguing differences between them. Figure 13 (next page) shows the average age at first birth versus IQ percentile group of the subjects in both studies.

This shows that for the subjects of the Young Women study who were 18 years of age in 1967, the age at the birth of the first child was actually modestly higher in the lowest and second lowest 10th percentile group than in the third percentile group, the difference between the extremes of the lowest 10th percentile and the upper 10th percentile was only 1.8 years, and the average age at the birth of the first child was 23.4 years. By contrast, for the NLSY women who were 18 years of age in 1978, there was a much greater correlation between IQ and the age at the birth of the first child. Here, those in the lowest 10th percentile had their first child at age 19.9 years. There was then a progressive increase in IQ to those in the highest 10th percentile who were having their first child at age 26.7 years. Here the difference between the lowest and the highest IQ was 6.8 years, and the average age at the birth of the first child was 22.5 years.

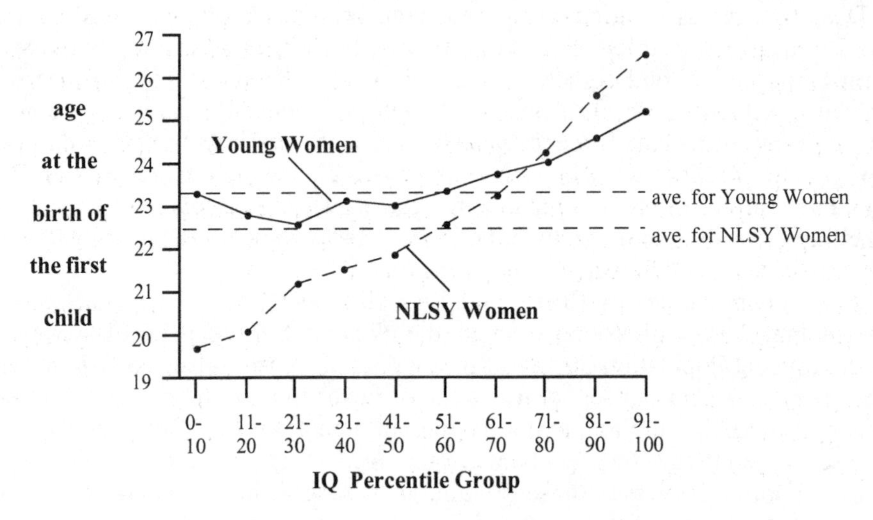

Figure 13. Age at the birth of the first child by IQ percentile group for women in the NLS Young Women database and the NLSY database. The results for the NLSY study given here are presented slightly differently than in earlier chapters in that the ten IQ percentile groups were chosen to contain equal numbers of subjects. The was done to maximize the comparison with the Young Women study where the IQ percentile groups also contained equal numbers.

One could argue that in the earlier parts of the twentieth century, the minimal differences in the age at the birth of the first child for those with lower IQs versus those with higher IQs, and the possible tendency for those in the very lowest IQ ranges to actually have children later than average, might account for the above "paradox" and suggest that, for the major part of the twentieth century, IQs were not dropping. The above graph also suggests that in the latter part of the twentieth century, with the majority of those with a higher IQ attending college and delaying childbirth and those with a lower IQ dropping out of school and initiating childbirth at a younger age, finally there really was a selection for genes associated with a lower IQ, learning disabilities, and disruptive behaviors.

Summary – In attempting to understand the apparent increase in disruptive behavioral disorders since World War II, one of the changes during this period was a dramatic increase in percentage of the population completing a high school or college education. In the 1920s only 2% of the population completed college. This increased modestly to 6% prior to WW II, and then exploded to between 18% and 35% in the post-war period. There was a direct, positive relationship between IQ and the percentage of students both starting and completing college.

There was an even more dramatic increase in the number of 17-year-olds obtaining a high school diploma. This increased from less than 8% in the 1910s to over 70% by the 1970s. Virtually all of the modern high school dropouts cluster in the lower IQ ranges and also show a significantly higher frequency of delinquent and illegal behaviors.

Data from three different National Longitudinal Studies showed that for the women who were of college age around 1947, 12.1% attended college, and for the whole group, the average age at the birth of their first child ranged from 18.9 to 22.6 years, a span of only 4.0 years. By contrast, for the women who were of college age around 1978, 36.2% attended college, and for the whole group, the average age at the birth of their first child ranged from 18.7 to 27.6 years, a span of 8.9 years. Combining both factors, the percent attending college and the delay in age of initiating childbirth for those continuing their education shows a fortyfold increase in this selective force from 1920 to 1990.

These results suggest that the dramatic increase in availability and attainment of a high school and college education during the twentieth century, in conjunction with a high degree of correlation between a lack of such education with learning, disruptive, and addictive behaviors and the early initiation of childbearing, provides the single greatest driving force for the differential selection of the genes contributing to these problem behaviors.

Chapter 39

Birth Control

In the Introduction, I suggested that the availability of efficient methods of birth control in increasingly complex societies was one of the factors leading to the selection of genes for disruptive and addictive behaviors, and learning disorders. By contrast, in 1940 Frederick Osborn[209] proposed in his Eugenic Hypothesis that "the distribution of births in industrial welfare-state democracy would become more eugenic as the environment improved with respect to health, educational, and occupational opportunities, and particularly with respect to the spread of birth control to the point where freedom of parenthood became a reality for all citizens." In a subsequent report he and Bajema[210] stated the "evidence indicates that the reproductive trend toward an increase in the frequency of genes associated with higher IQ discernible by some populations in the most recent generation of Americans will probably continue in the foreseeable future in the Unites States and will be found also in other industrial welfare-state democracies." This would "produce childbearing patterns that appear to bring about an increase in the frequency of genes in human populations which upon interaction with environment lead to the development of above average intelligence." In summary, they felt that the dysgenic trend of a negative correlation between IQ and family size would stop and then reverse itself when everyone, including those with a lower IQ, had total access to effective birth control methods. This would result in couples only having children when they could afford them, and this, in turn, would result in those with a higher IQ, more education, and higher socioeconomic status once again having the most children.

This idea was also voiced by Coale in 1965:[49] "Our best estimate is that, quite likely, there will be a positive association between socioeconomic class and natality within another decade or so. This is so partly because everybody has access now to means of controlling family size, and even an unskilled, uneducated person will be able to control it very well."

In a similar vein, Kirk[150] stated, "with the spread of birth control to most segments of the population...by the 1960s there was not much difference in completed family size by education and hence little opportunity for dysgenic effect from this differential." Finally, Cecil Carter[36,267] suggested that "the encouraging evidence with western European culture that within those sections of the population who plan their families, the trend is for average family size to become positively correlated with the intelligence and educational level of the parents."

These are hopeful and optimistic hypotheses, and I would vastly prefer to be presenting such a positive outlook on the future. Unfortunately, all of the facts presented in this book suggest that the very increase in education and greater availability of effective contraception that Osborn predicted would have a eugenic effect, instead has produced a dysgenic effect leading to the selection of genes for learning disorders and disruptive behaviors. Thus, the effective use of birth control can be either a help or a hindrance to the problem of the selection of genes for disruptive behaviors, depending upon who uses it. The problem is, these hypotheses, which focus entirely on IQ, fail to consider the impulsive, compulsive, disruptive behavioral disorders. As shown in Chapter 24, there is a strong tendency for those with the highest frequency of such behaviors to use contraception the least.

In addition, these hypotheses totally fail to consider the issue of the age at the birth of the first child. The significant jump in the span of years between the age at the birth of their first child for those dropping out of school versus those attending college or graduate school, and seen for the women who were 18 years old in 1947 compared to the women who were 18 in 1978 (see page 232) is probably due, in large part, to the introduction of the "pill" in the 1960s. This is one of the reasons I have implicated "technologically advanced societies" in the subtitle of this book as part of the problem. This contributes to the selection for genes associated with disruptive behavioral disorders, because those who struggle the most in school, have learning disorders, substance abuse,and other behavioral problems tend to drop out of school earlier, become sexually active earlier, use contraception less, have children earlier, and have more children. By contrast those without these problems are more responsible in the use of contraception, remain in school longer, delay the onset of having children and eventually have fewer children, all through the effective use of contraception. Thus, the availability of effective contraception, instead of having a leveling and eugenic effect, has a strong dysgenic effect by increasing the span in years for the age at the birth of the first child between those who drop out of school early versus those who remain in school.

Summary – A number of writers, worried about the inverse correlation between IQ and family size, optimistically predicted that the widespread availability of effective birth control methods would eventually reverse this trend, since only those who could truly afford children would have them. In reality, the widespread availability of birth control has simply made the problem worse, in that those with disruptive behaviors and learning disorders struggle the most in school, drop out of school earlier, become sexually active earlier, use contraception less, have children earlier, and have more children. By contrast, those without these problems are more responsible in the use of contraception, remain in school longer, delay the onset of having children, and have fewer children. Thus, instead of helping, the availability of the pill has been a major contributor to the gene selection problem.

Chapter 40

Other Causes

Illegitimacy

One of the most explosive changes occurring in the post WWII period in the United States was the increase in illegitimacy. This is shown in Figure 1.

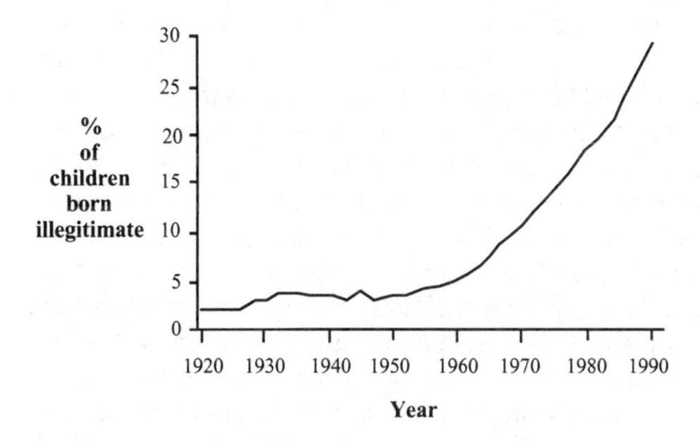

Figure 1. Percentage of children born illegitimate from 1920 to 1990. From Ventura et al.[265a] National Center for Health Statistics

Illegitimacy in and of itself has no dysgenic effect. For this to occur requires that individuals having illegitimate children have a higher frequency of learning disorders and disruptive and addictive behaviors. Chapters 25 and 32 show that this is the case.

Aid to Families of Dependent Children

In the mid-1930s a welfare program called Aid to Families with Dependent Children, or AFDC, was created. The original idea was to provide support for married women whose husbands had died and left them with minor children to raise. In this context it was a compassionate and popular program. This would allow the mothers to remain at home and provide support for their young children. While families where the husband abandoned the family were also eligible, these women were also married. As shown in Figure 2 (next page), in this form the proportion of families receiving AFDC remained between 0.5 and 2% until the mid-1960s.

Figure 2. The percentage of families on AFDC from 1935 to 1992. Redrawn from R. J. Herrnstein and C. Murphy, *The Bell Curve*, The Free Press, 1994, p 192.

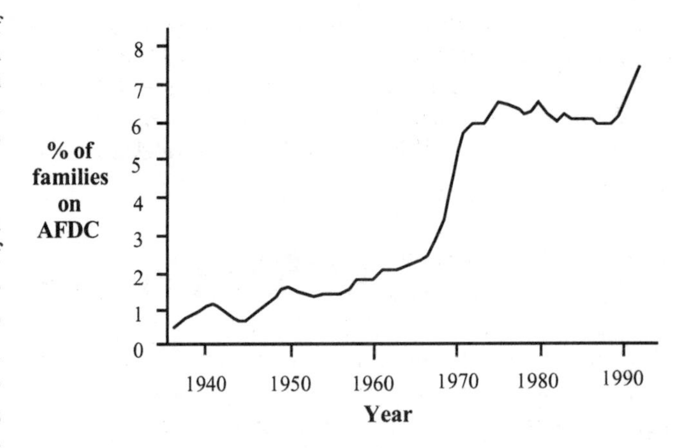

Between 1965 and 1975 the percentage of families on AFDC dramatically increased to 6-7%, leveled off until 1990, then began to rise again. This increase was due to the use of the AFDC program to support unwed mothers where the fathers provided no support for their children. Based on lengthy interviews with more than one hundred teenage mothers, Dash[88] stated:

> "A teenager may see any amount of money as a source of independence, since it enables her to set up her own household free of parental control and supervision...In addition, the availability of welfare may be a deciding factor in whether an unwed mother keeps her child or gives the child up for adoption."

A number of aspects of the AFDC program actually contributed to its expansion and abuse. For example, if a social worker made an unannounced visit to the family and found the father or any other man living with the mother, the funds could be cut off. In addition, if the mother had additional children, each new child was also supported with increased funding to the "family." These two features both discouraged having the father or any man in the house to help financially support the family, and encouraged the mother to have additional illegitimate children since she was, in essence, being paid to do so. This contributed to a whole culture of young teenagers getting pregnant, dropping out of school, and having children and more children. The AFDC program supported and encouraged teenagers having children.

To the extent that such women tended to have increased rates of conduct disorder, drug abuse, learning problems, and low IQ (Chapter 26), this provided an additional driving force for the selection of these genes.

Divorce and Single Parent Families

In their report on the increasing prevalence of learning and psychological disorders in the U.S., Zill and Schoenborn[286] reviewed a number of possible causes. These included the increasing numbers of children with divorced parents or from unmarried mothers. This was supported by the fact that 29.6% of children with a stepfather had had a development delay, learning or behavioral disorder, compared to only 14.6% of children from intact families. In single mother families, 24.8% of the children had one of these three disorders. While this is most often interpreted as indicating that environmental factors associated with parental divorce or single parenthood contribute to these problems, the inverse may be of equal or greater importance – namely, parents or women carrying genes for disruptive and addictive behaviors or learning disorders are more likely to divorce or have children outside of marriage and then pass these genes on to their children.

The Feminist and Women's Movement

The rise of the feminist movement for equality in jobs, opportunity, and income for women in all areas of employment began in the 1960s and coincided with the general availability of the "pill." This freedom for women from the usual encumbrances of pregnancy and childrearing played a major role in releasing women to voluntarily seek careers in areas other than only motherhood. The potentially dysgenic effect of the feminist movement is that competition with men for the more prestigious and higher-paying jobs required increased levels of higher education and longer periods of time spent in a commitment to the job to be competitive. Both of these require selecting out the brighter and most talented, and those least likely to be involved in substance abuse, mental, personality, or learning disorders. As shown in Chapter 20, this is associated with a later onset of childbearing, a decrease in the number of children, and increase in the frequency of childlessness. This trend is verified by Bloom and coworkers who commented on the growing trend to childlessness, especially among the better educated.[18,19]

Attitudes About Having Children

Somewhat related to the women's movement is the dramatic change in attitudes about having children.[217] Between 1957 and 1976 the percentage of adults who felt positive about parenthood – that is, who viewed parenthood as a role that could fulfill their major values – dropped from 58 to 44.[266] Between 1970 and 1983 the percentage of women who felt having a child was a high priority dropped from 53 to 26%.[217] Between 1962 and 1980 the percentage of couples that felt it was important to have children dropped from 84% to 43%.[258,261] These trends would only become dysgenic if such attitudes were more prevalent in those individuals in the higher IQ percentile. The trends shown in Chapter 20, Figure 6, indicate this is the case.

Summary – Several factors other than education and birth control may also contribute to the selection of genes for learning disorders and disruptive behaviors. These include the explosion in illegitimate births, aid to dependent children, divorce, and childlessness – all showing dramatic changes in the second half of the twentieth century. These are potentially dysgenic because those with illegitimate births and on AFDC tend to have their children earlier and are more likely to have learning and disruptive behaviors, while those deciding to have fewer or no children disproportionately come from those with the least number of these problems.

Part V

Proving or Disproving the Hypothesis

Chapter 41

Criticisms of the Hypothesis

In the following pages I have attempted to list a number of potential criticisms of the hypothesis and responses to these criticisms. While I strongly suspect others may find more objections to add to the list, or feel the responses are inadequate, I have tried to be as inclusive and candid as possible.

Criticism #1: The hypothesis is racist.

Almost anytime that the subject of genes and behavior comes up, some will complain that the ideas are racist. This carries with it the implication that some of the genes in question occur more often in some races than in others, and, in this case, that the selective forces are occurring more rapidly in one race, especially minorities, than another. In fact, as far as this hypothesis is concerned, the reverse may well be the case since many lines of evidence suggest that the selective forces discussed are occurring to a greater degree in Whites than in Blacks. Some of the many examples of this are as follows: a) As shown in Chapter 17, Figure 3, a much higher proportion of Whites who became unwed mothers had behavior problems than Blacks who became unwed mothers. b) As shown in Chapter 22 on drug abuse, there tended to be a greater disparity in age at the birth of the first child and in IQ for Whites using drugs than for the total group that included minorities. In these examples, the selection for genes associated with disruptive behaviors would actually be greater in Whites than in Blacks or other minorities.

Thus, in regards to the hypothesis, it is not the absolute rates of teen pregnancy, or of having children earlier, or of having large families, or problems with disruptive behaviors and drug use that is important, it is the relative degree to which these occur in those predominantly carrying the relevant genes versus those who do not carry the genes that count. In most areas where this information is available, this ratio is greater in Whites than in Blacks. I would thus argue that not only is this not a greater problem in Blacks or other minorities than in Whites, it is in fact a greater problem in Whites than in Blacks.

The other issue is whether these genes are more common in one race than others. Scientists who specialize in the study of races repeatedly point out that the entire concept of race has little meaning in a genetic sense, since the genetic variation within races is far greater than the genetic variation between races.[41] In addition, I have repeatedly pointed out that the mechanism of inheritance of the disorders discussed in this book is polygenic. Dozens of different genes are involved. Even though there is some variation in the frequency of some genes in different "races" while some are more common in one "race," others are less common, and the differences soon cancel each other out. All the evidence that I am aware of suggests that environmental and socioeconomic factors play a greater role in racial differences in the frequency of these disorders, than to differences in gene frequencies.

Criticism #2. The hypothesis will encourage discredited coercive Nazi-like eugenic programs.

This entire subject is so important that it is discussed in much more detail in *Part VI. What to Do*. However, the major issue in regard to eugenics will be briefly discussed here, and that relates to what might best be termed "voluntary" versus "involuntary" eugenics, which I would equate to "acceptable" versus "evil" or "Nazi-type" eugenics. Voluntary eugenics refers to voluntary reproductive decisions made by couples based on their own religious and ethical values and on information they have obtained about their own health or the genes they carry. A classic example would be the decision of a person whose parent had Huntington's disease to not have children herself because the risk of having a child with Huntington's disease would be so high, i.e. 25% if she had no knowledge about whether she carried the mutant *Huntington* gene, 50% if she knew she carried the gene, and 100% if she knew her unborn baby carried the gene. This is the parents' own voluntary decision. No one is forcing the decision on them. No one is forcing them to be sterilized. No one is sending them to a death camp because they may carry the gene. By contrast, involuntary, evil, or Nazi-type eugenics is eugenics using coercion and often state-controlled programs that force individuals to do what the state believes is in the best interest of the state. Such approaches, I will categorically state, need to be assiduously avoided.

Some have categorized any mention of any kind of eugenics as "starting down the slippery slope" to coercive eugenics. As long as all approaches are purely voluntary and no approaches are coercive, forced, mandated, legalized, or involuntary, the slippery slope can be avoided.

I recognize that this does not entirely eliminate the potential controversy over eugenics because many would argue against even the voluntary eugenic approach, since in some cases it may require the use of a therapeutic abortion. However, as suggested above, since that is part of a given couple's religious and ethical values, such couples would voluntarily not endorse such a program, and this is their choice. Thus, the objection based on a pro-life view, in fact, verifies the strength of the voluntary approach to eugenics. Neither group should force their views or approach on the other.

Criticism #3. All studies of genetic factors in behavior are inherently evil because they encourage abuse by bigots, despots, and totalitarian governments.

The cause of various behavioral disorders is often considered to be environmental or genetic. In truth, they are both. Behavior is the classic example of the interaction between one's genes and one's environment. One simple example of such an interaction is scurvy, or vitamin C deficiency. While some animals are able to make vitamin C, humans have lost these genes and thus have a genetic inborn disorder of vitamin C metabolism; since vitamin C is present in many foods, this defect usually causes no trouble. However, if you were a sailor embarked upon a long sea journey and you ran out of fresh fruit or vitamin C tablets, you would soon come down with scurvy. Some would argue scurvy is a genetic disorder, and others could argue it is an environmental disorder. The truth is that it is both – a genetic predisposition to scurvy (being a human being) requires a vitamin C-deprived environment (lack of fresh fruit) to be expressed. Many behavioral disorders work the same way – a given set of genes might only cause serious problems in the presence of the right predisposing environmental setting. To claim that studies of the genetic component of behavior is potentially evil, while the study of the environmental component is not, is ludicrous. While Hitler abused millions on the basis of genetic theories of behavior, Stalin abused an equal or even greater number on the basis of environmental theories of behavior. Stalin supported Lysinko, who believed (correctly) that improved environment could improve the phenotype (physical and mental characteristics), but also believed (incorrectly) that the improved phenotype would be passed on genetically to subsequent generations. This supported Stalin's communist theories, and he murdered millions of Russians who opposed him. The point is, legitimate scientific inquiry does not murder and abuse people, people murder and abuse people, and will adopt the flimsiest of excuses to do so.

Criticism #4. The hypothesis advocates genocide.

Genocide refers to the selective killing of a specific racial, ethnic, or religious group, often by a totalitarian government, as in Nazi Germany or Stalinist Russia, or by the act of war between or within countries. Most genocide is based upon perceived differences between those committing the genocide and the victims of genocide.

Scientists who work in the field of behavioral genetics are sometimes accused of promoting genocide, not because they have found that some behavioral genes are more common in one ethnic group than another, but because of the fear that such differences might be found. As discussed above, behavioral disorders are polygenic and multifactorial, with environment playing as much a role as genes. The mutant genes involved are common, with 5 to over 50% of the general population carrying each gene. It is the chance coming together of a certain number of these genes that causes behavioral problems, not the presence of any single gene.

While the frequency of these genes may vary from one ethnic group to another, they average out such that no one group carries a disproportionate number of all the genes involved. Thus, it is extraordinarily likely that the average number of such genes will be the same in groups carrying out the genocide as in the groups they are victimizing. Environmental, social, economic, cultural, religious, and political factors play a far more important causative role in genocide than genetic factors ever will.

It could also be suggested that the hypothesis seeks to prevent some groups of individuals from having children and thus promotes reproductive genocide. In fact, the hypothesis places far more emphasis on the age of onset of childbearing, rather than the number of children. The recommendation is to change the age of birth of the first child, not to change the number of children. The recommendation that everyone at least complete a high school education before having children is not genocide. Social studies show this is beneficial to everyone – the child, the mother, the family, and society in general. The hypothesis simply suggests it also makes good genetic sense.

Criticism #5. Gene selection is not significant for polygenic disorders.

Most of the examples of rapid selection of genes in man, such as the selection for sickle cell trait in Africa because it protected against falciparum malaria, involved examples of single genes. This could lead some to believe that a comparable degree of selection would not occur in polygenic disorders. More specifically, if a given gene mutation is present in only 45% of affected individuals, and present in 20% of unaffected individuals, the selective forces would be dramatically less than if a given gene mutation was present in 100% of affected individuals and less than 1% of unaffected individuals. While this is obviously correct, it fails to take into consideration the fact that if an individual with an impulse and learning disorder due to the presence of ten different mutant genes has her first child at age 15, while another individual without any impulse or learning disorder has only five such genes, stays in school, and has her first baby at age 26, there will be a selection, on average, for five of her ten impulse-learning disorder genes. Differences in the age of birth of the first child and total number of children will result in the selection for both single gene and polygenic disorders. Support for this concept comes from Sewel Wright, the father of population genetics. He proposed that evolution acts through the selection of multiple interacting genes rather than on single genes.[274a]

Criticism #6. Even if there was selection for genes for learning, disruptive, and addictive disorders, a new equilibrium in gene frequencies will quickly be reached, after which no further selection will occur.

One of the fundamental tenets of population genetics is that gene selection rarely works in only one direction. The example of sickle cell anemia illustrates the point. Carriers of the sickle cell gene are protected from death by falciparum malaria. In those parts of Africa where falciparum malaria was endemic, the frequency of the sickle cell trait quickly rose to the point that 40% of the population carried the sickle cell gene. It did not rise to 100% because individuals who carried both genes and had sickle cell disease were at great risk of dying before they had children. An equilibrium between the two opposing forces soon set in, and the frequency of carriers leveled off at 40%.

One could argue that since severe cocaine addicts have impaired libido, IV drug users get and die of AIDS or other infectious diseases, severe alcoholics die of alcohol-related disorders, severe depressives commit suicide, severe criminals end up in jail, etc., that the selection of the genes associated with these disorders would soon come to equilibrium and stop increasing. This is true. The genes involved would probably never continue to increase in frequency until 100% of the population carried them all. However, my concern is that if selection for these genes is occurring, the process may have just begun in earnest since WWII, and we may still have a long way to go before equilibrium is reached. The species could begin to self-destruct long before equilibrium was reached, especially if the disparity between the age at birth of the first child and the number of couples who opt to have no children continues to grow and show a disproportionate representation among those with, versus those without, these genes.

Criticism #7. Not all of these genes produce negative behaviors.

I have previously pointed out that the genes for Tourette syndrome, ADHD and obsessive-compulsive disorder are not all negative.[52] Many such individuals can be very creative, talented musicians and very bright, workaholic, moderately aggressive, and perfectionist. This indicates that the same genes that can be detrimental in some combinations can lead to very desirable behaviors in other combinations. This is well-illustrated in Kay Jamison's book, *Touched with Fire*,[139] documenting the high degree of creativity present in individuals with manic-depressive disorder. It has been suggested that, of the immigrants who packed up and left their home countries to immigrate to the U.S., a disproportionate number were restless, sensation-seeking, easily bored, workaholic, and individualistic – a collection of traits that are not all bad. Such traits could lead to a society full of creative entrepreneurs. To the extent that this is true, it suggests that a modest number of these genes can be good, but too many is bad. This is diagrammed in Figure 1.

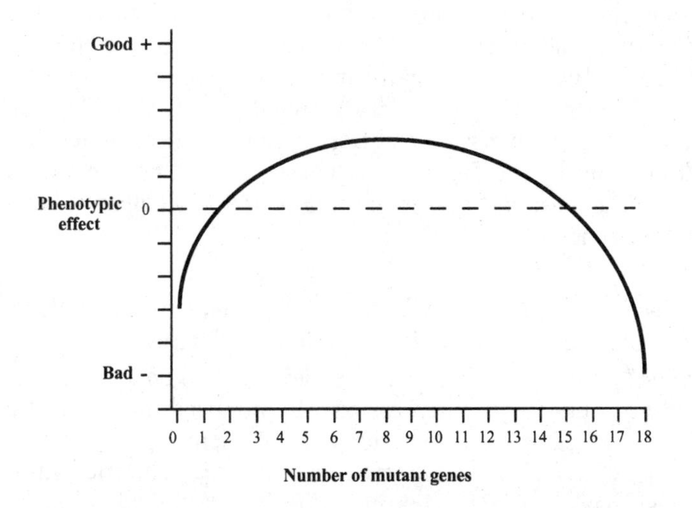

Figure 1. Hypothetical relationship between the number of mutant behavioral genes and effect on behavior (phenotype). The number of genes at the different points will vary as more direct molecular information becomes available.

I am concerned that the present selective forces may be sending us into the direction of the downward slope. When an individual inherits too many of the genes in question, the result is largely negative.

Criticism #8. Since eugenic programs have a trivial effect on gene frequencies they should all be avoided, good or bad.

The backbone of population genetics is the Hardy-Weinburg equation, where p2 + 2pq + q2, p = the frequency of the normal gene and q = the frequency of the mutant gene, and p + q = 1. For genes causing recessive disorders, where a double dose of q is required, the frequency of the disease is q2. For genes causing dominant diseases, the frequency of the disorder is 2pq + q2, or approximately twice the product of p x q. Imagine, for example, that disease X was a recessive disorder that caused aggressive behavior and learning disabilities and was present at a frequency of 1 in 100 people in the population.

An enthusiastic eugenicist might suggest that if everyone with disease X were sterilized, the disorder would rapidly disappear. Not so. At a population frequency of 1 in 100, q2 = 1/100, or .01, and q, the frequency of the disease X gene would be the square root of .01, or .1, and the frequency of X disease gene carriers would be 2 x .9 x .1, or .18, or 18%, or 1 in 5.5. Thus, it can be seen that the sterilization of a small portion of individuals with X disease will have a trivial effect on the frequency of the mutant gene, and sterilization of 100% of individuals with X disease would have to go on for many generations or hundreds of years to have any effect of the frequency of the mutant gene.

The point is, for recessive diseases, eugenic programs are inefficient. However, they are considerably more efficient for dominant disorders. For rare disorders, each affected individual who decides to not have children (or ensures the children are normal by amniocentesis and therapeutic abortion) has a much greater effect on reducing the frequency of the disease. Here, if every affected person (or potentially affected person if the age of onset of the disease was after the beginning of the reproductive period) in the population avoided passing the gene on to the next generation, the disease would disappear in one generation.

The efficiency of controlling polygenically inherited disorders would be somewhat intermediate between these two. Every affected person who decided not to pass on the mutant genes would have an effect on a number of genes, but the effect would be diluted out by the fact that not every affected person would carry all of the responsible genes, and that many unaffected individuals would also carry the same mutant genes.

Criticism #9. The earlier age of onset of a number of disorders simply reflects the presence of expandable repeats, not an increase in gene frequency.

Tim Crow in England[85] suggested an exclusively genetic explanation for the secular trends of increasing frequency of depression. He proposed that a genetic element mutated between generations by alterations in a tandem repeat and this caused earlier age of onset of symptoms and increased risk for psychiatric symptoms. A similar hypothesis has recently been proposed for the earlier age of onset and increased severity in younger subjects in a large manic-depressive disorder pedigree[185,201] and schizophrenia pedigree.[15] In view of the recent discoveries that this type of mechanism occurs in fragile-X syndrome, myotonic dystrophy, and other neurological disorders[37], such a hypothesis is appealing and may be involved in some families but not others.[11] However, while this mechanism can easily explain the earlier age of onset and increased severity of symptoms in the relatives of individuals with these disorders, it does not easily explain the dramatic increase in frequency and earlier age of onset of these diseases across the whole population. The selection hypothesis would explain the increase in frequency, the earlier age of onset, and the increased severity in both the general population and within families.

Recently, molecular genetic studies that actually tested for repeats showed an increase in the frequency of such sequences in manic-depressive disorder and schizophrenia.[202] However, rather than finding a single gene, a number of such sites were found. This is consistent with a polygenic mode of inheritance in which a number of genes are defective because of the presence of increased lengths of such repeats. Thus, in the ultimate analysis, both mechanisms may play a role if there was a selective increase in the prevalence of a number of mutant genes, some of which were due to repeat sequences which increased in size in subsequent generations.

Criticism #10. Since both genetic and environmental factors play a role in learning, disruptive, and addictive disorders, the increase in their prevalence does not prove the frequency of the genes is increasing, only that the environmental causes are increasing.

While there is a great deal of truth to this, I believe that an increase in both the environmental and genetic factors is involved. For example, it would be crazy to conclude the explosive increase in the use of, and addiction to, crack cocaine over a very short period of time was due to an increase in the frequency of genes associated with drug addiction. Clearly, the major factor was the sudden widespread availability of crack, its highly addictive nature, and the availability of pushers to sell and profit from the sale of crack, especially in areas of lower socioeconomic status. However, the fact that genetic factors are involved in the learning, disruptive, and addictive disorders, the observation that individuals with these problems have children earlier, have more children, and come from larger families demonstrates that the means for the selection of these genes is present. This is the source of concern. The maximally effective approach to these problems is one that considers all of the factors involved and targets both genetic and environmental causes.

Criticism #11. The concern about the IQ and fertility paradox fails to consider multiple IQs.

In many parts of this book I have discussed the issue of the IQ paradox, namely that many studies in the twentieth century have documented a negative correlation between IQ and fertility, suggesting that IQ should be dropping by 1 to 2 points per generation, and yet a number of studies where large scale IQ testing was performed at different times suggest these IQs are actually increasing. As reviewed in Chapter 45, this may in part be a result of the Flynn Effect, which is either an improvement in test-taking skills, or an improvement in the phenotypic IQ through improved environmental factors such as nutrition and education, without an improvement in genotypic IQ, the part due to genes. In fact, the phenotypic IQ could be increasing while the genotypic IQ is decreasing, but at a slower rate.

However, there are some other possible explanations of the IQ paradox. One is the proposal that there are multiple IQs,[284] and only a small number of them are being tested in most of these studies. If one of the most important IQs was not even being tested, then the "important kind" of intelligence may not be decreasing. Against this is the fact that in the NLSY data, multiple different measures of IQ were used, and yet these studies provide some of the strongest evidence for a negative correlation between a wide range of behaviors (alcoholism, drug abuse, delinquency, conviction rate, socioeconomic status, jobs, years of school completed, etc.) and IQ.

Criticism #12. The concern about the IQ and fertility paradox fails to consider the role of plagues.

Another potential explanation of the IQ paradox may be that a simple correlation between IQ and fertility fails to take into consideration the problem of death due to infectious diseases and plagues. As thoroughly documented by Laurie Garrett in her book *The Coming Plague*,[138] the death rate from a variety of infectious diseases, both indigenous and in plagues, is much higher in densely packed, low income regions of the world. This could provide a leveling factor not considered in the usual studies of IQ versus the number of births. However, this is basically a question of whether the IQ is inversely proportional to the number of children surviving to reproductive age, not just the number of children born. There are some studies of this as well, including studies before the onset of modern antibiotic therapy.

One of these, based on occupation rank, is shown in Figure 2. This shows that the number of surviving children was actually significantly lower for those in the higher ranking professions than for unskilled and agricultural laborers in 1911.

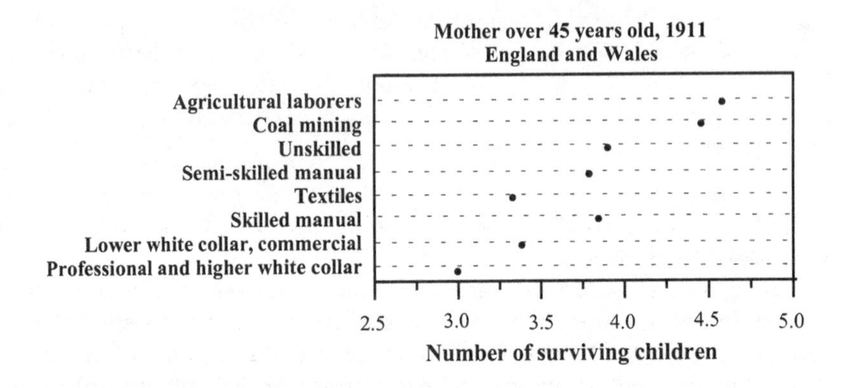

Figure 2. Correlation between number of surviving children and occupation rank in England in 1911 for mothers over 45 years of age. From Vining, D.R.Jr. *Behavioral and Brain Sciences* 9:167-219, 1986.

Criticism #13. The increase in crime and other disruptive behaviors in the past forty years can all be explained by neurotoxicity due to trace elements like lead and manganese.

Masters et al.[179a] have suggested that the increase in violent and disruptive behaviors in the past several decades can be traced to the increase in environmental lead and manganese. They observed a significant correlation between the levels of these trace elements and the rate of violent crime by county in the United

States. While this may explain some of the increases, especially those in violent crime, it is not clear whether it could also explain the worldwide increases in depression and other psychiatric disorders. My suspicion is that a genetic model is best able to explain the increases in a wide spectrum of disorders in many countries.

Criticism #14. The recommendations are a Brave New World attempt to build a perfect master race.

I have no interest in building a master race or eugenically improving on the present status of *Homo sapiens*. I personally believe the species, despite its imperfections, is fine as it is. My concern is to keep it that way, not make it better.

Criticism #15. The book is a thinly veiled cover for a hidden conservative or liberal, political agenda.

I am apolitical. I have voted for both Republican and Democratic candidates based on my perception of their qualifications for the job. I am not a member of any politically oriented organization and don't work for "think tanks" of any persuasion. My entire purpose for writing this book is a concern growing out of my involvement in the field of genetics of human behavior. As described in Part VI, concerning what to do if the hypothesis is correct, the different approaches involve suggestions that range all over the political map. I have no hidden agenda other than an open discussion of the hypothesis.

Summary – A number of potential criticisms of the hypothesis and responses to those criticisms are given. The hypothesis is not racist or anti-minority, since it is not the absolute frequencies for various behaviors that are important, but rather the relative frequencies for those having children early versus those having children later, or having more children, that produces the selective drive. In most cases, this is more pronounced in Whites than in other races. There is no evidence that the average frequency of the many genes involved is any different in different ethnic groups.

The hypothesis does not call for Nazi-like coercive eugenics policies. As reviewed in the following chapter, all proposals are entirely voluntary and up to the individual couples. The hypothesis does not encourage genocide. In addition, the recommendation is to change the timing of onset of childbearing, not to eliminate it.

Despite being polygenic disorders, gene selection still occurs. Although a new equilibrium of gene frequencies will eventually occur, that may be at a far higher frequency than exists presently. Although the genes in question can produce positive as well as negative behaviors, when an individual inherits too many of these genes, the result is largely negative. While it is difficult to alter the frequencies of recessive genes through voluntary eugenic programs, it may be easier with polygenic disorders. While the expandable repeat theory could explain the earlier age of onset and greater severity that occurs in later generations in some of these disorders, it cannot explain the increase in frequency on the general population. It is clear that environmental factors are also very important. The disorders in question are due to an interaction between the genes and environment. The presence of multiple types of IQ is unlikely to explain away the IQ-fertility paradox, since, in different studies, a wide range of IQs and behaviors have been examined, and all show the same effect. Some object that many of the studies examined fertility on the basis of birth rate rather than number of surviving children. However, when numbers of surviving children are examined instead, the results are the same.

The concern is not to build a master race or species. It is an attempt to keep the species as it is. Finally, the author has no hidden political agenda. The issues raised should be of concern to all members of the human race regardless of ethnic, religious, or political background.

Chapter 42

Proving the Hypothesis

The only reliable way to prove the hypothesis of this book is to identify the specific genes involved and perform the studies necessary to show if these genes are either increasing in frequency or at risk to increase in frequency. I will address each of these separately.

Identification of the Genes Involved

This task would had been virtually impossible as recently as the mid-1980s. However, since that time there has been a virtual explosion in the field of molecular genetics with the discovery and development of restriction fragment length polymorphisms (RFLPs) and the polymerase chain reaction (PCR). These two techniques[60] allow the identification of the vast variation between individuals in the base sequence of DNA. These and other techniques have led to the identification and localization of many thousands of markers up and down all the human chromosomes.

The second revolutionary development has been the cloning and sequencing of many thousands of the different human genes. This has been so extensive that virtually all of the genes that play a role in the synthesis, use, and breakdown of all the neurotransmitters (dopamine, serotonin, norepinephrine, GABA, acetylcholine, and others) have been cloned and sequenced. These genes include the receptors, transporters, vesicular transporters, and enzymes that make and break down the neurotransmitters. When the RFLP and PCR polymorphisms are present in or close to these genes, association studies allow the detection of their role in a variety of behaviors, even when the effect is fairly subtle. As reviewed in the companion book, *Search for the Tourette Syndrome and Human Behavior Genes*,[60] and in Part II, polymorphisms at the dopamine D_2 and D_3 receptors, the *dopamine β-hydroxylase*, *dopamine transporter*, *tryptophan 2,3-dioxygenase*, *monoamine oxidase,* and other genes, have shown a significant correlation with various behaviors.

However, many more studies need to be done, since less than 5% of the reasonable candidate genes have been studied. Investigations of alcoholism, drug addiction, ADHD, Tourette syndrome, conduct disorder, manic-depressive disorder, and other behaviors at a molecular level have just begun. Thus, the first part of what needs to be done is to continue, or hopefully increase, this research.

Studies Necessary to Verify or Refute the Hypothesis

The most reliable method of determining if the genes involved in these disruptive, addictive, and learning disorders are increasing in frequency would be to study, over the next fifty years, the frequency of these genes in cohorts of same aged individuals. For example, DNA from the umbilical cord blood of at least five hundred subjects from an epidemiologically sound sample, in each of a number of racial groups, could be studied at five-year intervals. The problem with this approach is that if the hypothesis is true, by the time it was proven, many years of valuable time would have been lost. An obvious preference is to perform a study that would give the same answer in only a few years. Four approaches seem reasonable.

The first would be to examine the frequency of these genes in an epidemiologically sound sample of five hundred individuals in each of several racial groups in different age groups, such as newborns and older individuals at ten-year intervals, i.e. twenty, thirty, forty, etc. While this would produce a dramatic improvement in the time required to get an answer, it carries the potential problem of bias if the mutant genes examined had a differential effect on the death rate. For example, if a mutation of a serotonin receptor was examined that increased the risk of depression and hypertension, it could be associated with an increase in the frequency of suicide and death due to heart disease. As such, the older individuals in the population would have a lower frequency of the mutation, giving the false impression that the mutant gene was increasing in frequency because it was more common in the younger members. However, if detailed records were taken on the psychological and physical health status of the subjects in every age group, it would be possible to identify associations with such disease entities. Such information would allow the frequencies in the older age groups to be corrected for the rates of death due to such disorders in each age group.

A second approach would be to examine the frequency of the genes in question in a group of women of the same racial background who were giving birth to their first child at less than 18, 19 to 24, and more than 25 years of age. A progressive, significant decrease in the frequency of the mutant genes being tested with increasing age at the birth of the first child, would be consistent with the gene selection hypothesis.

Thirdly, the hypothesis could be tested in relation to the total number of children a women gave birth to, by determining if women who were past their childbearing years (40 years of age or older) and had three or more children carried more of the genes in question than similar women with two, one, or no children.

Finally, because of the enormous data available, a uniquely valuable source of subjects would be the NLSY individuals themselves. In regards to verifying the hypothesis, it would be extremely valuable to obtain blood samples from the NLSY subjects before it is too late.

While many of the relevant genes have yet to be identified, enough have been suggested to allow such studies to be started. Many health risks, such as cancer, heart disease, and others, while tragic individually, do not pose a threat to the species as a whole. However, if the hypothesis is true, the longer we delay in performing the necessary research the greater the *Homo sapiens* species will have changed.

I would argue that such research not only needs to be done, but needs to be given the highest priority. In commenting on the politically motivated cancellation of a conference examining the possible role of biological and genetic factors in aggression and violence, Wright stated[274] "Again and again, when genetic research turns toward human nature...politics swamps the discussion and often sinks the research." Perhaps it is time that politicians and governments realized that one of their most urgent priorities might be to ensure that the research into these issues be encouraged and funded. The survival of our species may depend on it.

Summary – Testing the hypothesis will require the replication of studies already done, and the identification of additional genes important in learning disorders and the addictive, aggressive, and disruptive behaviors. Three tests of the hypothesis that would give answers relatively quickly would be to a) determine the frequency of these genes in individuals of different age groups, b) determine if the frequency of these genes was higher in women having their first child in the teen years versus those having their first child later, c) determine if women who have had three or more children carry more of the mutant genes in question than women who have had two, one, or no children, and d) examine gene frequencies in the NLSY sample. At the present time, the only hindrance to such studies is motivation to do the research and the funding to carry it out.

[Since this chapter was written we have identified at a molecular level alleles of several genes that were associated with both, a) one of more of the behaviors discussed in this book (ADHD, subtance abuse, learning disabilities, others), and b) a significantly earlier age of first intercourse, earlier age at the birth of the first child, and/or greater parity (having more children). These studies are in progress.]

Part VI

What to Do

Chapter 43

What to Do – Without Genetic Testing

The following discusses what can be done to attempt to reverse the proposed selection for genes associated with impulsive, compulsive, addictive, and learning disorders. While this chapter makes the assumption that the hypothesis is true, i.e. that genes are involved in these behaviors being selected for, I would argue that most of the suggestions are reasonable even if the hypothesis takes years to prove, or even if it is shown not to be true. Throughout the book, I have argued that one of the major selective factors for these genes is the strong tendency for affected individuals with these conditions to have children earlier, resulting in a more rapid turnover and thus an increase in frequency of the genes involved. Thus, "what to do" obviously relates to reversing this trend. This is illustrated as follows (Figure 1):

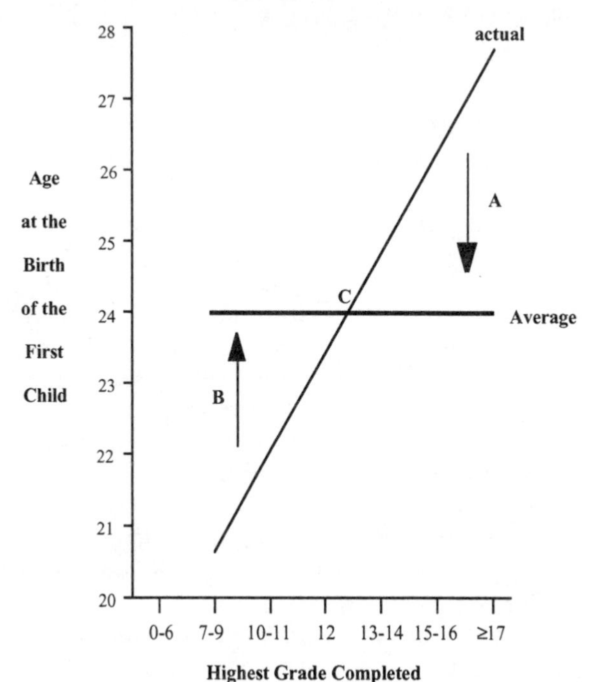

Figure 1. Diagram of the problem of the selection for genes for learning and disruptive behaviors, and approaches to the solution.

The line labeled "actual" is the observed correlation between the age at the birth of the first child and the highest grade completed, based on the NLSY database (see Chapter 28). The line labeled "average" is the correlation that would result in no selection for the genes in individuals who drop out of school early, and no selection against the genes in individuals who remain in school and go to college and graduate school. "A" refers to those approaches that would encourage individuals remaining in school to start having children earlier, and "B" refers to those that would encourage individuals usually dropping out of school to have children later. "C" refers to the numerically most common group – those who have graduated from high school and may have attended a few years of college. No program is recommended for them since, from a gene selection point of view, they are neutral.

A second major factor is whether testing for the variant genes is, or is not, carried out. Programs based on genetic testing are discussed in the next chapter.

None of these solutions are based on actual genetic testing of the genes involved, since such genes are either unknown or at a very early stage of investigation. In addition, the whole concept of genetic testing is fraught with emotional, ethical, legal, and moral issues. While few would argue with the genetic testing that is already being carried out, i.e. testing for PKU and other treatable genetic disorders at birth, and testing for cystic fibrosis, sickle cell, Tay-Sachs, Huntington disease, and other genes in at-risk individuals, even this testing has raised numerous issues relating to problems with insurance, stigmatization, privacy of records, and related problems. To bypass these issues, this chapter applies to programs that could be carried out in the absence of any such testing.

Solution "A"

I will first address part "A" of the above diagram – things that could be done to encourage individuals in college or graduate school to start having children earlier or to have children at all.

I. Support for married students attending college and graduate school.

These couples are usually on such a tight budget that they can ill afford to begin having children. There are a number of simple programs that might encourage such couples to begin a family while still in college. The following are some examples.

a) Day care centers in colleges. While it is very difficult to attend college and raise children at the same time, the availability of free day care centers at colleges would be of considerable help. These could be set up and run at a minimal cost by requiring that one or both spouses volunteer to help staff the center an average of one day a week. Since the scheduling of most college and graduate programs allow several free mornings, afternoons or whole days, a significant amount of the staffing could be provided by the couples who actually utilize the services.

b) Tax incentives. Minor changes in the tax code could double or triple the deduction for children with parents who are college or graduate students. The tax incentives should also include the parents who are supporting children who are attending college and who themselves have children.

c) Medical care. Most colleges and universities provide clinics and inexpensive medical insurance for their students. This should be extended to include their children.

d). Direct aid. As discussed elsewhere, the Aid to Families of Dependent Children (AFDC) tends to contribute to the problem by encouraging high school dropouts to have children. A comparable program, such as Aid to College Students with Children (ACSC), would have the opposite effect and encourage the selection of genes associated with the skills required to remain and succeed in school.

II. Encouraging college graduates to have children.

The solution to the problem does not end when students graduate. It is clear that there is also a strong tendency for such individuals to further delay having children after they graduate, or to decide to have no children at all, preferring to concentrate on a career. Some may even be concerned with overpopulation in the world and mistakenly feel they are contributing most to society by having no children. The following are some potential ways to address this problem.

a) Awareness of the issue. Simply making individuals aware of the problem might help. That is one of the reasons I decided to write this book. If highly educated couples believe they are contributing to the improvement of society by having no children, they need to become aware that this may not be the case. Although I am also greatly concerned with the population explosion worldwide, there has been a progressive, continuing drop in fertility rates in the most technologically advanced countries, to the point that in such countries the issue of gene selection may be rapidly replacing the concern of overpopulation. As with any issue, from the burning of the rain forests to the depletion of the ozone layer, the first step to its solution is an awareness that the problem exists.

b) Day care centers at work. Just as the availability of day care centers in college would help those who decided to have children while still in school, the similar availability of day care centers in the workplace could play a role in encouraging career-minded mothers to realize that having children is not incompatible with those careers. Just as awareness of the problem could help couples decide to have children, the same awareness could help employers be more sympathetic to women in the workplace who also wish to be mothers.

c) Economics. One of the important reasons many couples delay having children is economics. If both are working and barely making ends meet, they may delay childbearing until their economic situation improves. This factor involves complex issues of the national and world economy and is unlikely to change a great deal.

Solution "B"

The second half of the equation is the identification of programs that could encourage individuals who drop out of school, or might drop out, to delay childbearing. For those who drop out of school and then have children, it is clear that programs to prevent dropping out of school would be beneficial.

I. Prevention of school dropouts.

One of the most common reasons that some students drop out of school is that the academic curriculum is difficult. Many of these students have learning disorders (LD) and would remain in school if the appropriate programs were available. It has been my experience that programs for LD children are much more readily available in the grade school years than in high school. I see a tendency for LD children 14-18 years old to "fall through the cracks." If they remain in school, they are getting D's or F's and barely receiving enough credits to progress. They are often largely ignored, assumed to be unmotivated or lazy, and encouraged to drop out and get a job which is, more often than not, impossible, since they are ill-prepared for all but the most menial of jobs. Unfortunately, the one job they tend to be good at is having children.

Many of these students are very good with verbal skills, but poor in performance skills and have poor handwriting. Some simple programs to increase the self-esteem and performance of such individuals include allowing them to use calculators in math classes, use tape recorders to take notes, to use the notes of a peer with better handwriting, be given oral rather than written tests, and be involved in hands-on experience rather than rote memorization.[52,96] Parents have to care enough to ask for an IEP, or individualized educational program, to ensure they are provided these things.

a) Medical treatment. Many children and especially adolescents who do poorly in school and have associated behavioral problems are suffering from undiagnosed attention deficit disorder or Tourette syndrome. These disorders are often very easily treated. As I have detailed elsewhere,[52] in some cases, the solution can be as simple as placing a small patch containing clonidine on their back once a week. I have hundreds of children whose ADHD has improved, grades have increased from D's and F's to A's and B's, and whose oppositional, defiant, and aggressive behavior has disappeared with such treatment. Other medications can be equally effective, usually with no side effects. Increasing knowledge of the genetic basis of these disorders[60] provides a rational support for why these medications are so effective.

I am particularly concerned with the number of adolescents I see who have performed well in grade school, only to begin to show poor performance, lack of motivation, and failing grades when they enter junior or senior high school. In many cases, these individuals have been found to have ADHD. They often have a history of other family members with childhood or adult ADHD. Appropriate treatment of these individuals often totally reverses this downward trend. Instead of becoming high school dropouts, some continue on and even complete college. It would be ideal if every student who seems to be at risk for dropping out of school could be screened by a health professional skilled in assessing and treating behavioral problems, and that appropriate treatment be available and affordable.

b) Alternative schools. Throughout the United States some 2,500 alternative schools, consisting of special programs to improve school performance, have been established. Some of these are specifically designed for pregnant teenagers who find it difficult to return to regular classes.[129] Many are located in separate facilities or even in the workplace. In addition to serving pregnant teenagers, they are responsive to the special needs of at-risk students, particularly those who are below grade level and are experiencing behavior and attendance problems. They stress individualized learning, counseling, social supports, and remedial education, and they often include work-study arrangements.

The potential disadvantage of alternative schools is that in some areas they may be viewed by educators as a place to put problem students to get them out of the regular schools. As such, they naturally have a high concentration of individuals who have severe behavioral and learning disorders, and if not staffed by very dedicated teachers and appropriate programs, they may produce an environment that is far from ideal for the already disabled student.

c) Vocational education. Many of the students who drop out of high school, or are on the verge of dropping out, might perform better in a vocational-type school and receive training for a specific trade. However, by their very nature, there are few or no scholarships for such programs, and the time required for these programs tends to be much shorter than for college. As a result, they are less likely to lead to a delay in the onset of childbearing, but are much better than dropping out of school entirely. Financial programs that upgrade the quality of these schools, provide remedial help, and increase the duration of such programs would be useful. In addition, scholarships directed less toward academic performance and more toward vocational skills performance for both men and women would be of further benefit. Accreditation of such programs and the development of a two-year degree, such as a VA, instead of an AA, would provide some prestige. Funding of such programs could come in part from the private sector, since graduates could provide a valuable source of skilled and semi-skilled workers for a number of industries.

In some places students are not allowed to enter vocational schools without a high school diploma. This is counterproductive, since students who do poorly in standard academic subjects and may flunk out of regular school excel in vocational training.

II. Prevention of teen pregnancies

Programs for the prevention of teenage pregnancies must address many of the different factors involved. The following are some examples.

a) Birth control information. One out of every ten women aged 15 to 19 becomes pregnant each year in the United States.[263] The vast majority of unmarried adolescents who become pregnant do not intend to do so.[282] In a 1979 survey, 58% of premaritally sexually active adolescents who did not use a method at their last intercourse, even though they did not want to become pregnant, thought they could not conceive.[282] In fact, half of all first pregnancies among adolescents occur within six months of the first intercourse, and 20% occur in the first month alone.[283] Thus, while sex education, including the teaching of methods of birth control, is often controversial, it is clearly a very important part of the prevention of teenage pregnancies. Many high school students have very naive ideas about sex, such as "You can't get pregnant the first time," "It won't happen to me," "Withdrawal is an effective birth control method," and many others. The name that best describes such women is "mothers."

Among those sexually active teenage girls who never practice contraception, 39% become pregnant within six months of their first sexual encounter, and 66% do so within two years. By contrast, among those who say they always use a method, only 7% get pregnant in the first six months, and 11% in the first two years.[280] Clearly, the availability, use, and understanding of the importance of contraception can make a huge difference in the incidence of teenage pregnancy.

The type of sex education is critical. When methods of contraception are not discussed, the simple presentation of the general facts of sexuality may have little effect on the use of contraception of sexually active teenagers.[89] Factors such as church attendance, parental education, and race were often more important than sex education in the prevention of teenage pregnancy.[179]

b) Programs discouraging teen pregnancies. A number of programs have been developed at the national and local level to attempt to discourage teen pregnancies. These have been well-reviewed in the book *Risking the Future.*[129] The major problem with such efforts to date is that there have not been enough of them.

c) Hot lines. Hot lines for sexual issues can be very useful because of their easy availability and confidentiality. Over a three-year period, one hot line received 32,000 calls. If even a modest portion of these resulted in referrals that eventuated in the effective use of contraception, they would be well worth the effort.

d) Birth control devices. It does little good for students to be well-informed about the methods of birth control if the means of obtaining them are unavailable. Again, while clearly controversial in some areas, high school students should have confidential access to counselors, nurses, or physicians who can provide them not only with the appropriate information but also with the means of preventing unwanted pregnancies. For males, this can be incorporated into AIDS prevention programs which make free condoms available.

While I understand the concern of parents who worry that such programs encourage the early initiation of sexual activity, the pubertal surge of hormonal activity is overwhelmingly more responsible for this than the easy availability of contraceptives. This is supported by the data in Chapter 24, showing that intercourse without the use of contraceptives, and the early initiation of intercourse, occurs significantly more often in subjects with problems with substance abuse, delinquent behavior, and learning disorders. It is also supported by studies that show that, on average, a teenager's first visit to a clinic or a doctor for contraception occurs eleven months after they first had sexual intercourse,[285] and 1 in 5 first attend a clinic for pregnancy testing.[42] Finally, no significant association has been found between the availability of family planning services and the probability that an adolescent girl would initiate sexual intercourse, independent of other risk factors.[121,129,284]

While the pill is one of the most effective methods of birth control, in the United States women are heavily influenced by myths about cancer and other side effects.[151] Based on experience with successful programs in Europe where contraceptive services to adolescents are heavily subsidized, providing such services in the U.S. could potentially be a very cost effective approach to the control of teenage pregnancies.[263]

Any program that makes contraception readily available to sexually active teenagers will help to delay the onset of childbearing in these subjects. The data suggests that these individuals tend to be sexually active whether contraceptive devices are available or not, so it makes sense to make them maximally available.

e) Family planning clinics. While information on the effectiveness of sex education programs in the prevention of teenage pregnancies has been difficult to obtain[129] the data available clearly support the effectiveness of the availability of family planning clinics. When available, these provide the critical elements of being easily accessible, inexpensive and confidential. They are especially important because they serve the population that needs them the most, such as those initiating sexual activity at an early age and who are financially unable to utilize a private physician.[285] Data from the National Surveys of Young Women showed that such clinics were the source of contraception for half of unmarried teenagers who used contraceptives.[285] Confidentiality is critical since the majority of sexually active teenagers indicated they would not come to a family planning clinic if parental consent was required.[262]

The availability of family planning clinics is especially important, since studies show that prescription forms of contraception, such as the pill or diaphragm, are much more effective in preventing pregnancy than non-prescription techniques such as condoms, foam, rhythm, or withdrawal.[159,207] In regard to the safety of the pill, the mortality risk associated with pill use is significantly lower than that associated with pregnancy and childbearing.[207]

Attendance to a family planning clinic has a positive effect on contraceptive behavior, a negative effect on birth rates, and presumably decreases the incidence of unintended pregnancy.[116] Such clinic patients are more likely than their counterparts who are not in family planning programs to use more reliable methods and less likely to use no method at all.[116]

f) School-based clinics. Clinics physically located in the school have many advantages. Since their general purpose is to improve the overall physical and mental health of teenagers, they include many services in addition to the reduction of teenage pregnancy. These include school physicals, treatment for minor illnesses and injuries, immunizations, testing for sexually transmitted diseases, treatment for drug and alcohol abuse, nutrition and weight reduction, as well as family planning.

> "School-based clinics are intended to capitalize on many of the features that existing research has shown are associated with teenagers' attendance at family planning clinics, including convenience, comfort, confidentiality, and cost. Located within the school buildings or on school grounds, clinics are accessible. Students don't have to take a bus or drive to another part of town or request parents' assistance in getting them to the services. Most clinics operate during school hours and do not require appointments. Because they are visible entities in the school, clinic staff become familiar to students and vice versa. In addition, because the programs are geared to the needs of adolescents, and students are aware that their friends use the services, school-based clinics seem more approachable to many young people than doctors' offices, hospitals, or freestanding adult clinics. Most, if not all, school-based clinics require written consent from parents before students can receive medical services. Generally parents are asked to sign a blanket permission form at the beginning of the academic year, but are not informed when stu-

> dents come to the clinic for services. Nor are patient records accessible to teachers or school officials. Moreover, because the clinics provide a wide range of services, the reason for an individual's visit cannot be automatically assumed. In most clinics, services are provided free to registered students, although several charge a nominal annual fee."[129]

Most important, school-based clinics seem to be quite effective in reducing unwanted teenage pregnancies. In one study, the fertility rate in schools with clinics dropped substantially, from 79 births per 1,000 in 1973 to 26 births per 1,000 in 1983-1984.[98] The latter compares well with the national average of 45 pregnancies per 1,000 students. In another evaluation, over the period of the study, pregnancies increased dramatically from 32 to 51% in the control schools, but dropped 26% in the schools with clinics.[279]

The disadvantages of in-school clinics is that they are not open on weekends or during the summer months, and do not service school dropouts. In addition, those that do not fill prescriptions force students to go elsewhere to obtain birth control pills.

g) Consent Laws. By 1987, twenty-nine states had laws that specifically authorized minors to give their own consent for family planning services, i.e. their parents did not have to be notified. In the other twenty-one states, such laws do not exist or the issue is ambiguous. Since private physicians in states that do not have explicit consent laws are significantly less likely to serve unmarried teenagers,[208] such laws should be passed in those states that do not have them.

III. Revision of AFDC statutes.

In an article entitled "The AFDC Conundrum: A New Look at an Old Institution," Lieberman[168] stated that changes are needed because, "the current policies are a colossal failure." He felt that a major defect in the present AFDC system is that simply giving money to unwed mothers fails to provide the structure, education, and support necessary to keep the mothers from simply continuing to have more and more illegitimate children. For example, Furstenberg and colleagues[120] called attention to the need for "a facility for pregnant teenagers that stressed the importance of finishing school and delaying subsequent births" They found that graduates of the Poe School in Baltimore, a nonresidential institution, "were over twice as likely to have been using birth control a year after their first child was born and were only one-third as likely to have been receiving public assistance 17 years later."

Lieberman's suggestions included the following:

• Immediate implementation of the 1988 amendment authorizing states to require AFDC mothers under age 18 to live with parents or in maternity homes, by prompt issuance of federal regulations and by active promotion within state departments. This age might even be increased to 21.

• Promotion of state laws supporting maternity home care for at least six weeks following birth. The benefits would include improved prenatal and postnatal care, provision of adoption services, education in birth control methods not in conflict with the mother's religion, assistance in ensuring adequate educational and employment plans, and discouragement of further out-of-wedlock births.

• More vigorous use of the provisions of the AFDC statute allowing protective payments, that is, direct payments to providers in the case of young mothers, as a means of circumscribing the liberty gained by out-of-wedlock parenthood.
• A reorientation of the entire AFDC program.

I suspect that, of all the programs listed above, the most effective approach to solution "B" would be the encouragement of private, local, state, and federal support for the widespread availability of in-school clinics and family planning clinics in general, and changes in the AFDC system.

IV. The Issue of Politics.

Conservative values. It is clear that the U.S. society, as well as societies in other parts of the world, tends to stratify along a continuum of conservative to liberal values. At the most conservative end of the spectrum there are strong sentiments for pro-life and anti-abortion, pro-family, and, if not against sex education, at least against programs making contraceptives available to unmarried teenagers. As far as the present hypothesis is concerned, in regard to gene selection, some aspects of conservative programs can be very helpful, while others can be counter-productive. The institution of family values, programs against drug use, a preference for sexual activity only among married couples, the instillation of religious instruction, encouragement of a middle class work ethic, and respect for education are all very positive approaches to the gene selection problem. However, opposition to sex education or the ready availability of contraception will discourage rather encourage solution "B." In addition, part of the pro-life view is that unwanted babies or babies from teen parents should be allowed to proceed to term, and then, if appropriate care is not available, be placed for adoption. However, this encourages rather than discourages solution "B," since the genes present in the parents involved are passed onto the adopted child.

Clinicians who treat children with impulsive behavioral disorders (ADHD, Tourette syndrome, conduct disorder) observe a disproportionately high frequency of adopted children in their clinics. In addition, surveys of adopted children show a higher prevalence of ADHD, aggressive, oppositional, and other behavioral disorders than in non-adopted children.[25] Since many of these children are adopted at birth and display these behaviors early in life – often before they know they have been adopted, the most likely explanation is that the genes that contributed to the parents acting impulsively and producing a child they were unprepared to care for, are passed on to the adopted child, where they contribute to a similar group of impulse disorders. To the extent that conservatives are particularly concerned with the increasing rates of crime, drug abuse, school drop-outs, and teenage pregnancy, they should at least be aware of the possibility that some parts of their agenda, such as opposing sex education in school, opposing the ready availability of contraception to teenagers, opposing abortion, and assuming that adoption can solve the problems of unwanted pregnancies – from a gene selection point of view may be making the problems worse instead of better.

Liberal values. Just as a conservative agenda contains elements that are both positive and negative from the view point of the gene selection problem, the same can be said of those with a more liberal agenda. There are, of course, a number of universal truths or values that everyone tends to share, whether liberal or conservative. These include valuing education and a work ethic, and being against crime and drug abuse. As has been repeatedly emphasized in this book, a continuing education is perhaps the single most important element in postponing the onset of childbearing. On the positive side, liberals are more likely to support programs that encourage sex education in high school, that provide for in-school and family planning clinics, and that support the right of women to choose whether they wish to continue a pregnancy or not. On the other hand, to the extent that they support an open-ended expansion of programs such as Aid to Families of Dependent Children (AFDC), as discussed elsewhere, without some limitations, this program may tend to encourage teenage pregnancy and dependence on welfare support. While such programs are critical to the support of needy children, some limitations need to be incorporated into them to discourage multigenerational, single-parent dependence on welfare.

Summary – Approaches to the solution of the gene selection problem that are not based on genetic testing can be divided into those that encourage college and post-graduate students to both have children, and have them earlier, and those that encourage high school dropouts or potential dropouts to not drop out and to have children later. The majority of individuals actually fall into neither group and require no program, since, from a selection point of view, they are neutral. The single program that is most likely to be effective is ensuring the availability of family planning clinic services in schools or out, for all teenagers who need it.

Chapter 44

What to Do – With Genetic Testing

Genetic tests are now available for a very wide range of disorders, including Huntington's disease, cystic fibrosis, neurofibromatosis, polycystic disease of the kidneys, Tay-Sachs disease, fragile-X-syndrome, hemophilia, and many, many others. In countries such as the United States, where abortion is legal for any purpose, these tests can help to ensure that couples who have already had one affected child, or who know the disease in question is present in their family, will have an unaffected child. This is possible through the use of tests of a single cell of an embryo fertilized in the test tube, or tests in the first 6 to 12 weeks using chorionic villus biopsy, or tests at 14 to 16 weeks using amniocentesis. The DNA from these cells can be tested for any known mutation using the ultra-sensitive PCR technique.

When the disease in question is life-threatening or results in severe mental retardation or physical disability, there are few who would question the use of these remarkable procedures. However, the milder the disorder, the easier it is to treat, and the later its age of onset, the more controversial these techniques become. For example, the genetic causes of Alzheimer's disease are beginning to be understood. If a specific genetic test became available and your father had died of Alzheimer's disease, and testing showed you were a carrier, would you terminate the pregnancy of your unborn son if he also carried the gene? One could argue that since the genetic form of Alzheimer's disease usually has an onset after age 60, and since your son could thus have sixty years of a normal life, or even die of some other cause before the onset of the disease, the pregnancy should not be terminated. On the other hand, if you personally witnessed your father's mental status deteriorate over a period of fifteen years, and both you and your mother had great difficulty dealing with his chronic illness, there might be no way you would want your son and his family to be subjected to this, even if it occurred late in his life. The point is, these are very personal decisions. What is an obvious decision to one person may be an equally obvious decision, but in the opposite direction, to another person.

A further example actually occurs in practice. Amniocentesis is routinely performed in pregnant women who are over 35 years of age to rule out Down's syndrome, a chromosome disorder which is associated with mental retardation and occurs much more often in older mothers. In the vast majority of cases, when the result is positive, the parents terminate the pregnancy – but not always. Some couples prefer raising a Down's syndrome child to terminating the pregnancy. This is their choice.

At the other extreme, after the announcement of the discovery of a gene causing obesity in mice with the possibility that a comparable gene might be found in humans, a TV station carried out a survey which indicated that the majority of people questioned would seriously consider terminating the pregnancy if their unborn child had a gene for severe obesity. As frivolous as this may seem to some, this is also their choice.

With this background, I will address the issue of "What to Do – With Genetic Testing." First, it is necessary to explore the problems of testing for polygenic disorders.

Testing for Polygenic Disorders. There is presently enough information to illustrate some of the problems with testing for polygenic disorders. For example, as shown in Chapter 10, three genes affecting dopamine metabolism were found to show a highly significant correlation with ADHD, stuttering, conduct disorder, oppositional defiant, and other behaviors. Despite this, statistical tests indicated that these three genes accounted for less than 12% of the variation in the ADHD score in the several hundred subjects tested. Thus, it was not surprising that some individuals carrying all three gene markers had no ADHD, and some who carried none of the markers had severe ADHD.

In other words, since only a portion of the genes and factors responsible for ADHD had been identified, the genetic tests could not be used to diagnose ADHD, could not be used to identify everyone at risk to develop ADHD, and certainly could not be used in prenatal diagnosis. However, this does not mean that other genes that will prove to play a role in ADHD, conduct disorder, depression, impulsivity, alcoholism, drug abuse, and other behavioral disorders will not be found. Given the rapid rate of discovery of genetic factors in disease, it is very likely that in five to ten years the genes will be identified that explain over 90% of the variation in these disorders. Testing for fifty different genes may identify individuals with a two-, five-, ten-, twenty-, or fiftyfold increased risk for developing one or more of the impulsive, compulsive, addictive, and learning disorders discussed in this book. In other words, in ten years the genes responsible for the trends reported will have all been identified. It is also likely that if the hypothesis of this book turns out to be correct, given the necessary funding, the genes that are undergoing the most rapid rate of selection will also have been identified.

Into the future. In the not-too-distant future I predict that a) the programs outlined in the previous chapter, based on "no genetic testing," while helpful, will be shown to be inadequate to the task of decreasing the rate of selection of the genes in question, and b) the selective forces identified, such as the differential age at the birth of the first child, will still be as dramatic, or even more so, than they are now. In other words, the programs suggested in the previous chapter can slow the rate of gene selection but cannot stop it, and certainly cannot reverse it.

So, now what do we do? There are several possibilities.

1. Do nothing. Although I cannot predict the future, I will rule this out as being unlikely. It is not in character for the human race to identify a major threat that could result in its eventual destruction, and just sit by without making some attempts to eliminate the problem.

2. Compulsory testing and compulsory eugenic programs. This I also categorically rule out. One episode of Nazism per millennium is one episode too many.

3. Voluntary testing. I see this as both the only viable option and also a very effective one. Since procedures for the simultaneous testing of multiple mutations are already available, it is clear that in the future it will be possible to test for all of the mutant genes involved in these disorders with a single, inexpensive blood test. This means it will also be possible to perform prenatal tests for these genes. This would allow virtually every couple the voluntary option of being tested to determine which of the relevant mutant genes they carry. Couples with a family history of learning disabilities, aggressive or abusive behaviors, ADHD, Tourette syndrome, autism, severe depression, suicide, or related disorders, may be especially motivated to have themselves tested. This would give some indication about their risk of having affected children. If the risk was very high, or if they already had a severely affected child, some may voluntarily choose to have no more children. If they wanted children they would next have the voluntary option to have prenatal diagnosis performed.

It is abundantly obvious that prenatal diagnosis in polygenic disorders is different than prenatal diagnosis in single gene disorders. In the latter, if the gene is present, the disorder will be present. However, in polygenic inheritance, we are reduced to talking about relative risks or odds ratios. Since such information is less precise, how couples act on the information will vary widely. One couple may find their unborn child has a twentyfold increased risk of having an impulsive, aggressive conduct disorder, and yet choose to carry the pregnancy to term, while another couple may choose to terminate a pregnancy where their child has only a fivefold increased risk for the same disorder. Yet another couple may be having a child with a fiftyfold increased risk and yet they either choose not to be tested, or don't care, or don't know testing is available. These are the natural variations that occur with a voluntary, optional, free-choice system. The reason this approach may be successful in eventually causing the selection of these genes to level off, or even reverse itself, is that almost no one, or relatively few couples, will choose to terminate pregnancies where the risk for these conditions is less than average, average, or increased only slightly. Thus, if enough couples choose to be tested, even if many choose not to terminate a pregnancy despite a high risk, many will. Over a period of time, this will allow a totally voluntary program to be successful.

None of this would be possible, however, and the hypothesized gene selection could continue unabated, if the problem is ignored or simply wished away. It will be important to periodically initiate new sets of studies such as the NLSY and include obtaining blood samples on the subjects studied. It will be important to fund the research needed to identify the genes involved and determine if they are increasing in frequency.

Summary – In the near future, all the genes involved in increasing one's risk of developing alcoholism, drug abuse, ADHD, aggressive and impulsive conduct and related disorders, will have been identified, and the relative risk of developing these disorders for individuals with varying combinations of the mutant genes will be known. If the programs based on "no genetic testing" have been unsuccessful, and if research shows that continued gene selection is occurring, doing nothing is not a viable course of action. The opposite extreme of compulsory testing and compulsory eugenics is even less acceptable. This leaves a middle ground consisting of a voluntary program in which each couple has the option to have themselves tested and, if they still wish to have children, utilize prenatal diagnosis. If sufficient numbers of couples opt for testing, over time this approach will be successful, despite wide variations in how couples act on that information, since, on average, more low risk pregnancies will be continued than high risk pregnancies.

Summary

Many different studies have documented an increase in the frequency and a decrease in the age of onset of a wide range behavioral disorders, including depression, suicide, alcohol and drug abuse, anxiety, ADHD, conduct disorder, autism, and learning disorders in the second half of the twentieth century. All of these disorders have a strong genetic component. The usual explanation of these trends has been that they are the result of an increasingly fast-paced and technologically complex society. I have suggested that converse is true – that the increasingly complex society is selecting for the genes causing these behaviors.

The major source of this selective force comes from the increasing numbers of individuals attending college and graduate school in the latter half of the twentieth century. There is a very high positive correlation between IQ, the number of years of attending school, and the age at the birth of the first child. Individuals with the highest IQs continue in school the longest, have their children at older ages, have fewer children, and have the lowest frequency of learning disorders and impulsive, compulsive, aggressive, and addictive behaviors. By contrast, those who drop out of high school have children earlier, have more children, have lower IQs, more learning disorders, and a higher frequency of impulsive, compulsive, aggressive, and addictive behaviors. To the extent that each of these conditions is genetically controlled, this differential in the age at the birth of the first child will exert a powerful selective advantage on the genes involved. The magnitude of this selective force has increased dramatically in the second half of the twentieth century.

Because of their marked effect on reproductive behavior, the learning disorders and other impulsive, compulsive, aggressive, and addictive disorders have the ability to result in progressive and permanent changes in the frequency of the associated genes – potentially leading to the genetic meltdown of the species. A number of approaches to slowing, stopping, or reversing the selection for these genes is suggested. An important priority is funding the research into identifying the genes involved. A number of straight-forward approaches to determining if these genes are actually increasing in frequency, are proposed.

It is pointed out that the hypothesis is not racist, since, if anything, the selective forces seem to be accelerating more rapidly for Whites than minorities. Nazi-type coercive policies are to be assiduously avoided. It is suggested the ultimate solution will come from more research to provide a precise identification and effective treatment of the problems, and from a totally voluntary approach by which individuals take charge of their own reproduction.

One of the most dangerous aspects of this "hidden" epidemic is that the effects are slow and subtle. Like the gradually tightening coils of a python, the danger may not be appreciated until it is too late.

References

1 16th Annual Report to Congress on the Implementation of The Individuals with Disabilities Education Act. (1994). U.S. Dept. Education, Washington, D.C.

2 Abrahamse AF, Morrison PA, and Waite LJ (1988). *Beyond Stereotypes: Who Becomes a Single Teenage Mother?* RAND Corp, Santa Monica, CA.

3 Abrahamse AF, Morrison PA, and Waite LJ (1988). Teenagers willing to consider single parenthood: Who is at greatest risk? *Family Planning Perspectives* 20:13-18.

4 Achenbach TM (1991). Manual for the Teacher's Report Form and 1991 Profile. University of Vermont Dept. Psychiatry, Burlington, VT.

5 Achenbach TM (1991). *Manual for the Child Behavior Checklist/4-18 and 1991 Profile.* University of Vermont Dept.Psychiatry, Burlington, VT.

6 Achenbach TM and Howell CT (1993). Are American children's problems getting worse? A 13-year comparison. *J Am Acad Child Adolesc Psychiatry* 32:1145-1154.

7 Allebeck P and Allgulander C (1990). Psychiatric diagnoses as predictors of suicide. A comparison of diagnoses at conscription and in psychiatric care in a cohort of 50,465 young men. *Br J Psychiatry* 157:339-344.

8 Allebeck P and Allgulander C (1990). Suicide among young men: Psychiatric illness, deviant behavior and substance abuse. *Acta Psychiatr Scand* 81:565-570.

9 Anastasi A (1956). Intelligence and family size. *Psychological Bulletin* 53:187-209.

10 Angst J, Degonda M, and Ernst C (1992). The Zurich Study: XV. Suicide attempts in a cohort from age 20 to 30. *Eur Arch Psychiatry Clin Neurosci* 242:135-141.

11 Asherson P, Walsh C, Williams J, Sargent M, Talor C, Clements A, Gill M, Owen M, and McGuffin P (1994). Imprinting and anticipation. Are they relevant to genetic studies of schizophrenia? *Br J Psychiatry* 164:619-624.

12 Bachman JG, Johnston LD, and O'Malley PM (1981). Smoking, drinking, and drug use among American high school students: Correlates and trends, 1975-1979. *Am J Public Health* 71:59-69.

13 Baldwin W (1983). Trends and consequences of adolescent childbearing in the United States. Statement prepared for the Select Committee on Children, Youth and Families, July 20.

14 Barkley RA, Fischer M, Edelbrock CS, and Smallish L (1990). The adolescent outcome of hyperactive children diagnosed by research criteria: I. An 8-year prospective follow-up study. *J Am Acad Child Adolesc Psychiatry* 29:546-557.

15 Bassett AS and Honer WG (1994). Evidence of anticipation in schizophrenia. *Am J Hum Genet* 54:864-870.

15a Benjamin J, Patterson C, Greenberg B, Murphy DL, and Hamer D (1995). *Dopamine D_4* receptor gene association with normal personality traits. *Psychiat Genet* 5:S36.

16 Biederman J, Faraone SV, Spencer T, Wilens T, Norman D, Lapey KA, Mick E, Lehman BK, and Doyle A (1993). Patterns of psychiatric comorbidity, cognition, and psychosocial functioning in adults with attention deficit hyperactivity disorder. *Am J Psychiatry* 150:1792-1798.

17 Biederman J, Newcorn J, and Sprich S (1991). Comorbidity of attention deficit hyperactivity disorder with conduct, depressive, anxiety, and other disorders. *Am J Psychiatry* 148:564-577.

18 Bloom DE and Pebley AR (1982). Voluntary childlessness: A review of the evidence and implications. *Population Research and Policy Review* 1:203-224.

19 Bloom DE and Trussell J (1984). What are the determinants of delayed childbearing and permanent childlessness in the United States? *Demography* 21:591-611.

20 Blum K, Noble EP, Sheridan PJ, Montgomery A, Ritchie T, Jadadeeswaran P, Nogami H, Briggs AH, and Cohn JB (1990). Allelic association of human *dopamine D_2* receptor gene in alcoholism. *J Am Med Assn* 263:2055-2059.

21 Bohman M, Cloninger CR, Sigvardsson S, and von Knorring AL (1982). Predisposition to

petty criminality in Swedish adoptees. I. Genetic and environmental heterogeneity. *Arch Gen Psychiatry* 39:1233-1241.

22 Bouchard TJJr (1981). Familial studies of intelligence: A review. *Science* 212:1055-1059.

23 Bouchard TJJr, Lykken DT, McGue M, Segal NL, and Tellegen A (1990). Sources of human psychological differences: The Minnesota study of twins reared apart. *Science* 250:223-228.

24 Brennan PA, Mednick BR, and Mednick SA (1993). Parental psychopathology, congenital factors, and violence. In: *Mental Disorder and Crime*, edited by Hodgins S, Sage Publications, Newbury Park, CA. pp. 244-261.

25 Brodzinsky DM and Steiger C (1991). Prevalence of adoptees among special education populations. *J Learn Disabi* 24:484-489.

26 Bulusu L and Alderson M (1984). Suicides 1950-1982. *Popul Trends* 35:11-17.

27 Bumpass LL, Rindfuss RR, and Janosik RB (1978). Age and marital status at first birth and the pace of subsequent fertility. *Demography* 15:75-86.

28 Burd L, Fisher WW, Kerbeshian J, and Arnold ME (1987). Is development of Tourette disorder a marker for improvement in patients with autism and other pervasive developmental disorders. *J Am Acad Child Adolesc Psychiatry* 26:162-165.

29 Burd L and Kerbeshian J (1988). Familial pervasive development disorder, Tourette disorder, and hyperlexia. *Neurosci Biobehav Rev* 12:233-234.

30 Burke KC, Burke JD, Rae DS, and Reiger DA (1991). Comparing age at onset of major depression and other psychiatric disorders by birth cohorts in five US community populations. *Arch Gen Psychiatry* 48:789-795.

30a Cadoret BJ, Yates WR, Troughton E, Woodsworth G, and Stewart MA (1995). Genetic-environmental interaction in the genesis of aggressivity and conduct disorders. *Arch Gen Psychiatry* 52: 916-925.

31 Campbell SB and Werry JS (1986). Attention deficit disorder (hyperactivity). In: *Psychopathic Disorders of Childhood*, edited by Quay HC and Werry JS. Wiley, New York. pp. 1-35.

32 Canino GJ, Bird HR, Shrout PE, Rubio-Stipec M, Bravo M, Martinez R, Sesman M, and Guevara LM (1987). The prevalence of specific psychiatric disorders in Puerto Rico. *Arch Gen Psychiatry* 44:727-735.

33 Cantwell DP (1972). Psychiatric illness in the families of hyperactive children. *Arch Gen Psychiatry* 27:414-417.

34 Cardon LR, Smith SD, Fullker DW, Kimberling WJ, Pennington BF, and DeFries JC (1994). Quantitative trait locus for reading disability on chromosome 6. *Science* 266:276-279.

35 Carmelli D, Swan GE, Robinette D, and Fabsitz R (1992). Genetic influence on smoking - A study of male twins. *New Eng J Med* 327:829-833.

36 Carter C (1961). Promising families: Some conclusions. *Eugenics Rev* 52:197-200.

37 Caskey, CT, Pizzuti, A, Fu, YH, Fenwick, RG,Jr. and Nelson, DL (1992). Triplet repeat mutations in human disease. *Science* 256:784-789.

38 Cattell RB (1936). Is our national intelligence declining? *Eugenics Rev* 28:181-203.

39 Cattell RB (1937). *The Flight for Our National Intelligence*. King & Sons, London.

40 Cattell RB (1974). Differential fertility and normal selection for IQ: Some required conditions in their investigation. *Social Biology* 21:168-177.

41 Cavalli-Sforza LL, Menozzi P, and Piazza A (1993). *The History and Geography of Human Genes*. Princeton University Press, Princeton, N.J.

42 Chamie M, Eisman JD, Forrest JD, Orr M, and Torres A (1982). Factors affecting adolescents' use of family planning clinics. *Family Planning Perspectives* 14:126-139.

43 Chess S and Hassibi M (1970). Behavior deviations in mentally retarded children. *J Am Acad Child Psychiatry* 9:282-297.

44 Cloninger CR and Gottesman II (1987). Genetic and environmental factors in antisocial behavior disorders. In: *The Causes of Crime*, edited by Mednick SA, Moffitt TE, and Stack SA. Cambridge Univ. Press, New York. pp. 92-109.

45 Cloninger CR and Reich T (1983). Genetic heterogeneity in alcoholism and sociopathy. Res Publ Assoc Res *Nerv Ment Dis* 60:145-166.

46 Cloninger CR, Reich T, Sigvardsson S, von Knorring AL, and Bohman M (1988). Effects of changes in alcohol use between generations on the inheritance of alcohol abuse. In: *Alcoholism: Origins and Outcome*, edited by Rose RM and Barrett JE. Raven Press, New York. pp. 49-74.

47 Cloninger CR, Sigvardsson S, Bohman M, and vonKnorring AL (1982). Predisposition to petty criminality in Swedish adoptees. II. Cross-fostering analysis of gene-environment interaction. *Arch Gen Psychiatry* 39:1242-1247.

48 Cloninger CR, von Knorring AL, Sigvardsson S, and Bohman M (1986). Symptom patterns and causes of somatization in men: II. Genetic and environmental independence from somatization in women. *Genet Epidemiol* 3:171-185.

49 Coale A (1965). Comment. *Eugenics Quarterly* 12:58.

50 Comings BG and Comings DE (1987). A controlled study of Tourette syndrome. V. Depression and mania. *Am J Hum Genet* 41:804-821.

51 Comings DE (1987). A controlled study of Tourette syndrome. VII. Summary: a common genetic disorder causing disinhibition of the limbic system. *Am J Hum Genet* 41:839-866.

52 Comings DE (1990). *Tourette Syndrome and Human Behavior.* Hope Press, Duarte, CA. pp. 1-828.

53 Comings DE (1993). A genetic hypothesis for the secular increases in psychiatric disorders. *Psychiat Genet* 3:176.

54 Comings DE (1994). The role of genetic factors in human sexual behavior based on studies of Tourette syndrome and ADHD probands and their relatives. *Am J Med Gen (Neuropsych Genet)* 54:227-241.

55 Comings DE (1994). Genetic factors in substance abuse based on studies of Tourette syndrome and ADHD probands and relatives. II. Alcohol abuse. *Drug and Alcohol Dependence* 35:17-24.

56 Comings DE (1995). The role of genetic factors in conduct disorder based on studies of Tourette syndrome and ADHD probands and their relatives. *J Dev Behav Pediatr* 16:142-157.

57 Comings DE (1995). Genetic Mechanisms in Neuropsychiatric Disorders. In: *Handbook of Psychoneurogenetics*, edited by Blum K, Noble EP, Sparks RS, and Sheridan PJ. CRC Press, Boca Raton, FL. p (in press).

58 Comings DE (1995). Genetic factors in depression based on studies of Tourette syndrome and Attention Deficit Hyperactivity Disorder probands and relatives. *Am J Med Gen (Neuropsych Genet)* 60:111-121.

59 Comings DE (1995). Tourette syndrome: A hereditary neuropsychiatric spectrum disorder. *Ann Clin Psychiatry* 6:235-247.

60 Comings DE (1996). *Search for the Tourette Syndrome and Human Behavior Genes*. Hope Press.

61 Comings DE and Comings BG (1984). Tourette's syndrome and attention deficit disorder with hyperactivity: Are they genetically related? *J Am Acad Child Psychiatry* 23:138-146.

62 Comings DE and Comings BG (1987). A controlled study of Tourette syndrome. I. Attention deficit disorder, learning disorders, and school problems. *Am J Hum Genet* 41:701-741.

63 Comings DE and Comings BG (1987). A controlled study of Tourette syndrome. II. Conduct. *Am J Hum Genet* 41:742-760.

64 Comings DE and Comings BG (1987). A controlled study of Tourette syndrome. VI. Early development, sleep problems, allergies, and handedness. *Am J Hum Genet* 41:822-838.

65 Comings DE and Comings BG (1987). A controlled study of Tourette syndrome. IV. Obsessions, compulsions, and schizoid behaviors. *Am J Hum Genet* 41:782-803.

66 Comings DE and Comings BG (1987). A controlled study of Tourette syndrome. III. Phobias and panic attacks. *Am J Hum Genet* 41:761-781.

67 Comings DE and Comings BG (1988). Tourette's syndrome and attention deficit disorder. In: *Tourette's Syndrome and Tic Disorders: Clinical Understanding and Treatment*, edited by Cohen DJ, Bruun RD, and Leckman JF. John Wiley & Sons, New York. pp.120-135.

68 Comings DE and Comings BG (1990). A controlled family history study of Tourette syndrome. I. Attention deficit hyperactivity disorder, learning disorders, and dyslexia. *J Clin Psychiat* 51:275-280.

69 Comings DE and Comings BG (1990). A controlled family history study of Tourette syndrome. III. Other Psychiatric Disorders. *J Clin Psychiat* 51:288-291.

70 Comings DE and Comings BG (1990). A controlled family history study of Tourette syndrome. II. Alcoholism, drug abuse, and obesity. *J Clin Psychiat* 51:281-287.

71 Comings DE and Comings BG (1991). Clinical and genetic relationships between autism-PDD and Tourette syndrome: A study of 19 cases. *Am J Med Genet* 39:180-191.

72 Comings DE and Comings BG (1993). SIDS and Tourette syndrome: Is there an etiologic relationship? *J Dev Physical Disabil* 5:265-279.

73 Comings DE and Comings BG (1993). Comorbid Behavioral Disorders. In: *Tourette Syndrome and Related Disorders*, edited by Kurlan R. Marcel-Decker, New York. pp.111-147.

74 Comings DE, Comings BG, Devor EJ, and Cloninger CR (1984). Detection of major gene for Gilles de la Tourette syndrome. *Am J Hum Genet* 36:586-600.

75 Comings DE, Comings BG, Muhleman D, Dietz G, Shahbahrami B, Tast D, Knell E, Kocsis P, Baumgarten R, Kovacs BW, Levy DL, Smith M, Kane JM, Lieberman JA, Klein DN, MacMurray, Tosk J, Sverd J, Gysin R, and Flanagan S (1991). The dopamine D_2 receptor locus as a modifying gene in neuropsychiatric disorders. *J Am Med Assn* 266:1793-1800.

76 Comings DE, Ferry L, Bradshaw-Robinson S, Burchette R, Chiu C, and Muhleman D (1996). The *Dopamine D2* Receptor *(DRD2)* Gene: A Genetic Risk Factor in Smoking. *Pharmacogenetics. 6:73-79.*

77 Comings DE, Himes JA, and Comings BG (1990). An epidemiological study of Tourette syndrome in a single school district. *J Clin Psychiat* 51:463-469.

78 Comings DE, Muhleman D, Ahn C, Gysin R, and Flanagan SD (1994). The *dopamine D2* receptor gene: A genetic risk factor in substance abuse. *Drug and Alcohol Dependence* 34:175-180.

79 Comings DE, Rosenthal RJ, Lesieur HR, Rugle L, Muhleman D, Chiu C, Dietz G, and Gade R (1996). A study of the *dopamine D2* receptor gene in pathological gambling. *Pharmacogenetics (in press).*

80-81 Comings DE, Wu H, Chiu C, Ring RH, Gade R, Ahn C, MacMurray JP, Dietz G, and Muhleman D (1996). Polygenic inheritance of Tourette syndrome, stuttering, attention deficit hyperactivity, conduct and oppositional defiant disorders: The additive and subtractive effect of three dopaminergic genes - *DRD2, DβH,* and *DAT1. Am J Med Gen (Neuropsych Genet)* 67:264-288.

82 Cook RC (1951). *Human Fertility: The Modern Dilemma.* Sloane, New York.

83 Coryell W, Endicott J, and Keller M (1992). Major depression in a nonclinical sample: demographic and clinical risk factors for first onset. *Arch Gen Psychiatry* 49:117-125.

84 Cross-National Collaborative Group, (1992). The Changing Rate of Major Depression. *J Am Med Assn* 268:3098-3105.

85 Crow TJ (1986). Secular changes in affective disorder and variations in the psychosis gene. *Arch Gen Psychiatry* 43:1013-1014.

86 Crowe RR (1972). The adopted offspring of women criminal offenders: A study of their arrest records. *Arch Gen Psychiatry* 27:600-603.

87 Crowe RR (1974). An adoption study of antisocial personality. *Arch Gen Psychiatry* 31:785-791.

88 Dash L (1989). *When Children Want Children: An Inside Look at the Crisis of Teenage Childbearing.* Morrow, New York.

89 Dawson DA (1986). The effects of sex education on adolescent behavior. *Family Planning Perspectives* 18:162.

90 Deb S and Pradad KBG (1994). The prevalence of autistic disorder among children with a learning disability. *Br J Psychiatry* 165:395-399.

91 *Diagnostic and Statistical Manual of the American Psychiatric Assn. IV.* American Psychiatric Assn, Washington, DC (1994).

92 Diamond J (1995). Easter's End. *Discover* 16:62-69.

93 Dinwiddie SH and Reich T (1991). Epidemiological perspectives on children of alcoholics. *Recent Dev Alcohol* 9:287-299.

94 Donovan JE and Jessor R (1985). Structure of problem behavior in adolescence and young adulthood. *J Consult Clin Psychol* 53:890-904.

95 Donovan JE, Jessor R, and Jessor L (1983). Problem drinking in adolescence and young adulthood: A follow-up study. *J Stud Alcohol* 44:109-137.

96 Dornbush MP and Pritt SK (1995). *Teaching the Tiger. A Handbook for Individuals Involved in the Education of Students with Attention Deficit Disorder, Tourette Syndrome or Obsessive-Compulsive Disorder.* Hope Press, Duarte, CA. pp.1-250.

97 Edelbrock C, Castello AJ, and Kessler MD (1984). Empirical corroboration of attention deficit disorder. *J Am Acad Child Psychiatry* 23:285-290.

98 Edwards L, Steinman M, Arnold K, and Hakanson E (1980). Adolescent pregnancy prevention services in high school clinics. *Family Planning Perspectives* 12:6-14.

99 Eibl-Eibesfeldt I (1986). Intelligence and selection. *Behav Brain Sci* 9:191-192.

100 Elliott D and Huizinga D (1984). *The Relationship between Delinquent Behavior and ADM Problems.* Behavioral Research Institute, Boulder, CO.

101 Elster AB, Ketterlinus R, and Lamb ME (1990). Association between parenthood and problem behavior in a national sample of adolescents. *Pediatrics* 85:1044-1050.

102 Elster AB, Lamb ME, Peters L, Kahn J, and Tavare J (1987). Judicial involvement and conduct problems of fathers of infants born to adolescent mothers. *Pediatrics* 79:230-234.

103 Elster AB, Lamb ME, and Tavare J (1987). Association between behavioral and school problems and fatherhood in a national sample of adolescent youths. *J Pediatr* 111:932-936.

104 Ensminger ME (1990). Sexual activity and problem behaviors among black urban adolescents. *Child Dev* 61:2032-2046.

105 Erenberg G, Cruse RP, and Rothner AD (1986). Tourette syndrome: an analysis of 200 pediatric and adolescent cases. *Clev.Clin Q* 53:127-131.

106 Erenberg G, Cruse RP, and Rothner AD (1987). The natural history of Tourette syndrome: a follow-up study. *Ann Neurol* 22:383-385.

107 Falconer DS (1981). *Introduction to Quantitative Genetics, 2nd Edition.* Longman, New York. pp.1-365.

108 Faraone SV, Biederman J, Chen WJ, Kritcher B, Keenan K, Moore C, Sprich S, and Tsuang MT (1992). Segregation analysis of attention deficit hyperactivity disorder. *Psychiat Genet* 2:257-275.

109 Faraone SV, Biederman J, Keenan K, and Tsuang MT (1991). Separation of DSM-III attention deficit disorder and conduct disorder – Evidence from a family genetic study of American child psychiatric patients. *Psychol Med* 21:109-121.

110 Farrington DP (1978). The family backgrounds of aggressive youths. In: *Aggression and Antisocial Behavior in Childhood and Adolescence*, edited by Hersov LA and Berger M. Pergamon Press, London. pp.73-94.

111 Farrington DP (1989). Early predictors of adolescent aggression and adult violence. *Violence Vict* 4:79-100.

112 Farrington DP, Loeber R, and van Kammen WB (1987). Long-term criminal outcomes of hyperactivity – impulsivity – attention deficit and conduct problems in childhood. Meetings of the Society for Life History Research, St. Louis, MO.

113 Ferry LH, Robbins AS, Scariti PD, Masterson A, Abby DD, Everett DD, and Burchette RJ (1993). Enhancement of smoking cessation using the antidepressant bupropion hydrochloride. *Circulation 84* (Abst supp):I-671.

114 Flynn JR (1984). The mean IQ of Americans: Massive gains 1932 to 1978. *Psychological Bulletin* 95:29-51.

115 Flynn JR (1987). Massive IQ gains in 14 nations: What IQ tests really measure. *Psychological Bulletin* 101:171-191.

116 Forrest JD, Hermalin A, and Henshaw S (1981). The impact of family planning clinic programs on adolescent pregnancy. *Family Planning Perspectives* 13:109-116.

117 Fowler RC, Rich CL, and Young D (1986). San Diego Suicide Study. II. Substance Abuse in Young Cases. *Arch Gen Psychiatry* 43:962-965.

118 Freedman DS, Shear CL, Burke GL, Srinivasan SR, Webber LS, Harsha DW, and Berenson GS (1987). Persistence of juvenile-onset obesity over eight years: The Bogalusa Heart Study. *Am J Public Health* 77:588-592.

119 Freedman DX (1991). Foreward. In: *Psychiatric Disorders in America*, edited by Robins LN and Regier DA. The Free Press, New York. pp.1-449.

120 Furstenberg F, Brooks-Gunn J, and Morgan J (1987). Adolescent mothers and their children in later life. *Family Planning Perspectives* 19:142-151.

121 Furstenberg FF, Moore KA, and Peterson JL (1985). Sex education and sexual experience among adolescents. *Am J Public Health* 75:1331-1332.

122 Garn SM (1985). Continuities and changes in fatness from infancy through adulthood. *Cur Probl Pediatr* 15:1-47.

122a Gelernter J (1995). DRD4 alleles are associated with Tourette's syndrome. *Psychiat Genet* 5:S20.

123 Gershon ES, Hamovit JH, and Guroff JJ (1987). Birth-cohort changes in manic and depressive disorders in relatives of bipolar and schizoaffective patients. *Arch Gen Psychiatry* 44:314-319.

124 Gillberg C, Steffenburg S, and Schaumann H (1991). Is autism more common now than ten years ago? *Br J Psychiatry* 158:403-409.

125 Gittelman R, Mannuzza S, Shenker R, and Bonagura N (1985). Hyperactive boys almost grown up. *Arch Gen Psychiatry* 42:937-947.

126 Goldney RD and Katsikitis M (1983). Cohort analysis of suicide rates in Australia. *Arch Gen Psychiatry* 40:71-74.

127 Hagin RA, Beecher R, Pagano G, and Kreeger H (1982). Effects of Tourette syndrome on learning. *Adv Neurol* 35:323-328.

128 Hallowell EM and Ratey JJ (1994). *Driven to Distraction.* Pantheon Books, New York, N.Y. pp.1-318.

129 Hayes CD (1987). *Risking the Future. Adolescent Sexuality, Pregnancy, and Childbearing.* National Academic Press, Washington, DC. pp.1-337.

130 Heath AC, Berg K, Eaves LJ, Solaas MH, Corey LA, Sundet J, Magnus P, and Nance WE (1995). Education policy and the heritability of educational attainment. *Nature* 314:734-736.

131 Herrnstein RJ and Murray C (1994). *The Bell Curve.* The Free Press, New York. pp.845.

132 Higgins JV, Reed EW, and Reed SC (1962). Intelligence and family size: A paradox resolved. *Social Biology* 9:84-90.

132a Hillman RE, Kanafani N, Bright JD, Spence MA, and Miles JH (1995). Changing prevalence of autism in Missouri: Effect of a comprehensive state autism project. *Am J Hum Genet* 57: A164.

133 Hinshaw SP (1992). Externalizing behavioral problems and academic underachievement in childhood and adolescence: Casual relationships and underlying mechanisms. *Psychological Bulletin* 111:127-155.

134 Hirschi T and Hindelang MJ (1977). Intelligence and delinquency: A revisionist review. *Am Socialog Rev* 42:571-587.

135 Howell DC, Huessy HR, and Hassuk B (1085). Fifteen-year follow-up of a behavioral history of attention deficit disorder. *Pediatrics* 76:185-190.

136 Huba GJ, Wingard JA, and Bentler PM (1981). A comparison of two latent variable causal models for adolescent drug use. *J Pers Soc Psychol* 40:180-193.

137 Hutchings B and Mednick SA (1975). Registered criminality in adoptive and biological parents of registered male criminal adoptees. In: *Genetic Research in Psychiatry*, edited by Fieve RR, Rosenthal D, and Brill H. Johns Hopkins University Press, Baltimore, MD. pp.105-116.

138 Jacobs BL, Gannon PJ, and Azmita EC (1984). Atlas of serotonergic cell bodies in the cat brainstem: an immunocytochemical analysis. *Brain Res Bull* 13:1-31.

139 Jamison KR (1993). *Touched with Fire. Manic-depressive illness and the artistic temperament.* The Free Press, New York. pp.1-370.

140 Jessor RL and Jessor SL (1977). *Problem behavior and psychological development: A longitudinal study of youth.* Academic Press, New York.

141 Johnston LD, O'Malley PM, and Eveland LK (1978). Drugs and delinquency: A search for casual connection. In: *Longitudinal Research on Drug Use: Empirical Findings and Methodological Issues*, edited by Kandel DB. Hemisphere Publishing, Washington, D.C.

142 Kaminer RK and Cohen HJ (1995). Introduction. *J Dev Behav Pediatr* 16 (suppl):S1-S2.

143 Kandel DB (1980). Drug and drinking behavior among youth. *Ann Rev Sociol* 6:235-285.

144 Kandel E, Knop J, Mednick S et al., (1988). IQ as a protective factor for subjects at high risk for antisocial behavior. *J Consult Clin Psychol* 52:224-226.

145 Kaplan HB (1975). *Self-attitudes and deviant behavior.* Goodyear, Pacific Palisades, CA.

146 Kaplan HB (1977). Antecedents of deviant responses: Predicting from a general theory of deviant behavior. *J Youth & Adolesc* 6:89-101.

147 Kaplan HB (1978). Self-attitudes and multiple modes of deviance. In: *Drugs and Suicide: When Other Coping Strategies Fail*, edited by Lettieri D. Sage, Beverly Hills, CA. pp.75-116.

148 Karno M, Hough RL, Burmam A, Escobar JI, Timbers DM, Santana F, and Boyd JH (1987). Lifetime prevalence of specific psychiatric disorder among Mexican-Americans and Non-Hispanic Whites in Los Angeles. *Arch Gen Psychiatry* 44:695-701.

149 Kessler RC, McGonagle KA, Zhao S, Nelson CB, Hughes M, Eshleman S, Wittchen H-U, and Kendler KS (1994). Lifetime and 12-month prevalence of DSM-III-R psychiatric disorders in

the United States. Results from the National Comorbidity Survey. *Arch Gen Psychiatry* 51:8-19.

150 Kirk D (1969). The genetic implications of family planning. *J Med Education* 44 (Supplement 2):80-83.

151 Kisker EE (1985). Teenagers talk about sex, pregnancy, and contraception. *Family Planning Perspectives* 17:83.

152 Klerman GL (1976). Age and clinical depression: Today's youth in the 21st century. *J Gerontology* 31:318-323.

153 Klerman GL (1978). Affective Disorders. In: *The Harvard Guide to Modern Psychiatry*, edited by Nicholi AM. Harvard University Press, Cambridge, Mass. pp.253-281.

154 Klerman GL (1979). The age of melancholy. *Psychology Today* 12:36-42.

155 Klerman GL (1988). The current age of youthful melancholia. Evidence for increase in depression among adolescents and young adults. *Br J Psychiatry* 152:4-14.

156 Klerman GL, Lavori PW, Rice J et. al. (1985). Birth cohort trends in rates of major depressive disorder among relatives of patients with affective disorder. *Arch Gen Psychiatry* 42:689-693.

157 Klerman GL and Weissman MM (1989). Increasing rates of depression. *J Am Med Assn* 261:2229-2235.

158 Knell E and Comings DE (1993). Tourette syndrome and attention deficit hyperactivity disorder: Evidence for a genetic relationship. *J Clin Psychiat* 54:331-337.

159 Koenig MA and Zelnik M (1982). Repeat pregnancies among metropolitan area teenagers: 1971-1979. *Family Planning Perspectives* 14:341-344.

160 Kominski R (1990). Estimating the national high school dropout rate. *Demography* 27:303-311.

161 Kovacs M, Krol RSM, and Voti L (1994). Early onset psychopathology and the risk of teenage pregnancy among clinically referred girls. *J Am Acad Child Adolesc Psychiatry* 33:106-113.

162 Kurlan R, Whitemore D, Irvine C, McDermott MP, and Como PG (1994). Tourette's syndrome in a special education population: A pilot study involving a single school district. *Neurology* 44:699-702.

163 Lahey BB, Strauss CC, and Frame CL (1984). Are attention deficit disorders with and without hyperactivity similar or dissimilar disorders? *J Am Acad Child Psychiatry* 23:302-309.

164 Lavori PW, Klerman GL, Keller MB, Reich T, Rice J, and Endicott J (1987). Age-period-cohort analysis of secular trends in onset of major depression: Findings in siblings of patients with major affective disorder. *J Psychiatr Res* 21:23-35.

165 Lee KC, Kowak YS, and Rhee H (1987). The national epidemiological study of mental disorders in Korea. *J Korean Med Sic* 2:19-34.

166 Lerer RJ (1987). Motor tics, Tourette syndrome, and learning disabilities. *J Learn Disabil* 20:266-267.

167 Lerner B (1983). Test scores as measures of human capital and forecasting tools. *J Social Econ Political Stud* 11:131-159.

168 Liebmann GW (1993). The AFDC conundrum: A new look at an old institution. *Social Work* 38:36-43.

169 Loeber R (1982). The stability of antisocial and delinquent child behavior: A review. *Child Dev* 53:1431-1446.

170 Loeber R, Green SM, Keenan K, and Lahey BB (1995). Which boys will fare worse? Early predictors of the onset of conduct disorder in a six-year longitudinal study. *J Am Acad Child Adolesc Psychiatry* 34:499-509.

171 Loeber R and Stouthamer-Loeber M (1987). Prediction. In: *Handbook of juvenile delinquency*, edited by Quay HC. Wiley, New York. pp.325-382.

172 Loewy AD (1981). Raphe pallidus and raphe obscurus projections to the intermediolateral column in the rat. *Brain Res* 222:129-132.

173 Loney J, Kramer J, and Milich R 1981. The hyperkinetic child grows up: Predictors of symptoms, delinquency, and achievement at follow-up. In: *Psychosocial Aspects of Drug Treatment for Hyperactivity*, edited by Gadow K and Loney J. Westview Press, Boulder, CO. pp.381-415.

174 Loney J, Whaley-Klahn M.A, Koiser T, and Conboy J (1983). Hyperactive boys and their brothers at 21: Predictors of aggressive and antisocial outcomes. In: *Prospective Studies of Crime*

and Delinquency, edited by VanDusen KT and Mednick SA. Kluwer Academic Publishers Group, Boston. pp.181-206.

174a Lyons MJ, True W-R, Elsen SA, Goldberg J, Meyer JM, Farone SV, Evans LJ, Tsuang MT. Differential heritability of adult and juvenile antisocial traits. *Arch Gen Psychiatry* 52: 906-915.

175 Lynn R and Hampson S (1986). Further evidence for secular increases in intelligence in Britain, Japan, and the United States. *Behav Brain Sci* 9:203-204.

176 Mannuzza S, Klein RG, Bonagura N, Malloy P, Giampino TL, and Addalli KA (1991). Hyperactive boys almost grown up. *Arch Gen Psychiatry* 48:77-83.

177 Mannuzza S, Klein RG, Konig PH, and Giampino TL (1989). Hyperactive boys almost grown up. IV. Criminality and its relationship to psychiatric status. *Arch Gen Psychiatry* 46:1073-1091.

178 Mao Y, Hasselback P, Davies JW, Nichol R, and Wigle DT (1990). Suicide in Canada: A epidemiological assessment. *Can J Public Health* 81:324-328.

179 Marsiglio W and Mott F (1986). The impact of sex education on sexual activity, contraceptive use, and premarital pregnancy among American teenagers. *Family Planning Perspectives* 18:151.

179a Masters RD, Greolotti DJ, Hone BT, Gonzalez D, and Jones DJr (1995). Neurotoxicity and violence. (submitted for publication).

180 Maughan B and Pickles A (1990). Adopted and illegitimate children growing up. In: *Straight and devious pathways from childhood to adulthood*, edited by Robins LN and Rutter M Cambridge Univ.Press, Cambridge, England. pp.36-61.

181 McCall RB and Clement ME (1994). Role of serotonin1A and serotonin2 receptors in the central regulation of the cardiovascular system. *Pharmacol Rev* 46:231-243.

182 McConaughly SH and Ritter DE (1986). Social consequences and behavioral problems of learning disabled boys aged 6 to 11. *J Learn Disabil* 19:39-45.

183 McGee R, Williams S, and Silva PA (1984). Behavioral and developmental characteristics of aggressive, hyperactive, and aggressive-hyperactive boys. *J Am Acad Child Psychiatry* 23:270-279.

184 McGue M, Bouchard TJJr, Iacono WG, and Lykken DT (1993). Behavior genetics of cognitive ability: A life-span perspective. In: *Nature, nurture and psychology*, edited by Plomin R and McClearn GE. American Psychological Assn, Washington, D.C. pp.59-76.

185 McInnis MG, McMahon GA, Chase GA, Simpson SG, Ross CA, and DePaulo JRJr (1993). Anticipation in bipolar affective disorder. *Am J Hum Genet* 53:385-390.

185a McLaughlin SA, Valdes MG, Jacobson RM, Wollan PC, Beard CM, Weisman L, and Jacobson S. (1995). Incidence of sudden infant death syndrome in Olmstead County, Minnesota: 1945 through 1992. *Mayo Clin Proc* 70,837-843.

186 Mednick S, Gabrielli WJr., and Hutchings B (1984). Genetic influences in criminal convictions: Evidence from an adoption cohort. *Science* 224:891-894.

187 Mednick SA, Gabrielli WF, and Hutchings B (1983). Genetic influences on criminal behavior: Some evidence from an adoption court. In: *Prospective Studies of Crime and Delinquency*, edited by VanDusen KT and Mednick SA. Kluuer-Nijhoff Pub., Boston.

188 Miller L (1988). Neuropsychological perspectives on delinquency. *Behav Sci Law* 6:409-428.

189 Miller PY and Simon W (1974). Adolescent sexual behavior: Context and change. *Social Problems* 22:58-76.

190, 191 Moffitt TE (1990). Juvenile delinquency and attention deficit disorder: Boys' developmental trajectories from age 3 to age 15. *Child Dev* 61:893-910.

192 Moffitt TE, Gabrielli WF, Mednick SA, and Schlusinger F (1981). Socioeconomic status, IQ, and delinquency. *J Abnormal Psychology* 90:152-156.

193 Moffitt TE and Henry (1989). Neuropsychological assessment of executive functions in self-reported delinquents. *Dev Psychopathology* 1:105-118.

194 Moffitt TE, Mednick SA, and Gabrielli WFJr (1989). Predicting careers of criminal violence: Descriptive data and predispositional factors. In: *Current Approaches to the Prediction of Violence*, edited by Brizer DA and Crowner M. American Psychiatric Press, Washington, D.C.

195 Moffitt TE and Silva PA (1988). Self-reported delinquency, neuropsychological deficit, and history of attention deficit disorder. *J Abnorm Child Psychol* 16:553-569.

196 Mott FL and Haurin RJ (1988). Linkages between sexual activity, and alcohol and drug use among American adolescents. *Family Planning Perspectives* 20:128-136.

197 Murphy GE and Wetzel RD (1980). Suicide risk by birth cohort in the United States, 1949 to 1974. *Arch Gen Psychiatry* 37:519-523.

197a Neuman R, Geller B, Rice J, and Tood R (1995). Differences between first-degree relatives of affectively disordered children and affectively disordered adults. *Psychiat Genet* 5:S28.

198 Noble EP, Blum K, Khalsa ME, Ritchie T, Montgomery A, Wood RC, Fitch RJ, Ozkaragoz T, Sheridan PJ, Anglin MD, Paredes A, Treiman LJ, and Sparkes RS (1993). Allelic association of the *D2 dopamine* receptor gene with cocaine dependence. *Drug Alcohol Depend* 33:271-285.

199 Noble EP, Jeor ST, Ritchie T, Syndulko K, Jeor SC, Fitch RJ, Brunner RL, and Sparkes RS (1994). *D2 dopamine* receptor gene and cigarette smoking: A reward gene? *Medical Hypothesis* 42:257-260.

200 Noble EP, Noble RE, Ritchie T, Grandy DK, and Sparkes RS (1993). Allelic association of the *D2 dopamine* receptor gene with obesity. *Am J Hum Genet* 53:A117.

200a Novick O, Ebstein R, Umansky R, Priel B, Osher Y, and Belmaker RH (1995). D-4 receptor polymorphism associated with personality variation in normals. *Psychiat Genet* 5:S36.

201 Nylander P-O, Engström C, Chotai J, Wahlström J, and Adolfsson R (1994). Anticipation in Swedish families with bipolar affective disorder. *J Med Genet* 31:686-689.

202 O'Donovan MC, Guy C, Craddock N, Murphy KC, Cardino AG, Jones LA, Owen MJ, and McGuffin P (1995). Expanded CAG repeats in schizophrenia and bipolar disorder. *Nature Genet* 10:380-381.

203 O'Quinn AN and Thompson RJJr (1980). Tourette's syndrome: an expanded view. *Pediatrics* 66:420-424.

204 Offord DR, Boyle MH, and Jones BR (1987). Psychiatric disorder and poor school performance among welfare children in Ontario. *Can J Psychiatry* 32:518-525.

205 Offord DR, Last JM, and Barrette PA (1985). A comparison of the school performance, emotional adjustment and skill development of poor and middle class children. *Can J Psychiatry* 76:174-178.

206 Ohishi M, Rakugi H, and Ogihara T (1994). Association between a deletion polymorphism of the angiotensin-converting-enzyme gene and left ventricular hypertrophy. *New Eng J Med* 331:1097.

207 Orr H, Forrest JD, and Lincoln R (1983). *Making choices: Evaluating the health risks and benefits of birth control methods*. Alan Guttmacher Institute, New York.

208 Orr M (1984). Private physicians and the provision of contraceptives to adolescents. *Family Planning Perspectives* 16:83-86.

209 Osborn F (1940). *Preface to Eugenics*. Harper, New York.

210 Osborn F and Bajema C (1972). The eugenics hypothesis. *Social Biology* 19:337-345.

211 Osborne RT (1975). Fertility, IQ, and school achievement. *Psychological Reports* 37:1067-1073.

212 Osgood DW, Johnston LD, O'Malley PM, and Bachman JG (1988). The generality of deviance in late adolescence and early adulthood. *Am Sociolog Rev* 53:81-93.

213 Pauls DL, Hurst CR, Kruger SD, Leckman JF, Kidd KK, and Cohen DJ (1986). Gilles de la Tourette's syndrome and attention deficit disorder with hyperactivity. Evidence against a genetic relationship. *Arch Gen Psychiatry* 43:1177-1179.

214 Pauls DL, Leckman JF, Raymond CL, Hurst CR, and Stevenson JM (1988). A family study of Tourette's syndrome: Evidence against the hypothesis of association with a wide range of psychiatric phenotypes. *Am J Hum Genet* 43:A64.

215 Pennington BF and Smith SD (1983). Genetic influences on learning disabilities and speech and language disorders. *Child Dev* 54:369-387.

216 Pipman EL, Offord DR, and Boyle MH (1994). Relation between economic disadvantage and psychosocial morbidity in children. *Can Med Assoc J* 151:431-437.

217 Popenoe D (1993). American family decline, 1960-1990: A review and appraisal. *J Marriage Family* 55:527-542.

218 Reber M (1992). Mental retardation: Review. *Psychiatr Clin North Am* 15:511-522.

218a Reich T, Begleiter H, Bucholoz K, Crowe R, Hesselbrock V, Nurnberger J, Porjesz B, Schuckt M and Van Eerdewegh P (1995). Familial transmission of alcohol dependence in a contemporary American sample. *Psychiat Genet* 5:S23.

219 Reich T, Cloninger CR, Eerdewegh PV, Rice JP, and Mullaney J (1988). Secular trends in the familiar transmission of alcoholism. *Alcohol Clin Exp Res* 12:458-464.

220 Reiss AJJr and Roth JA (1993). *Understanding and Preventing Violence*. National Acade-

my Press, Washington, D.C. pp. 1-464.

221 Retherford RD and Sewell WH (1988). Intelligence and family size reconsidered. *Social Biology* 35:1-40.

222 Rindfluss RR, Bumpass L, and St John C (1980). Education and fertility: Implications for the roles women occupy. *Am Sociolog Rev* 45:431-447.

223 Rindfuss RR and Sweet JA (1977). *Post-war Fertility Trends and Differentials in the United States*. Academic Press, New York, NY. pp.1-225.

224 Rintahaka PJ and Hirvonen J (1986). The epidemiology of sudden death syndrome in Finland in 1969-1980. *Forensic Sci Int* 30:219-233.

225 Robins LN (1966). *Deviant Children Grown Up*. Williams & Wilkins, Baltimore.

226 Robins LN (1978). Sturdy childhood predictors of adult antisocial behavior: replications from longitudinal studies. *Psychol Med* 8:611-622.

227 Robins LN, Helzer J, Croughan J, and Ratclif KS (1981). National Institutes of Health diagnostic interview schedule. *Arch Gen Psychiatry* 38:381-389.

228 Robins LN, Helzer JE, Weissman MM, Orvaschel H, Gruenberg E, Burke JD, and Reigier DA (1984). Lifetime prevalence of specific psychiatric disorders in three sites. *Arch Gen Psychiatry* 41:949-958.

229 Rutter M, Graham P, and Yule W (1970). *A Neuropsychiatric Study in Childhood. Clinics in Developmental Medicine*. SIMP/Heinemann, London, England.

230 Rutter M, Tizard J, and Whitmore K (1970). *Education, Health and Behavior*. Longman Group Limited, London.

231 Satterfield JH, Hoppe CM, and Schell AM (1982). A prospective study of delinquency in 110 adolescent boys with attention deficit disorder and 88 normal adolescent boys. *Am J Psychiatry* 139:795-798.

232 Saudou F, Amara DA, Dierich A, LeMeur M, Ramboz S, Segu L, Buhot M-C, and Hen R (1994). Enhanced aggressive behavior in mice lacking 5-HT1B receptor. *Science* 265:1875-1878.

233 Schachar R, Rutter M, and Smith A (1981). The characteristics of situationally and perversively hyperactive children: implications for syndrome definition. *J Child Psychol Psychiatry* 22:375-392.

234 Schachter DC, Pless IB and Bruck M (1991). The prevalence and correlates of behavioral problems in learning disabled children. *Can J Psychiatry* 36:323-331.

235 Schunkert H, Hense HW, Holmer SR et. al. (1994). Association between a deletion polymorphism of the angiotensin-converting-enzyme gene and left ventricular hypertrophy. *N Engl J Med* 330:1634-1638.

236 Semrud-Clikeman M, Biederman J, Sprich-Buckminster S, Lehman BK, Faraone SV. and Norman D (1992). Comorbidity between ADDH and learning disability: A review and report in a clinically referred sample. *J Am Acad Child Adolesc Psychiatry* 31:439-448.

237 Shaffer D, Gould M, and Hicks RC (1994). Worsening suicide rate in Black teenagers. *Am J Psychiatry* 151:1810-1812.

238 Shaywitz SE, Escobar MD, Shaywitz BA, Fletcher JM, and Makuch R (1992). Evidence that dyslexia may represent the lower tail of a normal distribution of reading ability. *New Eng J Med* 326:145-150.

239 Shilts Randy (1987). *And the Band Played On: Politics, People and the AIDS Epidemic*. St.Martin, New York, NY pp 630.

240 Sigvardsson S, Bohman M, von K.norring AL and Cloninger CR (1986). Symptom patterns and causes of somatization in men: I. Differentiation of two discrete disorders. *Genet Epidemiol* 3:153-169.

241 Sigvardsson S, Cloninger CR, Bohman M, and von Knorring AL (1982). Predisposition to petty criminality in Swedish adoptees. III. Sex differences and validation of the male typology. *Arch Gen Psychiatry* 39:1248-1253.

242 Silver LB (1981). The relationship between learning disabilities, hyperactivity, distractibility, and behavioral problems. *J Am Acad Child Psychiatry* 20:385-397.

243 Simon GE and VonKorff M (1992). Re-evaluation of secular trends in depression rates. *Arch Gen Psychiatry* 37:511-513.

243a Simon GE, VonKorff M, Ustun TB, Gater R, Gureje O, Sartorius N (1995). Is the lifetime risk of depression actually increasing? *J Clin Epidemiol* 48:1109-1118.

244 Singer HS and Rosenberg LA (1989). Development of behavioral and emotional prob-

lems in Tourette syndrome. *Pediatr Neurol* 5:41-44.

245 Skegg K and Cox B (1991). Suicide in New Zealand 1957-1986: The influence of age, period, and birth cohort. *Aust N Z J Psychiatry* 25:181-190.

245a Slutske WS, Heath AC, Madden PAF, Bucholz KK, Dinwiddle SH, Dunne MP, Statham D, and Martin NG (1995). Sex differences in the etiology of DSM-III-R conduct disorder. *Psychiat Genet* 5:S80.

246 Snyderman M and Rothman S (1988). *The IQ Controversy. The Media and Public Policy.* Transaction Books, New Brunswick, NJ.

247 Solomon MI and Hellon CP (1980). Suicide and age in Alberta, Canada, 1951 to 1977. *Arch Gen Psychiatry* 37:511-513.

248 Stevenson J (1992). Evidence for a genetic etiology in hyperactivity. *Behav Genet* 22:337-344.

249 Stewart MA, Cummings C, Singer S, and Deblois CS (1981). The overlap between hyperactive and unsocialized aggressive children. *J Child Psychol Psychiatry* 22:35-45.

250 Stewart MA, Deblois CS, and Cummings C (1980). Psychiatric disorders in the parents of hyperactive boys and those with conduct disorder. *J Child Psychol Psychiatry* 21:283-292.

251 Stoff SM, Pollock L, Vitiello B, Behar D, and Bridger WH (1987). Reduction of (3H)-imipramine binding sites on platelets of conduct-disordered children. *Neuropsychopharmacol* 1:55-62.

252 Surtees PG and Duffy JC (1989). Suicide in England and Wales 1946-1985: An age-period-cohort analysis. *Acta Psychiatr Scand* 79:216-223.

253 Sverd J (1988). Tourette syndrome associated with pervasive developmental disorder: Is there an etiological relationship? *J Multihandicapped Person* 1:281-291.

254 Sverd J (1991). Tourette's syndrome and autistic spectrum disorders: A significant relationship. *Am J Med Genet* 39:173-179.

255 Sverd J and Montero G (1993). Is Tourette syndrome a cause of sudden infant death syndrome and childhood obstructive sleep apnea? *Am J Med Genet* 46:494-496.

256 Sverd J, Sheth R, Fuss J, and Levine J (1995). Prevalence of pervasive developmental disorder in a sample of psychiatrically hospitalized children and adolescents. *Child Psychiatry Human Devel* 25:221-240.

257 Swan GE, Carmelli D, and Rosenman RH (1990). Smoking and alcohol consumption in adult male twins: genetic heritability and shared environmental influences. *J Substance Abuse* 2:39-50.

258 Sweet JA and Bumpass LL (1987). *American Family and Households*. Russell Sage Foundation, New York.

259 Teachman JD and Polonko KA (1985). Timing of the transition to parenthood: A multidimensional birth-interval approach. *J Marriage Family* 47:867-879.

260 Teasdale TW and Owen DR (1989). Continuing secular increases in intelligence and a stable prevalence of high intelligence levels. *Intelligence* 13:255-262.

261 Thornton A (1982). Changing attitudes toward family issues in the United States. *Family Planning Perspectives* 14:297-303.

262 Torres A, Forrest JD, and Eisman S (1980). Telling parents: Clinic policies and adolescents' use of family planning and abortion services. *Family Planning Perspectives* 12:284-292.

263 Trussell J (1988). Teenage pregnancy in the United States. *Family Planning Perspectives* 20:262-272.

264 Van Court M and Bean FD (1985). Intelligence and fertility in the United States. *Intelligence* 9:23-32.

265 Vaughn B, Trussell J, Menken J, and Jones L (1977). Contraceptive failure among married women in the U.S. *Family Planning Perspectives* 9:251-257.

265a Ventura SJ, Martin JA, Taffel SM et al. (1993). *Advance Report of Final Natality Statistics*, National Center for Health Statistics.

266 Veroff J, Douvan E, and Kulka RA (1981). *The Inner American: A Self-portrait from 1957 to 1976.* Basic Books, New York.

267 Vining DRJr (1986). Social versus reproductive success: The central theoretical problem of human sociobiology. *Behav Brain Sci* 9:167-216.

268 Vining DRJr (1982). On the possibility of the re-emergence of a dysgenic trend with respect to intelligence in American fertility differentials. *Intelligence* 6:241-264.

269 Webster TG (1963). Problems of emotional development in young retarded children. *Am J Psychiatry* 120:37-43.

270 Weiss G, Hechtman L, Milroy T, and Perlman T (1985). Psychiatric status of hyperactives as adults: A controlled prospective 15-year follow-up of 63 hyperactive children. *J Am Acad Child Psychiatry* 24:211-220.

271 West DJ and Farrington DP (1973). *Who Becomes Delinquent?* Heinemann, London.

272 White RK (1982). The relation between socioeconomic status and academic achievement. *Psychological Bulletin* 91:461-481.

273 Wolfgang M, Figlio RM, and Selin T (1972). *Delinquency as a Birth Cohort.* University of Chicago Press, Chicago.

274 Wright L (1995). Double Mystery. *The New Yorker* Aug 7:45-62.

274a Wright S (1941). The physiology of the gene. *Physiological Reviews* 21:487-527.

275 Wrong D (1958). Trends in class fertility in Western Nations. *Can J Economics & Political Sci* 24:216-229.

276 Wrong D (1980). *Class fertility trends in Western nations.* Arno Press.

277 Yamaguchi K and Kandel D (1987). Drug use and other determinants of premarital pregnancy and its outcome: A dynamic analysis of competing life events. *J Marriage Family* 49:257-270.

278 Zabin LS, Hardy JB, Smith EA, and Hirsch MB (1986). Substance use and its relation to sexual activity among inner city adolescents. *J Adolescent Health Care* 7:320-331.

279 Zabin LS, Hirsch MB, Smith R, Streett R, and Hardy JB (1986). Evaluation of a pregnancy prevention program for urban teenagers. *Family Planning Perspectives* 18:119-126.

280 Zabin LS, Kantner JF, and Zelnik M (1979). The risk of adolescent pregnancy in the first months of intercourse. *Family Planning Perspectives* 11:215.

281 Zajonc R and Bargh J (1980). Birth orders, family size, and decline of SAT scores. *Amer Psychologist* 35:662-668.

282 Zelnik M and Kantner J (1980). Sexual activity, contraceptive use and pregnancy among metropolitan area teenagers: 1971-1979. *Family Planning Perspectives* 12:289.

283 Zelnik M and Kantner JF (1979). Reasons for nonuse of contraception by sexually active women ages 15 – 19. *Family Planning Perspectives* 11:289.

284 Zelnik M and Kim YJ (1982). Sex education and its association with teenage sexual activity, pregnancy and contraceptive use. *Family Planning Perspectives* 14:117-126.

285 Zelnik M, Koenig MA, and Kim YJ (1984). Source of prescription contraceptives and subsequent pregnancy among women. *Family Planning Perspectives* 16:6-13.

286 Zill N and Schooenborn CA (1990). Developmental, learning, and emotional problems. Health of our nation's children, United States, 1988. *Advance Data* 190:1-18.

287 Zur Nieden M (1951). The influence of constitution and environment upon development of adopted children. *J Psychol* 31:91-95.

Abbreviations, Definitions, Glossary

< Less than. For example p <.001 means the value was less than 0.001.

A, Adenine One of the four bases in DNA. It pairs with thymine, T.

ADHD Attention Deficit Hyperactivity Disorder, previously called hyperactivity or minimal brain damage (MBD). ADHD is the most common behavioral disorder of childhood.

agonist A chemical compound that stimulates a given receptor. For example, dopamine is an agonist for dopamine receptors.

allele Genes come in two copies, one from the mother and one from the father. The individual genes, maternal or paternal, are called alleles. If there are different forms of the gene due to mutations, each of these different forms is also called allele.

allele-specific PCR This is a technique that can identify the specific mutations in different alleles.

ALTE Apparent Life Threatening Event, or near sudden death in an infant.

amino acids Building blocks of protein. Tryptophan and phenylalanine are the names of two of the more than twenty amino acids.

antagonist A chemical compound that inhibits or blocks a given receptor. For example, haloperidol is an antagonist of dopamine D_2 receptors.

ascertainment The process of finding of cases or individuals with a given disorder.

ascertainment bias The tendency to have more severe cases of a given disorder, or individuals with two separate disorders, come to medical attention instead of milder cases.

base The individual building block of DNA. There are four bases: adenine, cytosine, guanine, and thymine.

base pairs The bases align themselves into specific pairs. Guanine pairs with cytosine and adenine pairs with thymine.

Behavioral Questionnaire The 31-page questionnaire, covering many different aspects of human behavior, that all new patients to the City of Hope Clinic must complete. A copy is included in the book *Tourette Syndrome and Human Behavior.*

bipolar depression Manic-depressive disorder.

C, Cytosine One of the four bases in DNA. It pairs with guanine, G.

case/cases Individual(s) in a study.

cDNA DNA representing the messenger RNA sequence of a gene (produced by an enzyme called reverse transcriptase).

chromosome Unit of a giant piece of DNA containing many genes. Chromosomes and their associated proteins are most easily seen during cell division (mitosis) when they are stained (*chromo* - colored, *some* - body).

clone (*noun*). A segment of DNA that is "grown up" into many copies by being inserted into primitive organism or "vector." When the vector reproduces to make many copies, the segment of DNA is also grown up or "cloned" into many copies. (*verb*) *to clone*. To group up a segment of DNA by growing it in a vector.

coprolalia Compulsive, involuntary swearing.

D_1 The dopamine D_1 receptor.

D_2 The dopamine D_2 receptor.

D_3 The dopamine D_3 receptor.

D_4 The dopamine D_4 receptor.

D_5 The dopamine D_5 receptor.

DAT1 The human *dopamine transporter* gene.

DβH Dopamine beta-hydroxylase, an enzyme that converts dopamine to norepinephrine.

DβH The dopamine β-hydroxylase gene (italicized).

Diagnostic Interview Schedule A structured interview used to make a diagnosis of the DSM psychiatric disorders.

dimorphic Of two different sizes. Often used to refer to different sizes of parts of the brain in males and females.

DIS see **Diagnostic Interview Schedule**

disinhibition Poorly inhibited. Removing influences that inhibit nerve cell activity. Inability to refrain from inappropriate behaviors.

DNA DexoyriboNucleic Acid, the genetic material of the cell.

dominant A trait caused by a gene in which only a single abnormal gene is required to produce the disorder. In strictly dominant conditions, the disorder produced by one abnormal gene is the same as when two abnormal genes are present.

dopamine A chemical that acts as a neurotransmitter in the brain. Dopamine neurons are especially involved in the control of muscle movements and have been implicated in schizophrenia, Tourette syndrome, and Parkinson's disease.

dopamine β-hydroxylase An enzyme that converts dopamine to norepinephrine.

dopamine transporter A protein that directs the re-uptake of dopamine into the neuron from which it was released.

DRD1 The *Dopamine Receptor D_1* gene.

DRD2 The *Dopamine Receptor D_2* gene.

DRD3 The *Dopamine Receptor D_3* gene.

DRD4 The *Dopamine Receptor D_4* gene.

DRD5 The *Dopamine Receptor D_5* gene.

DSM-III The Diagnostic and Statistical Manual of the American Psychiatric Association, 1980 edition. A manual listing the diagnostic criteria for all psychiatric diagnoses in the United States.

DSM-III-R Diagnostic and Statistical Manual of the American Psychiatric Association, Third Edition, Revised 1987.

DSM-IV Diagnostic and Statistical Manual of the American Psychiatric Association, Fourth Edition 1994.

dysgenic Any factor or factors that tend to select for less desirable genes and against more desirable genes. While most often used in terms of the genes for IQ, the term can apply to other characteristics as well.

***Eco*R1** A restriction endonuclease enzyme.

electrophoresis A technique by which chemicals are separated in an electrical field.

endonuclease An enzyme that cuts DNA fragments in the center regions, i.e. not at the ends, as in exonuclease.

enzyme A protein that serves as a catalyst for a specific chemical reaction.

et al. Refers to co-workers in the citation of a paper.

etiological Causal.

etiology Cause.

exon A segment of DNA within the gene that is read into the nuclear RNA, remains in the messenger RNA, and is translated into protein.

expression Refers to manner in which different symptoms of a genetic disorder may be present in an individual carrying the gene(s). In mild expression there are few or mild symptoms; in severe expression there are many or severe symptoms.

familial Occurring in two or more members of a family. May or may not be genetic.

frontal lobes Major lobes of the brain located behind the forehead.

G, Guanine One of the four bases in DNA. It pairs with cytosine, C.

G->A A mutation involving a change from a guanine to an adenine in DNA.

G->T A mutation involving a change from a guanine to a thymine in DNA.

gene The portion of DNA that carries the code for a specific protein or RNA.

genetics The study of hereditary disorders or mechanisms.

genome All the genes and DNA between the genes in an organism.

genomic Pertaining to the DNA of the genome.

genotype The results of a specific genetic test on both chromosomes, i.e. from the mother and the father. For example, for blood groups A, B, and O, if a person was type AB, that would be his genotype for that gene.

***Gts* gene** The mutant Gilles de la Tourette Syndrome gene. Gene names are always italicized.

***gts* gene** The normal counterpart to the Gts gene.

haplotype The results of two or more genetic tests for genes close to each other on one chromosome. For example, if a mutation occurred at point A, resulting in a mutant A' on one chromosome and a normal A on the other, and if another mutation occurred close by at point B, resulting in chromosomes B' or B, if we knew that A and B' were on the same chromosome and A' and B were on the other chromosome, AB' and A'B would be the haplotypes for that pair of mutations.

hereditary Inherited by genetic mechanisms.

heterozygote An individual in whom the gene from the mother and the father, at a given site, are different. For example, an individual who has both a normal *gts* and mutant *Gts* gene is a *Gts/gts* heterozygote.

heterozygous The state of having different genes at a given site.

homozygote An individual in whom the gene from the mother and the gene from the father, at a given site, are the same. For example, an individual with two *Gts* genes is a *Gts/Gts* homozygote.

homozygous The state of having the same gene at a given site.

hybridization A technique for annealing or zipping together a DNA sequence, usually made radioactive, to non-radioactive matching sequence.

hypothalamus A portion of the brain just above the pituitary gland that plays a major role in regulating many bodily functions.

intron A segment of DNA inside the gene that is read into the nuclear RNA but is excised from the messenger RNA.

JAMA The *Journal of the American Medical Association*.

learning disorder The definition or learning disorders, based on Public Law 94-142 (1976) of the Education for All Handicapped Children Act is as follows:

Specific learning disability means a disorder in one or more of the basic psychological processes involved in understanding or in using language, spoken or written, which may manifest itself in an imperfect ability to listen, think, speak, read, write, spell, or do mathematical calculations. The term includes such conditions as perceptual handicaps, brain injury, minimal brain dysfunction, dyslexia, and developmental aphasia. The term does not include children who have learning problems which are primarily the result of visual, hearing, or motor handicaps, of mental retardation, of emotional disturbance, or of environmental, cultural, or economic disadvantage.

limbic Pertains to the limbic system, the part of the brain controlling emotions. It is called limbic because it bends around the central part of the brain like a limb.

linkage When two traits or genetic markers travel together in families they are said to be linked.

linkage disequilibrium A situation in which two different mutations or markers are so close to each other that there is little opportunity for rearrangement over time. As a result, these two markers tend to appear together in different people.

messenger RNA see ***mRNA***.

motor Pertaining to muscle. A motor movement is a muscle movement.

mRNA Messenger RNA. The RNA produced by the genes that passes from the nucleus to the cytoplasm where it is translated into a specific protein or enzyme.

multifactorial A form of inheritance in which more than one type of gene is involved.

mutation A change in the structure of DNA that can be inherited.

N Number. Especially refers to number of cases.

N.S. Not Significant. A term used when statistical comparison shows no differences between two or more groups.

neurons Nerve cells.

neurotransmitters Chemicals that are involved in the communication between one nerve and another.

NIAAA National Institute of Alcohol Abuse and Alcoholism.

NIDA National Institute of Drug Abuse.

NIH National Institutes of Health.

NIMH National Institute of Mental Health.

non-proband A member of the family other than the individual who first sought medical care of a specific disorder.

norepinephrine A chemical that acts as a neurotransmitter in the brain. It is formed from dopamine.

OCD Obsessive-Compulsive Disorder.

oligomers Short synthetic pieces of DNA used in the amplification of DNA by the PCR technique.

oligos Abbreviation for oligomers.

oxidase An enzyme which uses ionic oxygen (0-) as a substrate.

oxygenase An enzyme which uses molecular oxygen (O_2) as a substrate.

p Probability. In statistical comparisons, if $p = 0.05$ it is usually considered that there are significant differences between the groups being compared. If $p = 0.01$, it is more significant. If $p = 0.001$, it is highly significant. If $p = 0.0001$ or less, it is very highly significant.

paradigm A model. In science the accepted paradigm refers to the prevailing and accepted model in a given field.

paradigm shift A change in the accepted model. A paradigm shift takes place when the accumulated scientific evidence becomes so strong that most scientists feel compelled to accept the new model.

PCR Polymerase Chain Reaction. Refers to a technique used to amplify small pieces of DNA millions of times. An extremely powerful technique widely used in molecular genetics.

pedigree A recorded family history often diagrammed for easy interpretation.

penetrance Refers to whether an abnormal gene has produced any symptoms. If the gene is non-penetrant, there are no symptoms. If it is penetrant, there are some symptoms and the number is described by the degree of expression.

polygenes Our term for the genes involved in polygenic inheritance.

polygenic Caused by multiple genes.

polymerase An enzyme used in the synthesis or replication of nucleic acids. For DNA – DNA polymerase. For RNA – RNA polymerase.

polymerase chain reaction See ***PCR***.

polymorphism A genetic variation present in some people but not others. For example, ABO blood groups are polymorphisms. Some people are AA, some AB, and some O.

primer A short piece of DNA (oligomer) used to start the synthesis of DNA. Primers are especially used in PCR reactions.

proband The first individual in a family who comes to see the doctor.

prospective Done before a diagnosis is made. For example, if a type of data is collected before a diagnosis is made or before the patient is placed in a specific group, several types of bias are prevented. Such a study is called a prospective study.

psychopathology Psychological abnormalities or disorders.

receptors Structures on the surface of nerve cells that tightly bind neurotransmitters. They define which nerve cells will respond to which neurotransmitters and whether the effect will be to stimulate or inhibit the nerve.

recessive A trait caused by a gene in which two abnormal genes are required to produce the disorder. Individuals who carry a single abnormal gene have no symptoms.

renature In DNA studies this refers to allowing individual single strands of DNA to go back to the double-stranded form.

restriction endonuclease An enzyme that cuts DNA at specific sites.

RFLP Restriction Fragment Length Polymorphism. A polymorphism due to variations between people in the distance between two sites on a DNA molecule cut by a restriction enzyme.

RNA RiboNucleic Acid. Consists of several types such as messenger RNA, transfer RNA, ribosomal RNA, and heterogeneous nuclear RNA.

segregation analysis A mathematical analysis of how disorders are inherited based on studies of many families.

sequence The sequence of bases A,T,G,C, in a given segment of DNA.

sequenced Indicates that the sequence of a gene or segment of DNA is known.

serotonin A chemical that acts as a neurotransmitter in the brain. It functions as an inhibiting neurotransmitter. The limbic system and frontal lobes are particularly rich in serotonin.

SIDS Sudden Infant Death Syndrome.

single strand DNA is composed to two strands of nucleotides. The individual strand are termed single strands.

somatiform Soma refers to body. Somatiform disorder or complaints are complaints about the body. This term is synonymous with the older term psychosomatic, implying the bodily pain or dysfunction is largely psychological in origin.

SSRI Selective Serotonin Reuptake Inhibitor. Antidepressant medications that act by inhibiting the reuptake or clearance of serotonin from the nerve synapse. This allows the brain to use serotonin more effectively.

syndrome A group of symptoms or signs that identifies a specific disorder, such as Tourette syndrome.

T, Thymine One of the four bases in DNA. It pairs with adenine, A.

tandem repeat A tandem repeat refers to a series of identical sets of the same DNA sequence aligned end to end. For example, CAT CAT CAT would be a 3 base pair tandem repeat.

tandem repeat polymorphism A situation where different people have different numbers of repeats in a tandem repeat. For example, CAT CAT in some people and CAT CAT CAT in others would represent a 2 versus a 3 tandem repeat polymorphism.

***Taq* I** A restriction endonuclease enzyme.

***TDO2* gene** The gene that makes TDO2 or tryptophan dioxygenase or tryptophan 2,3-dioxygenase.

TDO2 Tryptophan DiOxygenase. An enzyme that breaks down tryptophan. The 2 refers to the presence of two oxygen molecules.

tic An involuntary, sudden, rapid muscle movement or vocal noise.

Tourette syndrome A common hereditary syndrome consisting of chronic motor (muscle) and vocal tics and associated with a wide range of other behaviors.

transporters Proteins that direct the re-uptake into the neuron of various neurotransmitters. These are dopamine, serotonin, norepinephrine and other transporters.

tryptophan An amino acid required to make serotonin.

tryptophan 2,3-dioxygenase The same as tryptophan oxygenase and tryptophan dioxygenase.

TS Tourette Syndrome or Gilles de la Tourette Syndrome.

type I error A statistical error resulting from stating something is true when, in fact, it is not true.

type II error A statistical error resulting from stating something is not true when, in fact, it is true. This error often results from not studying enough subjects.

Index

A

B

C

D

E

F

G

H

I

J

K

L

M

N

O

P

R

S

T

U

V

W

Y

Z

Order Form

cut cut cut cut cut

1. Books: Quantity — Amount

Tourette Syndrome and Human Behavior
_______ 1S Softback $39.95 _______

Search for the Tourette Syndrome and Human Behavior Genes
_______ 8H Hardback $34.00 _______
_______ 8S Softback $29.95 _______

The Gene Bomb ***Does Higher Education and Advanced Technology Accelerate the Selection of Genes for Learning Disorders, ADHD, Addictive and Disruptive Behaviors?***
_______ 9H Hardback $29.95 _______
_______ 9S Softback $25.00 _______

RYAN — A Mother's Story of Her Hyperactive-Tourette Syndrome Child
_______ 2S Softback $9.95 _______

What Makes Ryan Tick? A Family's Triumph over TS and ADHD
_______ 10S Softback $14.95 _______

Hi, I'm Adam - A Child's Book about Tourette Syndrome
_______ 4A Softback $4.95 _______

Adam and the Magic Marble
_______ 4B Softback $6.95 _______

Hi, I'm Adam + Adam and the Magic Marble
_______ 4C Both together $11.50 _______

Echolalia - An Adult's Story of Tourette Syndrome
_______ 5A Softback $11.95 _______

Don't Think About Monkeys - Extraordinary Stories by People with Tourette Syndrome
_______ 6A Softback $12.95 _______

Teaching the Tiger - A Handbook for Individuals Involved in the Education of Students with Attention Deficit Disorder, Tourette Syndrome or Obsessive-Compulsive Disorder
_______ 7A Softback $35.00 _______

Subtotal for Books _______

2. Tax: **California residents please add 8.25% sales tax** _______

3. Mailing and Handling:
- ☐ Fourth Class: $4.00 lst item $1.00 each additional item
- ☐ U.P.S. Ground: $6.00 lst item $1.00 each additional item
- ☐ U.P.S. Air: $10.00 lst item $2.00 each additional item _______

Total _______

Name: _______________________
Address: _______________________
City: _______________ State: _______ Zip: _______
Country (if other than U.S.A.): _______________________
Check Enclosed _____ **or** Visa ___ Mastercard ___
CC# _______________________ Expiration Date _________

send to: **Hope Press P.O.Box 188, Duarte, CA 91009-0188**

or Fill out this form with credit card # and FAX it to 818-358-3520

or Order by phone **1-800-321-4039** — 24 hr service

[Foreign buyers outside North America please: a) send bank check in U.S. dollars, or b) order by credit card with charge in U.S. dollars, or c) FAX in the form. For surface mail add $6.00 shipping for first book and $1.00 for each additional and allow 4-6 weeks. For air mail add $25.00 shipping and $2.00 for each additonal book and allow 1 week.]

for more details on each book visit our web site: **http://www.hopepress/tourette**

Order Form

1. Books: Quantity — Amount

Tourette Syndrome and Human Behavior
_______ 1S Softback $39.95 _______

Search for the Tourette Syndrome and Human Behavior Genes
_______ 8H Hardback $34.00 _______
_______ 8S Softback $29.95 _______

The Gene Bomb** **Does Higher Education and Advanced Technology Accelerate the Selection of Genes for Learning Disorders, ADHD, Addictive and Disruptive Behaviors?
_______ 9H Hardback $29.95 _______
_______ 9S Softback $25.00 _______

RYAN — A Mother's Story of Her Hyperactive-Tourette Syndrome Child
_______ 2S Softback $9.95 _______

What Makes Ryan Tick? A Family's Triumph over TS and ADHD
_______ 10S Softback $14.95 _______

Hi, I'm Adam - A Child's Book about Tourette Syndrome
_______ 4A Softback $4.95 _______

Adam and the Magic Marble
_______ 4B Softback $6.95 _______

Hi, I'm Adam + Adam and the Magic Marble
_______ 4C Both together $11.50 _______

Echolalia - An Adult's Story of Tourette Syndrome
_______ 5A Softback $11.95 _______

Don't Think About Monkeys - Extraordinary Stories by People with Tourette Syndrome
_______ 6A Softback $12.95 _______

Teaching the Tiger - A Handbook for Individuals Involved in the Education of Students with Attention Deficit Disorder, Tourette Syndrome or Obsessive-Compulsive Disorder
_______ 7A Softback $35.00 _______

Subtotal for Books _______

2. Tax: **California residents please add 8.25% sales tax** _______

3. Mailing and Handling:
- ☐ Fourth Class: $4.00 lst item $1.00 each additional item
- ☐ U.P.S. Ground: $6.00 lst item $1.00 each additional item
- ☐ U.P.S. Air: $10.00 lst item $2.00 each additional item _______

Total _______

Name: _______________
Address: _______________
City: _______________ State: _______ Zip: _______
Country (if other than U.S.A.): _______________
Check Enclosed _____ **or** Visa ___ Mastercard ___
CC# _______________ Expiration Date _______

send to: **Hope Press P.O.Box 188, Duarte, CA 91009-0188**

or Fill out this form with credit card # and FAX it to 818-358-3520

or Order by phone **1-800-321-4039** — 24 hr service

[Foreign buyers outside North America please: a) send bank check in U.S. dollars, or b) order by credit card with charge in U.S. dollars, or c) FAX in the form. For surface mail add $6.00 shipping for first book and $1.00 for each additional and allow 4-6 weeks. For air mail add $25.00 shipping and $2.00 for each additonal book and allow 1 week.]

for more details on each book visit our web site: **http://www.hopepress/tourette**

cut

Order Form

1. Books: Quantity | Amount

Tourette Syndrome and Human Behavior
_______ 1S Softback $39.95 _______

Search for the Tourette Syndrome and Human Behavior Genes
_______ 8H Hardback $34.00 _______
_______ 8S Softback $29.95 _______

The Gene Bomb Does Higher Education and Advanced Technology Accelerate the Selection of Genes for Learning Disorders, ADHD, Addictive and Disruptive Behaviors?
_______ 9H Hardback $29.95 _______
_______ 9S Softback $25.00 _______

RYAN — A Mother's Story of Her Hyperactive-Tourette Syndrome Child
_______ 2S Softback $9.95 _______

What Makes Ryan Tick? A Family's Triumph over TS and ADHD
_______ 10S Softback $14.95 _______

Hi, I'm Adam - A Child's Book about Tourette Syndrome
_______ 4A Softback $4.95 _______

Adam and the Magic Marble
_______ 4B Softback $6.95 _______

Hi, I'm Adam + Adam and the Magic Marble
_______ 4C Both together $11.50 _______

Echolalia - An Adult's Story of Tourette Syndrome
_______ 5A Softback $11.95 _______

Don't Think About Monkeys - Extraordinary Stories by People with Tourette Syndrome
_______ 6A Softback $12.95 _______

Teaching the Tiger - A Handbook for Individuals Involved in the Education of Students with Attention Deficit Disorder, Tourette Syndrome or Obsessive-Compulsive Disorder
_______ 7A Softback $35.00 _______

Subtotal for Books _______

cut

2. Tax: **California residents please add 8.25% sales tax** _______

3. Mailing and Handling:
☐ Fourth Class: $4.00 lst item $1.00 each additional item
☐ U.P.S. Ground: $6.00 lst item $1.00 each additional item
☐ U.P.S. Air: $10.00 lst item $2.00 each additional item _______

Total _______

cut

Name: ____
Address: ____
City: ___________________________ State:________ Zip: _________
Country (if other than U.S.A.): _______________________________
Check Enclosed _____ **or** Visa ___ Mastercard ___
CC# _________________________________ Expiration Date _________

cut

send to: **Hope Press P.O.Box 188, Duarte, CA 91009-0188**

or Fill out this form with credit card # and FAX it to 818-358-3520

or Order by phone **1-800-321-4039** — 24 hr service

[Foreign buyers outside North America please: a) send bank check in U.S. dollars, or b) order by credit card with charge in U.S. dollars, or c) FAX in the form. For surface mail add $6.00 shipping for first book and $1.00 for each additional and allow 4-6 weeks. For air mail add $25.00 shipping and $2.00 for each additonal book and allow 1 week.]

for more details on each book visit our web site: **http://www.hopepress/tourette**

cut

cut

cut

cut

cut

Order Form

1. Books: Quantity — Amount

Tourette Syndrome and Human Behavior
_______ 1S Softback $39.95 _______

Search for the Tourette Syndrome and Human Behavior Genes
_______ 8H Hardback $34.00 _______
_______ 8S Softback $29.95 _______

The Gene Bomb Does Higher Education and Advanced Technology Accelerate the Selection of Genes for Learning Disorders, ADHD, Addictive and Disruptive Behaviors?
_______ 9H Hardback $29.95 _______
_______ 9S Softback $25.00 _______

RYAN — A Mother's Story of Her Hyperactive-Tourette Syndrome Child
_______ 2S Softback $9.95 _______

What Makes Ryan Tick? A Family's Triumph over TS and ADHD
_______ 10S Softback $14.95 _______

Hi, I'm Adam - A Child's Book about Tourette Syndrome
_______ 4A Softback $4.95 _______

Adam and the Magic Marble
_______ 4B Softback $6.95 _______

Hi, I'm Adam + Adam and the Magic Marble
_______ 4C Both together $11.50 _______

Echolalia - An Adult's Story of Tourette Syndrome
_______ 5A Softback $11.95 _______

Don't Think About Monkeys - Extraordinary Stories by People with Tourette Syndrome
_______ 6A Softback $12.95 _______

Teaching the Tiger - A Handbook for Individuals Involved in the Education of Students with Attention Deficit Disorder, Tourette Syndrome or Obsessive-Compulsive Disorder
_______ 7A Softback $35.00 _______

Subtotal for Books _______

2. Tax: **California residents please add 8.25% sales tax** _______

3. Mailing and Handling:
☐ Fourth Class: $4.00 lst item $1.00 each additional item
☐ U.P.S. Ground: $6.00 lst item $1.00 each additional item
☐ U.P.S. Air: $10.00 lst item $2.00 each additional item _______

Total _______

Name: ______________________________
Address: ______________________________
City: ____________________ State: ________ Zip: ________
Country (if other than U.S.A.): ______________________________
Check Enclosed _____ **or** Visa ___ Mastercard ___
CC# ______________________________ Expiration Date _________

send to: Hope Press P.O.Box 188, Duarte, CA 91009-0188

or Fill out this form with credit card # and FAX it to 818-358-3520

or Order by phone **1-800-321-4039** — 24 hr service

[Foreign buyers outside North America please: a) send bank check in U.S. dollars, or b) order by credit card with charge in U.S. dollars, or c) FAX in the form. For surface mail add $6.00 shipping for first book and $1.00 for each additional and allow 4-6 weeks. For air mail add $25.00 shipping and $2.00 for each additonal book and allow 1 week.]

for more details on each book visit our web site: **http://www.hopepress/tourette**